Introduction
to
Mathematics

5TH EDITION

Introduction to Mathematics

BRUCE E. MESERVE

Professor of Mathematics
University of Vermont

MAX A. SOBEL

Professor of Mathematics
Montclair State College

PRENTICE-HALL, INC. ENGLEWOOD CLIFFS, N.J. 07632

Library of Congress Cataloging in Publication Data

Meserve, Bruce Elwyn (date)
   Introduction to mathematics.

   Includes index.
   1. Mathematics—1961-   I. Sobel, Max A.
II. Title.

QA39.2.M49 1984   510   83-10915
ISBN 0-13-487348-3

© 1984, 1978, 1973, 1969, 1964
by Prentice-Hall, Inc.,
Englewood Cliffs, NJ 07632

10 9 8 7 6 5 4 3 2 1

Printed in the United States of America

Cover photo: *Mertan* by Vasarely. Courtesy of the
Vasarely Center, NY.
Designer: Janet Schmid
Production: Nicholas Romanelli
Page Layout: Meg Van Arsdale
Manufacturing Buyer: John Hall
Photo Research: Teri Leigh Stratford

ISBN 0-13-487348-3

Prentice-Hall International, Inc., *London*
Prentice-Hall of Australia Pty. Limited, *Sydney*
Editora Prentice-Hall do Brasil, Ltda, *Rio De Janeiro*
Prentice-Hall of Canada Inc., *Toronto*
Prentice-Hall of India Private Limited, *New Delhi*
Prentice-Hall of Japan, Inc., *Tokyo*
Prentice-Hall of Southeast Asia Pte. Ltd., *Singapore*
Whitehall Books Limited, Wellington, *New Zealand*

ILLUSTRATION CREDITS: Compuserve 46, Castle & Cooke,
Inc. and Dole Pineapple 342, Creative Publications 35,
Greek National Tourist Office 337, IBM 63, 416, Ken Karp
78, 422, Las Vegas News Bureau, Convention Center,
376, NASA 54, 264, North Carolina State University 312, Dr.
Malcolm T. Sanford 130, Scott Publishing Co. 55, 115, 176,
286, 289, Irene Springer 174, 233, Teri Leigh Stratford 87,
339, 342, 381, 409, US Air Force 130, US Department of
Agriculture 342, US Fish and Wildlife Service, photo by
David Klinger 130, Vector Graphic 123. Ideal Toy Corp.
25, Pulsar Time, Inc. 79, New York Public Library 132, 139,
222

Contents

7 An Introduction to Algebra   **264**

8 Elements of Geometry   **312**

9 An Introduction to Probability   **376**

10 An Introduction to Statistics

Answers to Odd-Numbered Exercises

Index

Preface

Mathematics can be fun! The authors subscribe to this idea and trust that this fifth edition of *Introduction to Mathematics* will prove as popular with both students and instructors as has been the case for their previous editions. Thus this text presents a variety of interesting and timely topics without placing an undue emphasis on abstract manipulations. Hopefully such an approach will leave the reader with a better picture of the true meaning and beauty of mathematics.

Many users of the four earlier editions of this text, teachers and students alike, have made numerous valuable suggestions that have been incorporated into this fifth edition. Among the important changes are those that were made in an effort to be consistent with contemporary trends and recommendations. Thus a new Chapter 1 has been introduced on problem solving, and the guidelines and strategies developed there are brought into focus in later chapters. Attention to consumer needs is recognized in Chapter 2 (Calculators and Computers) as well as in the new Chapter 3 (Mathematics for the Consumer).

The topics considered in this book play major roles in the activities of citizens in our technological society. Some secondary-school introduction to algebra and geometry is expected of the reader, but no extensive working knowledge of any of the skills normally taught in these subjects is assumed. The subject matter has been selected for the undergraduate college student who is not a mathematics major but who wishes to acquire a basic understanding of the nature of mathematics. Frequently, prospective elementary-school teachers will be among such students. Throughout the book the emphasis is on key concepts of mathematics without undue concern over the mechanical procedures.

Prior editions of *Introduction to Mathematics* have been cited for their special attention to features that assist student learning. These include the following distinctive features that have been further expanded in this edition.

Illustrative Examples   Each section of the text contains illustrative examples with detailed solutions that provide the student with a model for the correct solutions of problems.

Exercises: Each section of the text is followed by a comprehensive set of exercises, with answers to the odd-numbered problems given at the back of the book. In all, there are more than 2000 exercises in the text so that choices may be readily made for classes of varying abilities.

Explorations   Explorations are included at the end of every section of the text. These are often discovery exercises that allow the reader to pursue independently open-ended explorations that supplement knowledge of the material under discussion. They also serve to provide vistas of possible future extensions of such material.

Chapter Review   A Chapter Review appears at the end of every chapter of the text, keyed to sections of the chapter. Detailed solutions of these exercises may be found within the body of the text.

Chapter Test   The Chapter Review is followed by a Chapter Test that enables the student to demonstrate a comprehensive mastery of the basic concepts and skills of the chapter.

Prior editions of *Introduction to Mathematics* have been cited for the clarity of exposition presented. A conscientious effort has been made to have this fifth edition even more readable so that students may find this a text that they can truly read and study on their own, rather than use merely as a source of exercises. Its distinctive features make this a text that students should find nonthreatening. This is especially important in that many of the students using this text will have had only a minimal amount of previous mathematical experience. Furthermore, the "light approach" used throughout should help overcome the anxiety that a significant number of students seem to have toward mathematics. However, additional challenges for the above-average student have also been included in the exercises and in the explorations so that these capable students may be suitably motivated.

The authors wish to express their appreciation to all those who have submitted suggestions for the revision of this text, as well as to the many students throughout the country who have both directly and indirectly contributed to the conception and formulation of this material. We are particularly grateful to the reviewers for their detailed comments and suggestions: Charles L. Adie, Northern Essex Community College; David R. Duncan, University of Northern Iowa; George Feissner, SUNY at Cortland; Barbara Gilfillan, Rhode Island College; Martha C. Jordan, Okaloosa-Walton Junior College; Mary K. Malm, Pheonix College; Charles Nelson, University of Florida.

Special recognition is given to Dorothy Meserve for her careful analysis of the manuscript, her detailed work in checking answers, and her preparation of the Instructor's Manual that accompanies this text.

The famous French mathematician René Descartes concluded his famous *La Géométrie* with the statement: "I hope that posterity will judge me kindly, not only as to the things which I have explained, but also as to those which I have intentionally omitted so as to leave to others the pleasure of discovery." The authors have attempted to provide a great deal of exposition in this text. They have, however, left many opportunities for the reader to experience the true beauty and excitement of mathematics through discovery.

BRUCE E. MESERVE

MAX A. SOBEL

List of Selected Explorations

Introduction
to
Mathematics

An Introduction to Problem Solving

1

1-1
Problem-Solving Strategies

The use of mathematics in our daily activities consists primarily in applying mathematics that we study in school to solve problems. Yet we have traditionally been so concerned with computation that problem solving has not been a major concern in our studies in the past. With today's technology and the ready availability of calculators and computers, there appears to be less reason for excessive concern with computation and thus more time can be devoted to problem solving.

There really is only one successful way to develop expertise in problem solving, namely to solve problems. That is, one must solve many problems in order to develop problem-solving skills. This is similar to the standard saying that the only way to learn to swim is to get into the water! However, just as the swimming instructor can provide some broad generalizations on dry land, so can we offer some suggestions before you begin your attempt to solve problems. These hints can be summarized within the following four broad categories.

1. **Study the problem carefully**

   Read the problem carefully and be certain that you understand what is given and what is to be found. Extract the key ideas from the statement of the problem and summarize these in your own words or symbols. Note whether the problem gives more information than is needed, or whether there may be insufficient information to solve the problem.

   The following problem appeared recently on an examination as part of the National Assessment of Educational Progress.

   One rabbit eats 2 pounds of food each week. There are 52 weeks in a year. How much food will 5 rabbits eat in one week?

   At one age level, almost 25% of the respondents gave 520 pounds as the answer. They failed to recognize that the second sentence concerning the number of weeks in a year consisted of information that was not needed for the solution of the problem. Thus they apparently felt compelled to use the number 52 in their solution to the problem! The question involved *one* week; 5 rabbits will eat 10 pounds of food in one week.

2. **Plan a strategy**

   Decide on a tentative approach to the solution, and be

ready to set this aside and follow another path if your first attempt proves unsuccessful. There are a number of distinct possibilities that you may wish to consider, such as these:

(a) Search for a pattern or relationships.
(b) Make a list.
(c) Draw a diagram or picture if appropriate.
(d) Attempt to solve a simpler but related problem.
(e) Make an educated guess or conjecture.
(f) Work backward from the answer.
(g) Use trial-and-error procedures; guess and test.
(h) Try to look at the problem from a different point of view.

We will illustrate some of these strategies in the illustrative examples that follow; others will be considered in the exercises.

Although these are some of the most effective problem-solving strategies, the list is not a complete one. You may think of other approaches to solving problems that will work for you. Also, at times you may use several different strategies within the solution of a single problem.

3. **Solve the problem**

Use the strategies suggested in the preceding discussion to solve the given problem. This is the ultimate goal of any problem-solving activity.

4. **Check the solution**

Where possible, attempt to check the solution. Usually, you will be able to return to the original statement of the problem to confirm your solution. At times it will be obvious that your solution is correct. However, there may be times that you are reasonably certain that your solution is correct, but no apparent method for checking seems feasible. In such cases, you may wish to estimate the answer and compare your estimate with your solution.

An important extension of this last step in the problem-solving process is to attempt to generalize or to extend the solution when possible. Such extensions or generalizations are often helpful in providing insights into the solutions of other problems.

The examples that follow will illustrate problem-solving approaches that make use of a variety of strategies. In each case, the student is urged to attempt to solve the problem before studying the given solution.

Example 1   Find the sum

$$\frac{1}{1 \times 2} + \frac{1}{2 \times 3} + \frac{1}{3 \times 4} + \cdots + \frac{1}{9 \times 10}$$

We shall attempt to find the sum by considering simpler, but related problems. Thus, by arithmetic, we find these sums.

$$\frac{1}{1 \times 2} = \frac{1}{2}$$

$$\frac{1}{1 \times 2} + \frac{1}{2 \times 3} = \frac{1}{2} + \frac{1}{6} = \frac{4}{6} = \frac{2}{3}$$

$$\frac{1}{1 \times 2} + \frac{1}{2 \times 3} + \frac{1}{3 \times 4} = \frac{1}{2} + \frac{1}{6} + \frac{1}{12} = \frac{9}{12} = \frac{3}{4}$$

Now compare each of the answers on the right with the denominator of the last fraction in the given sums. Note, for example, that the third sum has 3×4 as the denominator of the last fraction, and that the sum of the fractions given is 3/4. At this point you may wish attempt a guess at the sum of the fractions in the given problem. The last fraction has denominator 9×10, and a good *guess* is that the sum is 9/10.

In Example 1, we *conjectured* that the sum was 9/10. However, this does not constitute a proof. (A proof is considered in Exercise 25 of this section.) However, you may have greater confidence in your guess by testing it on another example. Thus you may wish to use the procedure for your guess to find the following sum; then confirm your answer by addition.

$$\frac{1}{1 \times 2} + \frac{1}{2 \times 3} + \frac{1}{3 \times 4} + \frac{1}{4 \times 5}$$

We can extend the solution to Example 1 by considering other similar problems, such as finding the following sum.

$$\frac{1}{1 \times 2} + \frac{1}{2 \times 3} + \frac{1}{3 \times 4} + \cdots + \frac{1}{98 \times 99} + \frac{1}{99 \times 100}$$

Does it appear that the sum should be 99/100? Note that we are extending a solution that has been conjectured, but not proved.

The problem stated in Example 2 has been popular in recent articles on problem solving. The solution illustrates problem-solving techniques with a nonstandard problem that appears impossible of solution at first glance.

Example 2 At a party for mathematicians the host announces the following:

> I have three daughters. The product of their ages is 72. The sum of their ages is the same number as our house number. How old are my daughters?

The guests confer, go outside to look at the house number, and then return to say that there is insufficient information to solve the problem. Thereupon the host adds this statement:

My oldest daughter loves chocolate pudding.

With this new bit of information the guests are able to determine the ages of the three daughters. What are these ages?

Solution Although the problem seems complex, read it several times and abstract the key items of information given.

It is often helpful to simplify the problem in your own words by listing briefly all of the given information.

Note here that the information concerning the oldest daughter is needed for the solution.

Three daughters.
Product of ages is 72.
Sum of ages equals house number.
Oldest daughter likes chocolate pudding.

For the time being we shall ignore the last statement, which seems to be extraneous. We begin by listing all possible sets of three numbers whose product is 72, as well as their sums. This is really a trial-and-error process of exhaustion of possibilities.

STRATEGY: Make a list. Note that the list is completed in a systematic manner to avoid the possibility of omitting any set of ages.

PRODUCTS	SETS OF POSSIBLE AGES	SUMS OF AGES
$1 \times 1 \times 72$	1, 1, 72	74
$1 \times 2 \times 36$	1, 2, 36	38
$1 \times 3 \times 24$	1, 3, 24	28
$1 \times 4 \times 18$	1, 4, 18	23
$1 \times 6 \times 12$	1, 6, 12	19
$1 \times 8 \times 9$	1, 8, 9	18
$2 \times 2 \times 18$	2, 2, 18	22
$2 \times 3 \times 12$	2, 3, 12	17
$2 \times 4 \times 9$	2, 4, 9	15
$2 \times 6 \times 6$	2, 6, 6	14
$3 \times 4 \times 6$	3, 4, 6	13
$3 \times 3 \times 8$	3, 3, 8	14

The next piece of information given is that the sum of the ages is equal to the house number. Since this information is not sufficient there must be two sums equal to the house number. Looking at the set of sums we note that there are two possible sets of ages with the sum of 14. Thus we assume that one of these must represent the ages. We conclude that the ages of the daughters must be one of the following:

2   6   6   or   3   3   8

Finally, we are told that the *oldest* daughter loves chocolate pudding. In the case of the ages 2, 6, 6 there is no oldest daughter. Therefore we conclude that their ages must be 3, 3, and 8.

Example 3 How many squares of all sizes are there on a standard checkerboard?

Solution

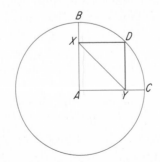

2 X 2 : 1
1 X 1 : 4
Total : 5

3 X 3 : 1
2 X 2 : 4
1 X 1 : 9
Total : 14

4 X 4 : 1
3 X 3 : 4
2 X 2 : 9
1 X 1 : 16
Total : 30

STRATEGY: Consider simpler but related problems. Then search for a pattern.

Example 4

STRATEGY: Look at a problem from a different point of view. Sometimes a period of *incubation* helps. That is, leave the problem and come back to it at a later time with a fresh viewpoint.

At first this problem might seem to be overwhelming because there are so many possible squares to count. For example, there are 64 squares that are 1 unit on each side. Then we must also consider squares that are 2 units on each side, and so forth.

To solve this problem we shall consider miniature checkerboards of different sizes, such as those shown. In each case, we count the number of possible squares of all sizes. Note the patterns for each case shown. We use these patterns to conclude that the sum for an 8 × 8 checkerboard will be

$$1 + 4 + 9 + 16 + 25 + 36 + 49 + 64 = 204$$

There are times that we are confronted with puzzle-type problems whose solution becomes obvious once it is known. It is very frustrating to find that some individuals can immediately see such solutions, whereas others have mental blocks that prevent the necessary insight. Example 4 is such a problem, and seems a suitable one to terminate this initial discussion of problem solving. We shall continue to consider the solution of a variety of problems throughout this text.

In the figure the radius *AB* of the circle is of length 5 cm and *AXDY* is a rectangle. Find the length of the line segment *XY*.

1 AN INTRODUCTION TO PROBLEM SOLVING

Solution It seems natural to consider XY as the hypotenuse of right triangle XAY, but we do not know the lengths of the legs of this triangle. The trick here is to draw the line segment AD. Then in rectangle $AXDY$ the diagonals are of equal length, so that $XY = AD$. But AD is a radius of the circle and all radii of a given circle are of equal length. Therefore $XY = AD = 5$ cm.

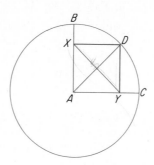

EXERCISES

Note that many of the problems considered here are included "just for fun." Do not feel discouraged if you are unable to solve some at first glance. Also, you should not feel disturbed if you discover that some are trick questions!

1. All of the following puzzles have logical answers, but they are not strictly mathematical. See how many you can answer.

    (a) How many two-cent stamps are there in a dozen?
    (b) How many telephone poles are needed in order to reach the moon?
    (c) How far can you walk into a forest?
    (d) Two American coins total 55¢ in value, yet one of them is not a nickel. Can you explain this?
    (e) How much dirt is there in a hole which is 3 feet wide, 4 feet long, and 2 feet deep?
    (f) There was a blind beggar who had a brother, but this brother had no brother. What was the relationship between the two?

2. A farmer has to get a fox, a goose, and a bag of corn across a river in a boat which is only large enough for him and one of these three items. Now if he leaves the fox alone with the goose, the fox will eat the goose. If he leaves the goose alone with the corn, the goose will eat the corn. How does he get all items across the river?

3. Three cannibals and three missionaries need to cross the river in a boat big enough only for two. The cannibals are fine if they are left alone or if they are with the same number or with a larger number of missionaries. They are dangerous if they are left alone in a situation where they outnumber the missionaries. How do they all get across the river without harm?

4. A bottle and cork cost $2.00 together. The bottle costs one dollar more than the cork. How much does each cost?

5. A cat is at the bottom of an 18-foot well. Each day it climbs up 3 feet; each night it slides back 2 feet. How long will it take for the cat to get out of the well?

6. If a cat and a half eats a rat and a half in a day and a half, how many days will it take for 50 cats to eat 50 rats?

7. Ten coins are arranged to form a triangle as shown in the accompanying figure. By rearranging only three of the coins, form a new triangle that points in the opposite direction from the one shown.

8. A woman goes to a well with three cans whose capacities are 3 liters, 5 liters, and 8 liters. Explain how she can obtain exactly 4 liters of water from the well.

9. Here is a mathematical trick you can try on a friend. Ask someone to place a penny in one of his hands, and a dime in the other. Then tell him to multiply the value of the coin in the right hand by 6, multiply the value of the coin held in the left hand by 3, and add. Ask for the result. If the number given is an even number, you then announce that the penny is in the right hand; if the result is an odd number, then the penny is in the left hand and the dime is in the right hand. Can you figure out why this trick works?

10. Four matchsticks are arranged to form a "cup" with a coin contained within three of the segments, as shown in the adjacent figure. By rearranging only two of the matchsticks, form a new figure that is congruent to the original one but with the coin no longer contained within any of the segments.

11. Three men enter a hotel and rent a room for $60. After they are taken to their room the manager discovers he overcharged them; the room rents for only $50. He thereupon sends a bellhop upstairs with the $10 change. The dishonest bellhop decides to keep $4 and returns only $6 to the men. Now the room originally cost $60, but the men had $6 returned to them. This means that they paid only $54 for the room. The bellhop kept $4. $54 + $4 = $58. What happened to the extra two dollars?

12. Find at least one way of using four 4's and ordinary arithmetic operations to write each of the numbers 1 through 10. Here are possible solutions for 1, 2, and 3.

$$\frac{44}{44} = 1 \qquad \frac{4}{4} + \frac{4}{4} = 2 \qquad \frac{4 + 4 + 4}{4} = 3$$

13. Use six matchsticks, all of the same size, to form four equilateral triangles.

14. Rearrange the right segments shown in the adjacent figure so as to form three congruent squares. Each of the four smaller segments is one-half of the length of one of the larger segments.

15. A sailor lands on an island inhabited by two types of people. The A's always lie, and the B's always tell the truth. The sailor meets three inhabitants on the beach and asks the first of these: "Are you an A or a B?" The man answers, but the sailor doesn't understand him and asks the second person what the first had said. The man replies: "He said that he was a B. He is, and so am I." The third inhabitant then says: "That's not true. The first man is an A and I'm a B." Can you tell who was lying and who was telling the truth?

16. Consider a house with six rooms and furniture arranged as in the accompanying figure. We wish to interchange the desk and the bookcase, but in such a way that there is never more than one piece of furniture in a room at a time. The other three pieces of furniture do not need to return to their original places. Can you do this? Try it using coins or other objects to represent the furniture.

Cabinet		Desk
Television set	Sofa	Bookcase

17. Arrange two pennies P and two dimes D as in the figure on the next page. Try to interchange the coins so that the pennies are at the right and the dimes at the left. A coin may be moved to an adjacent empty square. A coin may also be jumped over a single occupied square to an empty square. You may move only one coin at a time, and no two coins may occupy the same space at the same time.

moved only to the right, whereas dimes may be moved only to the left. What is the minimum number of moves required to complete the game?

18. Repeat Exercise 17 for three pennies and three dimes, using seven squares. What is the minimum number of moves required to complete the game?

19. Think of the decimal digits and identify the next two letters in the sequence: $O, T, T, F, F, S, S, \ldots$.

20. As in Exercise 19 identify the next two letters in the sequence: $E, F, F, N, O, S, S, \ldots$.

21. Place a half-dollar, a quarter, and a nickel in one position, A, as in the figure. Then try to move these coins, one at a time, to position C. Coins may also be placed in position B. At no time may a larger coin be placed on a smaller coin. This can be accomplished in $2^3 - 1$, that is, 7, moves.

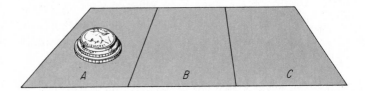

Next add a penny to the pile and try to make the change in $2^4 - 1$, that is, 15, moves.

This is an example of a famous problem called the **Tower of Hanoi**. The ancient Brahman priests were to move a pile of 64 such disks of decreasing size, after which the world would end. This would require $2^{64} - 1$ moves. Estimate how long this would take at the rate of one move per second.

22. Here is a game that must be played by two persons. Two players alternate in selecting one of the numbers 1, 2, 3, 4, 5, or 6. After each number is selected, it is added to the sum of those previously selected. For example, if player A selects 3 and player B selects 5, then the total is 8. If A selects 3 again, then the total is 11 and player B takes his

turn. The object of the game is to be the first one to reach 50. There is a way to win at all times if you are permitted to go first. See if you can discover this method for winning, and then try to play the game with a classmate.

23. Many tricks of magic have their basis in elementary mathematics and may be found in books on mathematical recreations. Here is one example of such a trick.

    Have someone place three dice on top of one another while you turn your back. Then instruct that person to look at and find the sum of the values shown on the two faces that touch each other for the top and middle dice, the two faces that touch each other for the middle and bottom dice, and the value of the bottom face of the bottom die. You then turn around and at a glance tell the sum. The trick is this: You merely subtract the value showing on the top face of the top die from 21. Stack a set of three dice in the manner described and try to figure out why the trick works as it does.

24. You are given a checkerboard and a set of dominoes. The size of each domino is such that it is able to cover two squares on the board. Can you arrange the dominoes in such a way that all of the board is covered with the exception of two squares in opposite corners? (That is, you are to leave uncovered the two squares marked **XX** in the adjacent figure.) Try to explain why you should or should not be able to arrange the dominoes in this way.

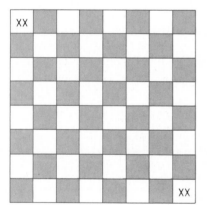

25. Note the following relationships.

$$\frac{1}{1 \times 2} = \frac{1}{1} - \frac{1}{2} \qquad \frac{1}{2 \times 3} = \frac{1}{2} - \frac{1}{3} \qquad \frac{1}{3 \times 4} = \frac{1}{3} - \frac{1}{4}$$

Use this pattern to confirm the answer found in Example 1 of this section.

26. Find the sum.

$$\frac{1}{2} + \frac{1}{2^2} + \frac{1}{2^3} + \frac{1}{2^4} + \cdots + \frac{1}{2^n}$$

To help you discover this sum, complete these partial sums and search for a pattern.

(a) $\dfrac{1}{2} + \dfrac{1}{2^2} = \dfrac{1}{2} + \dfrac{1}{4} = ?$

(b) $\dfrac{1}{2} + \dfrac{1}{2^2} + \dfrac{1}{2^3} = \dfrac{1}{2} + \dfrac{1}{4} + \dfrac{1}{8} = ?$

(c) $\dfrac{1}{2} + \dfrac{1}{2^2} + \dfrac{1}{2^3} + \dfrac{1}{2^4} = \dfrac{1}{2} + \dfrac{1}{4} + \dfrac{1}{8} + \dfrac{1}{16} = ?$

*27. Try to discover a rule that is being used in each case to obtain the answer given. For example, the given information

$$2, 5 \rightarrow 6 \qquad 3, 10 \rightarrow 12 \qquad 7, 8 \rightarrow 14 \qquad 5, 3 \rightarrow 7$$

should lead you to the rule $x, y \rightarrow x + y - 1$.

(a) $3, 4 \rightarrow 9$ $3, 3 \rightarrow 8$ $1, 5 \rightarrow 8$ $2, 8 \rightarrow 12$
(b) $2, 4 \rightarrow 9$ $3, 5 \rightarrow 16$ $1, 7 \rightarrow 8$ $3, 9 \rightarrow 28$
(c) $1, 5 \rightarrow 1$ $5, 2 \rightarrow 2$ $3, 9 \rightarrow 3$ $6, 5 \rightarrow 5$
(d) $4, 8 \rightarrow 4$ $5, 1 \rightarrow 5$ $6, 6 \rightarrow 6$ $8, 2 \rightarrow 8$
(e) $3, 4 \rightarrow 3$ $4, 4 \rightarrow 2$ $5, 5 \rightarrow 0$ $3, 2 \rightarrow 5$

28. In how many different ways can you make change for a 50-cent coin without using any pennies? Use the strategy of making a list.

29. From the adjacent figure remove three segments so as to form three squares of the same size. In this case, a trial-and-error procedure is appropriate.

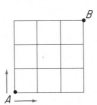

*30. In the diagram the segments represent one-way streets north (up) and east (right). How many different routes are possible for traveling from point A to point B? (*Hint:* Consider the strategy of reducing the situation to a simpler problem.)

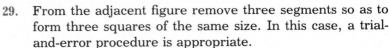

Explorations

Consider a triangular array of circles as shown in the figure. We are asked to place the numbers 1 through 9 in the circles so that the same sum is obtained for the four numbers along each side of the figure. In addition, we would like to obtain the smallest possible sum in figure (a) and the largest possible sum in figure (b).

To obtain the smallest sum it seems logical to place the smallest numbers (1, 2, 3) in the corner circles since they will each be added along two different directions. Similar reasoning

* An asterisk preceding an exercise indicates that the exercise is more challenging than the others.

        1 AN INTRODUCTION TO PROBLEM SOLVING

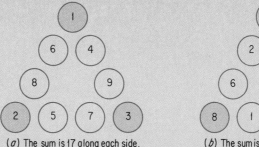

(a) The sum is 17 along each side.

(b) The sum is 23 along each side.

indicates that we should place the largest numbers (7, 8, 9) in the corner circles to obtain the largest sum. The remaining numbers are placed in the circles by a trial-and-error procedure to obtain the results

The following explorations provide guidelines for trying to obtain triangular arrays with sums 17, 18, 19, 20, 21, 22, and 23, that is, for all numbers from the smallest possible to the largest. As an aid to finding patterns that might be useful, copy and extend this table as you complete the explorations.

Sum along each side	17	18	19	20	21	22	23
Sum of numbers in corners	6						24

1. Construct a triangle with the sum of 19 along each side. (*Hint:* The sum of the numbers in the corners will be 12.)

2. Repeat Exploration 1 for the sum of 21. (*Hint:* The sum of the numbers in the corners will be 18.)

3. Study the entries in the table for the triangles completed thus far. What appears to be the pattern for the numbers that represent the sums of the corner numbers?

4. Look at the apparent pattern in the table. What do you conjecture for the sum of the corner numbers for a triangle whose sum will be 20 along each side? Construct such a triangle. Try to find more than one solution. Very often there may be several different solutions possible for a problem.

5. Unfortunately, it is impossible to construct triangles with the sums of 18 or 22 along each side. Experiment with a few cases to convince yourself that such is the case. Although this may shock some readers, it is well to note that not every problem in mathematics can be solved!

6. Read *Problem Solving in School Mathematics*, the 1980 Yearbook of the National Council of Teachers of Mathematics. Prepare a report on three of the chapters of the book.

1-2
Problem Solving with Arithmetic Patterns

A search for patterns is often a significant way to solve problems or to establish conjectures that can lead to the solution of problems. Furthermore, many individuals find the exploration of patterns to be one of the most fascinating aspects of the study of mathematics. Patterns appear in many branches of mathematics, especially in the study of arithmetic. We begin by exploring one of the first tables that a student encounters in elementary arithmetic, a table of *counting numbers* through 100.

Before reading on, see how many different patterns you can discover by yourself in this number table.

1	2	3	4	5	6	7	8	9	10
11	12	13	14	15	16	17	18	19	20
21	22	23	24	25	26	27	28	29	30
31	32	33	34	35	36	37	38	39	40
41	42	43	44	45	46	47	48	49	50
51	52	53	54	55	56	57	58	59	60
61	62	63	64	65	66	67	68	69	70
71	72	73	74	75	76	77	78	79	80
81	82	83	84	85	86	87	88	89	90
91	92	93	94	95	96	97	98	99	100

There are many different patterns that can be found in this table; we will explore a few of these and suggest others in the exercises at the end of this section. For example, the first ten counting numbers are on the horizontal line at the top (the first row). The *multiples* of 10 are on the vertical line at the right (the last column). Can you locate the multiples of 11? These begin with the first entry in the second row and continue down to the right:

11,   22,   33,   44,   55,   66,   77,   88,   99

STRATEGY: Search for a pattern.

Do you see why the multiples of 11 are on a line through the first entry in the second row and extending down diagonally to the right? As you move to a position one place to the right of a number in the table you are effectively adding one to the number; as you move one space down, you are adding 10.

14   1 AN INTRODUCTION TO PROBLEM SOLVING

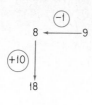

Next consider the pattern of multiples of 9. The multiples of 9 are on a line through 9 in the first row and extending down diagonally to the left:

$$9, \quad 18, \quad 27, \quad 36, \quad 45, \quad 54, \quad 63, \quad 72, \quad 81$$

Each movement one space to the left on the table has the effect of subtracting 1 from a number; each movement down adds 10 to a number. The adjacent diagrams show why the multiples of 9 are on the designated line.

Before leaving the number table, let us explore several more patterns. For example, consider any rectangular array of numbers and find the sums of the numbers in opposite corners. Here are a few examples.

⑥③	64	65	⑥⑥
73	74	75	76
⑧③	84	85	⑧⑥

$$63 + 86 = 149$$
$$66 + 83 = 149$$

④⑥	47	④⑧
56	57	58
⑥⑥	67	⑥⑧

$$46 + 68 = 114$$
$$48 + 66 = 114$$

STRATEGY: Make an educated guess or conjecture. To establish a proof, generalize by using $x, x + 1, x + 2$, and so on.

Do you think that the sums of the numbers in the opposite corners of such rectangular arrays will *always* be equal? Try a few more of your own and then conjecture your answer. Finally, see if you can offer a reasonable *proof* for a particular array that you select.

Earlier in this section we discovered a diagonal line in the number table that contained the multiples of 9. We now consider these multiples again in a vertical array.

Before reading on, see what patterns you can find in the column of multiples on the right.

$$1 \times 9 = 9$$
$$2 \times 9 = 18$$
$$3 \times 9 = 27$$
$$4 \times 9 = 36$$
$$5 \times 9 = 45$$
$$6 \times 9 = 54$$
$$7 \times 9 = 63$$
$$8 \times 9 = 72$$
$$9 \times 9 = 81$$

Most people speak of "adding digits," "subtracting from the units digit," and so on, as we have done. Many teachers are more precise in their terminology and recognize that digits are *numerals,* that is, symbols for numbers, rather than numbers. Numerals can be written. Only numbers can be added or subtracted. However, unless the more precise terminology is needed to avoid major confusion, we use the commonly accepted phraseology.

You may observe that the sum of the digits in each case is always 9. You should also observe that the units digit decreases (9, 8, 7, . . .), whereas the tens digit increases (1, 2, 3, . . .). What lies behind this pattern?

Consider the product

$$5 \times 9 = 45$$

This is another explanation for the appearance of the multiples of 9 along a diagonal line in the number table shown at the start of this section.

To find 6×9 we need to add 9 to 45. Instead of adding 9, we may add 10 and subtract 1.

$$
\begin{array}{rr}
45 & 55 \\
+\ 10 & -\ 1 \\
\hline
55 & 54 = 6 \times 9
\end{array}
$$

That is, by adding 1 to the tens digit, 4, of 45, we are really adding 10 to 45. We then subtract 1 from the units digit, 5, of 45 to obtain 54 as our product.

The number 9, incidentally, has other fascinating properties. Of special interest is a procedure for multiplying by 9 on one's fingers. For example, to multiply 9 by 3, place both hands together as in the adjacent figure, and bend the third finger from the left. The result is read as 27.

The next figure shows the procedure for finding the product 7×9. Note that the seventh finger from the left is bent, and the result is read in terms of the tens digit, to the left, and the units digit to the right of the bent finger. (Note that a thumb is considered to be a finger.) What number fact is shown in the figure on the right?

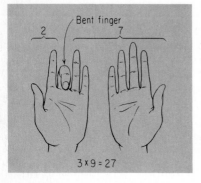

$3 \times 9 = 27$

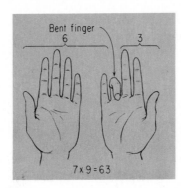

$7 \times 9 = 63$

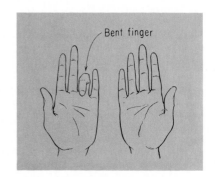

Here is one more pattern related to the number 9.

Use a calculator to verify that each of these statements is correct.

$$
\begin{array}{rl}
1 \times 9 + 2 = & 11 \\
12 \times 9 + 3 = & 111 \\
123 \times 9 + 4 = & 1{,}111 \\
1{,}234 \times 9 + 5 = & 11{,}111 \\
12{,}345 \times 9 + 6 = & 111{,}111
\end{array}
$$

Try to find a correspondence of the number of 1's in the number symbol on the right with one of the numbers used on the

1 AN INTRODUCTION TO PROBLEM SOLVING

left. Now see if, without computation, you can supply the answers to the following questions.

$$123{,}456 \times 9 + 7 = ?$$
$$1{,}234{,}567 \times 9 + 8 = ?$$

Let us see *why* this pattern works. To do so we shall examine just one of the statements. A similar explanation can be offered for each of the other statements. Consider the statement

$$12{,}345 \times 9 + 6 = 111{,}111$$

Note here that we discover a pattern, extend it to other cases, and then attempt to verify the pattern through use of a specific case.

We express 12,345 as a sum of five numbers.

11,111
1,111
111
11
1
12,345

We multiply each of these five numbers by 9.

$$11{,}111 \times 9 = 99{,}999$$
$$1{,}111 \times 9 = 9{,}999$$
$$111 \times 9 = 999$$
$$11 \times 9 = 99$$
$$1 \times 9 = 9$$

Finally, we add 6 by adding six 1's as in the following array, and find the total sum.

$$99{,}999 + 1 = 100{,}000$$
$$9{,}999 + 1 = 10{,}000$$
$$999 + 1 = 1{,}000$$
$$99 + 1 = 100$$
$$9 + 1 = 10$$
$$1 = 1$$
$$111{,}111$$

Study the pattern in the following table. Then try to write the next four lines without doing any computation. Finally, verify your entries by completing the necessary multiplications; use a calculator if one is available.

$$1 \times 1 = 1$$
$$11 \times 11 = 121$$
$$111 \times 111 = 12{,}321$$
$$1{,}111 \times 1{,}111 = 1{,}234{,}321$$
$$11{,}111 \times 11{,}111 = 123{,}454{,}321$$

Do you think that the pattern displayed will continue indefinitely? Compute the product $1{,}111{,}111{,}111 \times 1{,}111{,}111{,}111$ to help you answer this question.

Germany honored this famous mathematician by producing a stamp in his honor. Find out which mathematician has been pictured on a U.S. postage stamp.

Interesting discoveries can often be made by studying arithmetic patterns. A famous German mathematician by the name of Carl Gauss (1777–1855) is said to have been a precocious child who would often drive his teachers to despair. The story is told that on one occasion his teacher asked him to add a long column of figures, hoping to keep him suitably occupied for some time. Instead, young Gauss recognized a pattern, and gave the answer immediately. He is said to have found the sum of the first 100 counting numbers, as indicated in the following array.

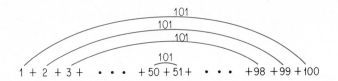

Gauss reasoned that there would be 50 pairs of numbers, each with a sum of 101 (consider $100 + 1$, $99 + 2$, $98 + 3$, . . . , $50 + 51$). Thus the sum is 50×101, that is, 5050.

Example   Use a method similar to that of Gauss to find the sum of the first 200 counting numbers.

Solution   Consider the sum $1 + 2 + 3 + \cdots + 198 + 199 + 200$. The sum of the first and last numbers is 201, and there will be 100 pairs of numbers, each with this sum. Thus the total sum is 100×201, that is, 20,100.

We conclude this section with the following problem. Suppose that you are offered a job for a month of 30 days with the option of being paid $1000 per day, or at the rate of 1¢ for the first day, 2¢ for the second day, then 4¢, 8¢, 16¢, and so forth. That is, your salary would double each day. Which option would you choose?

Of course, one could list the 30 daily salaries and add, but this would be a tedious job indeed. Let us, rather, rely on a very helpful tool used in problem solving: when possible, consider a similar but smaller task first. Thus we will consider total salaries in cents for five, for six, for seven, and for eight days first.

STRATEGY: Solve a simpler but related problem. This problem-solving strategy will often provide you with the clues to solve the more difficult problem.

FOR 5 DAYS	FOR 6 DAYS	FOR 7 DAYS	FOR 8 DAYS
1	1	1	1
2	2	2	2
4	4	4	4
8	8	8	8
<u>16</u>	16	16	16
31	<u>32</u>	32	32
	63	<u>64</u>	64
		127	<u>128</u>
			255

Do you see a pattern emerging? Compare the total for the first five days with the salary for the sixth day; compare the total for six days with the salary for the seventh day. Notice that the total for five days is one cent less than the salary for the sixth day; the total for six days is one cent less than the salary for the seventh day; and so forth. Thus your total salary for ten days will be one cent less than your salary for the eleventh day. That is,

$$1 + 2 + 4 + 8 + 16 + 32 + 64 + 128 + 256 + 512 = 2(512) - 1 = 1023$$

Your salary for the eleventh day would be 1024¢; in 10 days you will earn a *total* of 1023¢, or $10.23.

To use this approach to answer the original question will require a good deal more work, but will still be easier than the addition of 30 amounts. By a doubling process we first need to find your salary for the thirtieth day. Your total salary for all 30 days can then be found by doubling this amount (to find your salary for the thirty-first day), and subtracting 1.

To generalize, we note that the doubling process starting with 1 gives powers of 2.

$$2^0 = 1 \qquad 2^1 = 2 \qquad 2^2 = 4 \qquad 2^3 = 8 \qquad 2^4 = 16 \qquad 2^5 = 32 \qquad . . .$$

Then, by our discovery of a pattern, we may say that

$$2^0 + 2^1 + 2^2 + 2^3 + \cdots + 2^n = 2^{n+1} - 1$$

When $n = 5$, we have

$$2^0 + 2^1 + 2^2 + 2^3 + 2^4 + 2^5 = 1 + 2 + 4 + 8 + 16 + 32 = 2^6 - 1 = 63$$

To find your total salary for 30 working days, we need to find the sum

$$2^0 \ + \ 2^1 \ + \ 2^2 + \ \cdots \ + 2^{29}$$

1st day   2nd day   3rd day   30th day

From the preceding discussion we know that this sum is equal to $2^{30} - 1$. We can use a calculator, can compute 2^{30} by a doubling process, or can use a shortcut such as the following.

$$2^{10} = 2 \times 2^9 = 2 \times 512 = 1024$$

Alternatively, we may say that

$$2^{10} = 2^5 \times 2^5 = 32 \times 32 = 1024$$

In a similar manner we compute

$$2^{30} = 2^{10} \times 2^{10} \times 2^{10} = 1024 \times 1024 \times 1024 = 1,073,741,824$$

and thus $2^{30} - 1 = 1,073,741,823$. In 30 days you would earn a total of \$10,737,418.23, which is substantially more than the \$30,000 that you would earn at \$1000 per day.

EXERCISES

In the exercises the reader will have opportunities to solve problems and make discoveries through careful explorations of patterns.

1. Verify that the process for finger multiplication shown in this section will work for each of the multiples of 9 from 1×9 through 9×9. For each product, count from the left to the bent finger and specify its position as first, second, third, and so on.

2. Follow the procedure outlined in this section and show that
$$1234 \times 9 + 5 = 11,111$$

3. Study the following pattern and use it to express the squares of 6, 7, 8, and 9 in the same manner.
$$1^2 = 1$$
$$2^2 = 1 + 2 + 1$$
$$3^2 = 1 + 2 + 3 + 2 + 1$$
$$4^2 = 1 + 2 + 3 + 4 + 3 + 2 + 1$$
$$5^2 = 1 + 2 + 3 + 4 + 5 + 4 + 3 + 2 + 1$$

4. Study the entries that follow and use the pattern that is exhibited to complete the last four rows.
$$1 + 3 = 4, \quad \text{that is,} \quad 2^2$$
$$1 + 3 + 5 = 9, \quad \text{that is,} \quad 3^2$$
$$1 + 3 + 5 + 7 = 16, \quad \text{that is,} \quad 4^2$$
$$1 + 3 + 5 + 7 + 9 = ?$$
$$1 + 3 + 5 + 7 + 9 + 11 = ?$$
$$1 + 3 + 5 + 7 + 9 + 11 + 13 = ?$$
$$1 + 3 + 5 + \cdots + (2n - 1) = ?$$

1 AN INTRODUCTION TO PROBLEM SOLVING

5. An addition problem can be checked by a process called **casting out nines.** To do this, you first find the sum of the digits of each of the addends (numbers that are added), divide by 9, and record the remainder. Digits may be added again and again until a one-digit remainder is obtained. The sum of these remainders is then divided by 9 to find a final remainder. This should be equal to the remainder found by considering the sum of the addends (the answer), adding its digits, dividing the sum of these digits by 9, and finding the remainder. Here is an example.

ADDENDS	SUM OF DIGITS	REMAINDERS
4,378	22	4
2,160	9	0
3,872	20	2
1,085	14	5
11,495		11

When the sum of the remainders is divided by 9, the final remainder is 2. This corresponds to the remainder obtained by dividing the sum of the digits in the answer (1 + 1 + 4 + 9 + 5 = 20) by 9.

Try this procedure for several other examples and verify that it works in each case.

6. Try to discover a procedure for checking multiplication by casting out nines. Verify for several cases that this procedure works.

7. There is a procedure for multiplying a two-digit number by 9 on one's fingers provided that the tens digit is smaller than the units digit. The accompanying diagram shows how to multiply 28 by 9. Reading from the left, put a space after the second finger and bend the eighth finger. Read the product in groups of fingers as 252. Use this procedure to find

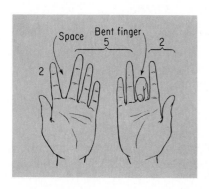

(a) 9 × 47   (b) 9 × 39   (c) 9 × 18   (d) 9 × 27. Check each of the numerical answers that you have obtained.

8. John offered to work for 1¢ the first day, 2¢ the second day, 4¢ the third day, and so on, doubling the amount each day. Bill offered to work for $30 a day. Which boy would receive more money for a job that lasted
(a) 10 days   (b) 15 days   (c) 16 days?

9. Using a method similar to that of Gauss, find
(a) The sum of the first 90 counting numbers.
(b) The sum of the first 300 counting numbers.
(c) The sum of all the odd numbers from 1 through 39.
(d) The sum of all the odd numbers from 1 through 299.
(e) The sum of all the even numbers from 2 through 500.

10. Use the results obtained in Exercise 9 and try to find a formula for the sum of
    (a) The first n counting numbers, that is,
$$1 + 2 + 3 + \cdots + (n - 1) + n$$

    (b) The first n odd numbers, that is,
$$1 + 3 + 5 + \cdots + (2n - 3) + (2n - 1)$$

For Exercises 11 *through* 15 *use the number table given at the beginning of this section.*

11. Consider any square array of nine numbers. Find the sum of the numbers and compare this with the number in the center of the array. Form a conjecture and test this by using several other such square arrays.

12. Draw a rectangle around any four numbers in a row. Compare the sum of the two outer numbers with the sum of the two inner numbers. State what you notice about these sums and check your conjecture with several other sets of four numbers in a row.

13. Repeat Exercise 12 for four numbers in a column.

14. Repeat Exercise 12 for four numbers along any diagonal.

15. Make a new table by replacing each of the two-digit numbers in the given table, except those that are multiples of 10, by the fraction that is the ratio of the first digit to the second. For example, the third row becomes

$$\frac{3}{1} \quad \frac{3}{2} \quad \frac{3}{3} \quad \frac{3}{4} \quad \frac{3}{5} \quad \frac{3}{6} \quad \frac{3}{7} \quad \frac{3}{8} \quad \frac{3}{9}$$

Consider these fractions and test the following conjectures for any rectangular array of such fractions.
    (a) The sums of the fractions in opposite corners are equal.
    (b) The products of the fractions in opposite corners are equal.

16. Write any three-digit number, such as 123. Repeat the digits to form a six-digit number, such as 123,123. Now divide this number successively by 7, by 11, and by 13. Repeat this procedure for several other three-digit numbers and state a discovery that you find. Then try to show why this works. (*Hint:* Find the product $7 \times 11 \times 13$.)

17. Find the number of different ways that you can add four odd counting numbers to obtain a sum of 10. For this exercise, a sum such as $7 + 1 + 1 + 1$ is considered to be

the same as the sum $1 + 1 + 7 + 1$. Use the strategy of making a list.

PERFECT SQUARES	PERFECT CUBES
$1 \times 1 = 1$	$1 \times 1 \times 1 = 1$
$2 \times 2 = 4$	$2 \times 2 \times 2 = 8$
$3 \times 3 = 9$	$3 \times 3 \times 3 = 27$
. . . .	. . . .

1	1	1
3	3	3
5	5	5
7	7	7

*18. Find the smallest counting number such that when that number is divided by 2 the result is a perfect square and when it is divided by 3 the result is a perfect cube. (A *perfect square* is a number such as 1, 4, 9, 16, . . .; a *perfect cube* is a number such as 1, 8, 27, 64, ...) For this exercise use a judicious trial-and-error approach.

*19. In the array shown in the margin, circle three numbers whose sum is 22.

*20. Obtain a true statement by changing the position of one digit only: $102 + 1 = 101$. (No other symbols are to be moved.)

Explorations

Consider the following table of multiplication facts from 1×1 through 9×9.

X	1	2	3	4	5	6	7	8	9
1	1	2	3	4	5	6	7	8	9
2	2	4	6	8	10	12	14	16	18
3	3	6	9	12	15	18	21	24	27
4	4	8	12	16	20	24	28	32	36
5	5	10	15	20	25	30	35	40	45
6	6	12	18	24	30	36	42	48	54
7	7	14	21	28	35	42	49	56	63
8	8	16	24	32	40	48	56	64	72
9	9	18	27	36	45	54	63	72	81

Try to find as many patterns as possible in the table. Here are several to get you started.

1. The entries in the diagonal from upper left to lower right are all perfect squares: 1, 4, 9, 16, 25, 36, 49, 64, 81. Explain why this is so.

2. The units digit for the multiples of 2 repeat: 2, 4, 6, 8, 0, 2, 4, 6, 8. Find other examples of such repetitions.

3. Note the sum of the digits for the multiples of 3: 3, 6, 9, 3, 6, 9, 3, 6, 9. Explore the sums of digits for other multiples in the table.

4. As in these two examples select any square array of four numbers from the given table of multiplication facts. Find the cross products for the selected array as in the adjacent figure. Explain why such cross products will always be equal.

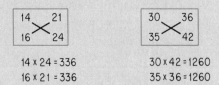

$$14 \times 24 = 336$$
$$16 \times 21 = 336$$
$$30 \times 42 = 1260$$
$$35 \times 36 = 1260$$

5. Extend the concept of Exploration 4 for the cross products of the corner entries in any rectangular array obtained from the given multiplication table.

6. For each column consider the respective entries in any two rows as the numerators and denominators of fractions. Then for rows 2 and 3 we obtain these fractions.

$$\frac{2}{3} \quad \frac{4}{6} \quad \frac{6}{9} \quad \frac{8}{12} \quad \frac{10}{15} \quad \frac{12}{18} \quad \frac{14}{21} \quad \frac{16}{24} \quad \frac{18}{27}$$

Show that for any two rows, not necessarily successive ones, a set of *equivalent fractions* is obtained. Explain why this is so.

7. Find the sum of all of the entries within the body of the table of multiplication facts. (*Hint:* Begin with several smaller tables first. Compare the sum of the entries within the table with the sum of the numbers across the top or side of the table, as in these two examples.)

This is another example of the problem-solving strategy of using a simpler but related problem.

X	1	2
1	1	2
2	2	4

Sum of entries = 9
Sum of numbers used as row (column) headings = 3

X	1	2	3
1	1	2	3
2	2	4	6
3	3	6	9

Sum of entries = 36
Sum of numbers used as row (column) headings = 6

1-3
Problem Solving with Geometric Patterns

For several years one of the most popular puzzles on the market has been Rubik's Cube, shown in the figure, an ingenious device of 27 cubes joined so that the set of nine cubes on each face may be rotated about its center. The object of the puzzle is to produce a cube with each of the six faces containing cubes of the same color. It is said that there are more than 43 quintillion possible positions! Thus trial- and-error procedures for solving this puzzle are not feasible. There are books that provide procedures for solving the puzzle, and some teenagers are able to produce a solution in less than a minute.

Although a trial-and-error approach to problem solving may be useful in forming conjectures, it is often more productive to use a small number of specific examples and search for some apparent clues or patterns. Consider, for instance, the problem of determining the number of triangles that can be formed from a given convex polygon by drawing all possible diagonals from a given vertex P. Any one of the vertices of the polygon may be selected as vertex P. First we draw some figures and present the results in tabular form.

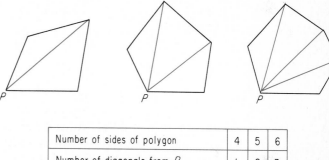

Number of sides of polygon	4	5	6
Number of diagonals from P	1	2	3
Number of triangles formed	2	3	4

From the pattern of entries in the table, it appears that the number of triangles formed is two less than the number of sides of the polygon. Thus we can expect that we can form ten triangles for a *dodecagon,* a polygon with 12 sides, by drawing all possible diagonals from any given vertex. Similarly, for a polygon with n sides, called an n-gon, we expect that we can form $n - 2$ triangles. This is reasoning by *induction.* We formed a generalization on the basis of some specific examples and an obvious pattern.

Note that the type of reasoning by induction shown here is a powerful method for making discoveries, but does *not* constitute a proof.

Next let us consider the problem of finding the total number of possible diagonals that may be drawn in a polygon. Again we first draw some figures and consider the results for possible patterns of answers.

 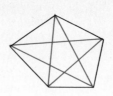 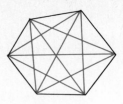

Number of sides of polygon	4	5	6
Total number of diagonals	2	5	9

No pattern appears to be evident. Therefore, we shall examine the number of diagonals in one of the figures, the hexagon, and then attempt to generalize our conclusions for a polygon of n sides.

From the vertex A of the hexagon $ABCDEF$ we can draw three diagonals, AC, AD, and AE as shown in the adjacent figure. We cannot draw diagonals from A to A, from A to B, or from A to F.

Do you see that three diagonals can be drawn from *each* vertex of the hexagon? This might lead us to conclude, *incorrectly*, that a total of 6×3, that is, 18, diagonals can be drawn. However, the diagonal drawn from A to C is the same line segment as the diagonal that can be drawn from C to A. In other words, if three diagonals are drawn from *each* vertex then each diagonal will be drawn twice. Thus the number of possible diagonals in the hexagon is $\frac{1}{2}(6 \times 3)$, that is, 9.

Let us try to generalize this discovery for an n-gon. Consider the vertex A_2 of a polygon $A_1A_2A_3A_4 \ldots A_n$ of the accompanying figure. From A_2 we *cannot* draw diagonals to A_2, to A_1, or to A_3.

Since an n-gon has n sides and n vertices, we draw only $n - 3$ diagonals from each vertex. That is, we can draw a diagonal from a given vertex to each vertex except the given one and its two adjacent vertices. Therefore, we may draw a total of $n \times (n - 3)$ diagonals, not necessarily distinct. As in the case of the hexagon, each diagonal will be drawn twice under this procedure. Therefore, the total number of possible diagonals in an n-gon is given by the formula

$$\frac{1}{2}n \times (n - 3), \quad \text{that is,} \quad \frac{n(n - 3)}{2}$$

If we test this formula for the case of the hexagon, $n = 6$, we have

$$\frac{6(6 - 3)}{2} = \frac{6(3)}{2} = 9$$

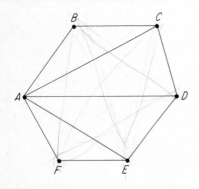

Problem solving in mathematics often consists of generalizing on the basis of several examples. Alternatively, we move from specific cases to a general situation. Here we explore the case for a hexagon, and then consider an n-gon, a polygon with n sides.

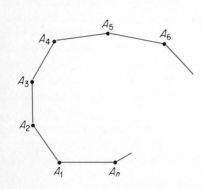

as originally stated. Does it appear that $n(n - 3)$ is always divisible by 2 for n greater than 3? Try to explain why this should be so.

The next figure shows a 24-sided polygon with all of its 252 diagonals. For $n = 24$ the formula for the number of diagonals is

$$\frac{24 \times (24 - 3)}{2} = 12 \times 21 = 252$$

Note that this figure, which has an appearance of including concentric circles, has been drawn with straight lines only. Draw a polygon with 24 sides by locating these points on a circle. Use a protractor and mark off arcs of 15° each; $360 \div 24 = 15$. Then connect the points and draw all of the diagonals to reproduce the figure shown.

Example 1 A basketball team consists of five members. Each member shakes hands with every other member of the team before the game starts. How many handshakes will there be in all?

Solution Surprisingly, this problem can be related to the preceding discussion of polygons. Consider a *pentagon* (a polygon of five sides) with each vertex representing a member of the team.

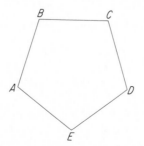 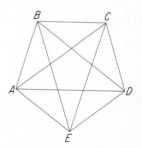

This problem is solved here through use of a diagram. It can also be solved by making a list. Thus the possible handshakes can be shown by listing all possible pairs of letters.

AB AC AD AE
BC BD BE
CD CE
DE

A handshake between two members of the team is represented by a line segment between two vertices, that is, either a diagonal or a side of the polygon. Thus there will be ten handshakes in all.

Example 2 Find the total number of diagonals and sides for a polygon of n sides.

Solution Let us consider two approaches to this problem.

(a) In the preceding discussion we found that the total number of diagonals in an n-gon is given by the formula $n(n - 3)/2$. To this we can add the number of sides, n, to obtain a formula that gives the total number of diagonals and sides as

$$\frac{n(n - 3)}{2} + n$$

(b) Consider an n-gon and note that from each of the n vertices we may draw segments to each of the remaining $n - 1$ vertices. Thus we may draw a total of $n \times (n - 1)$ segments, not necessarily distinct. Since each segment will be drawn twice under this procedure, the total number of distinct segments (diagonals and sides) may be given as

Try to show algebraically that
$$\frac{n(n - 3)}{2} + n = \frac{n(n - 1)}{2}$$

$$\frac{n(n - 1)}{2}$$

A convenient problem-solving approach is to use a general formula to solve a specific problem, as shown in Example 3.

Example 3 Find the total number of diagonals and sides for a polygon with ten sides.

Solution Use the formulas developed in Example 2. Note that both give the same result.

(a) Use $\dfrac{n(n - 3)}{2} + n$ for $n = 10$:

$$\frac{10(10 - 3)}{2} + 10 = \frac{70}{2} + 10 = 45$$

(b) Use $\dfrac{n(n - 1)}{2}$ for $n = 10$:

$$\frac{10(10 - 1)}{2} = \frac{90}{2} = 45$$

1 AN INTRODUCTION TO PROBLEM SOLVING

Patterns offer an opportunity to make reasonable guesses, but these conjectures need to be proved before they can be accepted with certainty. Consider, for example, the maximum number of nonoverlapping regions into which a circular region can be separated by line segments joining given points of a circle in all possible ways.

It is important to recognize that not all patterns lead to valid generalizations. There are times that an apparent pattern fails to continue beyond a certain point, as illustrated by the accompanying example.

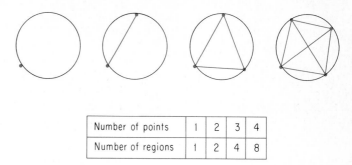

Number of points	1	2	3	4
Number of regions	1	2	4	8

Would you agree that a reasonable guess for the number of such regions derived from five points is 16? Draw a figure to confirm your conjecture. What is your guess for the maximum number of regions that can be derived from six points? Again, draw a figure to confirm your conjecture and count the number of regions formed. You may be in for a surprise!

EXERCISES

1. Take a piece of paper and fold the paper in half as in the figure. Then fold it in half again and cut off a corner that does not involve an edge of the original piece of paper.

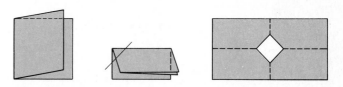

Your paper, when unfolded, should look like the preceding sketch. That is, with two folds we produced one hole. Repeat the same process but this time make three folds before cutting off an edge. Try to predict the number of holes that will be produced. How many holes will be produced with four folds? With n folds?

2. We wish to color each of the pyramids in the accompanying figure so that no two of the faces (sides and base) that have a common edge are of the same color.

Triangular
pyramid

Rectangular
pyramid

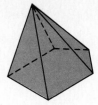

Pentagonal
pyramid

(a) What is the smallest number of colors required for each pyramid?

*(b) What is the relationship between the smallest number of colors required and the number of faces of a pyramid?

3. Consider the following set of figures. In each figure we count the number V of vertices, the number A of arcs, and the number R of nonoverlapping regions into which the figure separates the plane. A square, for example, has four vertices, four arcs, and separates the plane into two regions (inside and outside the square). See if you can discover a relationship between V, R, and A that holds for each case. Confirm your generalization by testing it on several other figures.

	V	R	A
□	4	2	4
◻	4	3	5
▨	5	5	8
☆	10	7	15

4. What is the largest number of pieces that can be obtained by cutting straight through an orange once? Twice? Three times? Four times?

5. Consider the problem of arranging sets of squares that are joined along their edges. In general, these figures are referred to as **polyominoes.** A single square, a **monomino,** can be arranged in only one way. Two squares, a **domino,** can also be arranged in only one way, since the position of the 2 × 1 rectangle does not affect the arrangement. Any two figures that have the same shape and size are **congruent figures.** There are two distinct arrangements for a **tromino,** that is, three squares.

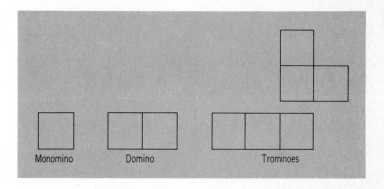

6. Next consider a **tetromino,** a set of four squares. Here we find that there are five possible arrangements such that no two are congruent. As in the case of the domino, we exclude any rearrangement that merely consists of a rotation which places the same squares in a congruent figure. Here are three of the five possible tetrominoes. Draw the two remaining tetrominoes.

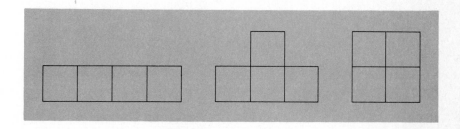

6. Consider a polyomino formed with five squares, called a **pentomino.** There are 12 such arrangements possible. Try to draw all 12. If necessary, cut out five square regions and make different arrangements of them.

7. Copy the given figure and form a figure consisting of two, triangles by removing three of the line segments.

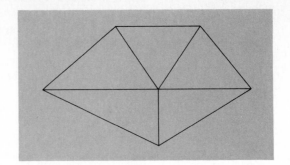

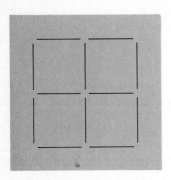

8. Twelve matchsticks are arranged to form the adjacent figure. By removing only two of the matchsticks form a figure that consists of two squares.

9. A square ornamental pool has a tree planted at each of its four corners. Show that the pool can be enlarged to twice its original size without replanting the trees, without having the trees surrounded by water, and without changing the shape of the pool.

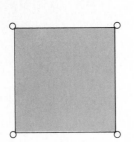

10. Twenty matchsticks are arranged to form the following figure. By rearranging only three matchsticks, form a new figure that consists of five congruent squares, that is, squares of exactly the same size.

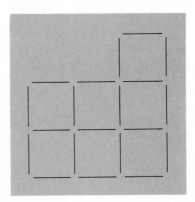

11. The **four-color problem** was proposed in 1853 by a student at University College, London: Can every map in the plane be colored with four colors? Any two countries with common

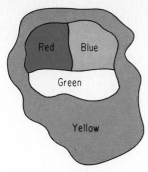

Four colors are required
for this map

boundaries must have different colors. Two countries with only single points in common may have the same color. Mathematicians searched for a solution for this problem for more than 100 years. In 1976 two mathematicians finally used computer analyses of nearly 2000 cases and 10 billion logical alternatives to decide that four colors would always be sufficient.

(a) Draw a map of five countries that requires only three colors.

(b) Draw a map of five countries that requires only two colors.

In Exercises 12 and 13 assume that you have a large collection of congruent circular disks, each painted one of the four colors red, yellow, blue, and green. The disks are to be placed flat (nonoverlapping) on a plane surface so that any two disks that touch are painted in different colors.

12. Sketch an arrangement of six disks such that each disk touches at least two others and only two colors are required.

13. Sketch an arrangement of (a) three disks, and (b) ten disks, for which three colors are required.

14. Copy the given figure and label each disk with one of the letters R, Y, B, G for the colors. Use as few colors as possible under the restriction that any two disks that touch must have different colors.

(a)

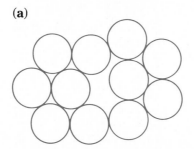

(b)

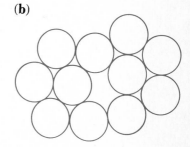

*15. Begin with a large sheet of newspaper and fold it in half to form two layers. Then fold it in half again to form four layers. Repeat this process to form eight layers; and so on.

(a) Try this experiment with different sizes of paper and determine the maximum number of times it is possible to make such folds.

(b) Assume that you could continue this process 50 times and that the original paper was 0.003 inch thick. Approximately how thick (or high) would the final pile be? First estimate the answer. Then use a calculator, if available, to obtain a better approximation.

16. A three-inch cube is painted red and then cut into 27 one-inch cubes. How many of these cubes will have red paint on **(a)** 0 faces **(b)** 1 face **(c)** 2 faces **(d)** 3 faces?

17. Repeat Exercise 16 for a four-inch cube cut into 64 one-inch cubes.

*18. Use the results of Exercises 16 and 17 to begin the following table. Extend it as shown and generalize for an n-inch cube.

Size of cube in inches	Number of cubes with red paint on :			
	0 faces	1 face	2 faces	3 faces
3				
4				
5				
6				
n				

*19. You have an unlimited supply of cubes, a pail of blue paint, and a pail of red paint. Each face of each cube is to be painted either all blue or all red. How many distinguishable (different-appearing) cubes can be painted under these conditions?

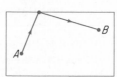

*20. One ball is on a pool table in position A of the adjacent figure, and another is in position B. Find approximately the point on the edge e of the table at which the ball at A must hit in order to rebound and hit the ball at B. To answer this question, note that the desired path from A to the edge of the table to B will be the shortest possible such path. Thus solve this problem by using a ruler and measuring different paths, such as those suggested by these figures.

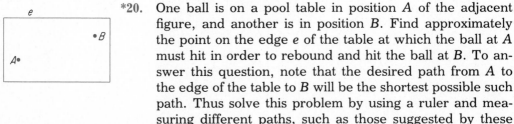

*21. What is the largest number of regions into which a circle can be divided using five chords? To solve this problem, form a table that answers this question for one, two, three, and four chords and conjecture a pattern for obtaining the largest number of regions. As an example of the possibilities, here are several diagrams that show regions in a circle for three chords.

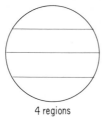

4 regions

6 regions

7 regions

Explorations

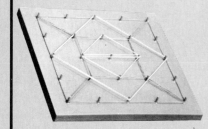

A **geoboard** is a square array of nails or pegs around which rubber bands may be stretched. Effective use can be made of clear plastic models on the overhead projector for demonstration purposes. If there are an insufficient number of geoboards available for classroom use, students may use dot paper or graph paper. There is a formula, known as **Pick's formula,** that relates the area of an enclosed region to the number of points involved. See if you can discover the formula from the following explorations.

1. For each of the following figures, count the number of points b in the boundary of the figure. Then compute the area A in terms of square units as in the given table. Finally, conjecture a formula for A in terms of b.

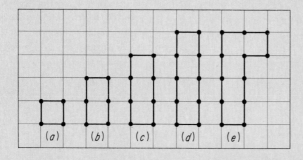

(a)   (b)   (c)   (d)   (e)

	(a)	(b)	(c)	(d)	(e)
Number (b) of boundary points	4	6	8	10	12
Area (A)	1	2			

2. Each of the figures that follow have boundary points b and interior points i. Compute the area A of each figure in

terms of square units as in the given table. Finally, conjecture a formula for A in terms of b and i.

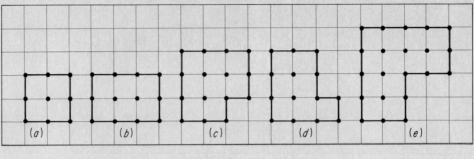

	(a)	(b)	(c)	(d)	(e)
Number (b) of boundary points	8	10	12		
Number (i) of interior points	1	2	3		
Area (A)	4	6			

1-4
Problem Solving with Algebraic Patterns

The use of algebra to generalize specific patterns is a powerful approach to problem solving. Consider, for example, the table of counting numbers that was introduced in Section 1-2. We can use algebra to show why the sums of the numbers in opposite corners of a rectangular array will always be equal. Consider any rectangular array in the table and use n to represent the first number shown. If we use an array of three rows and four columns, then the numbers in the first row can be represented as n, $n + 1$, $n + 2$, and $n + 3$. Also, each number in the second row will be 10 more than the corresponding number in the first row, and each number in the third row will be 20 more than the corresponding number in the first row.

n	$n + 1$	$n + 2$	$n + 3$
$n + 10$	$n + 11$	$n + 12$	$n + 13$
$n + 20$	$n + 21$	$n + 22$	$n + 23$

Note that we have used algebraic facts here to establish that a particular pattern will always hold. That is, we have used algebra as a means of proof.

Now consider the sum of the numbers in opposite corners.

$$n + (n + 23) = 2n + 23 \qquad (n + 3) + (n + 20) = 2n + 23$$

Since these sums are equal, we have *proved* that the sums of numbers in opposite corners will *always* be equal for any rectangular array of three rows and four columns. Try to prove that such sums will be equal for any rectangular array of four rows and three columns.

Example

Consider a square array of nine numbers from the table of counting numbers in Section 1-2. Show that the positive difference of the products of the numbers in opposite corners is always constant, and state what that constant is.

Solution

First we use a specific case to form a conjecture.

(23)	24	(25)	$23 \times 45 = 1035$
33	34	35	$25 \times 43 = 1075$
(43)	44	(45)	$1075 - 1035 = 40$

Note that a single *counterexample* is sufficient to show that a pattern does not hold, whereas any number of specific examples is insufficient to prove that a particular pattern is always true.

At this point we conjecture that the difference will always be 40. It is now important to test this conjecture on another sample; if the difference found is *not* 40, then we have a *counterexample* showing our conjecture to be false and there is no need to test that conjecture further. The reader should show that the difference of products is 40 for another square array. Finally, we can prove our conjecture by generalizing and finding the difference of products for this general case.

n	$n + 1$	$n + 2$	$n(n + 22) = n^2 + 22n$
$n + 10$	$n + 11$	$n + 12$	$(n + 2)(n + 20) = n^2 + 22n + 40$
$n + 20$	$n + 21$	$n + 22$	$(n^2 + 22n + 40) - (n^2 + 22n) = 40$

Clearly the difference is 40, and will *always* be 40 for such square arrays of nine numbers. What will the difference be for a rectangular array of four rows and three columns?

Many magic tricks and "mind-reading" activities can be explained through the use of simple algebraic techniques. Consider as an example the "think of a number" type of mathemagical trick.

Think of a number.

Add 3 to this number.

Multiply your answer by 2.

Subtract 4 from your answer.

Divide by 2.

Subtract the number with which you started.

If you follow these instructions carefully, your answer will always be 1, regardless of the number with which you start. We can explain why this trick works by using algebraic symbols or by drawing pictures, as shown below.

Think of a number:	n	☐	(Number of coins in a box)
Add 3:	$n+3$	☐ ○ ○ ○	(Number of original coins plus three)
Multiply by 2:	$2n+6$	☐ ○ ○ ○ ☐ ○ ○ ○	(Two boxes of coins plus six)
Subtract 4:	$2n+2$	☐ ○ ☐ ○	(Two boxes of coins plus two)
Divide by 2:	$n+1$	☐ ○	(One box of coins plus one)
Subtract the original number, n:	$(n+1)-n=1$	○	(One coin is left)

Try to make up a similar trick of your own and use algebra to explain why the trick works.

Now let us try to "build" a trick together. We begin by forming a square array and placing any six numbers in the surrounding spaces as in the figure.

+	3	4	1
7			
2			
5			

The reader should try to reproduce this trick, using sets of numbers different from the ones shown.

The numbers 3, 4, 1, 7, 2, and 5 are chosen arbitrarily. Next find the sum of each pair of numbers as in a regular addition table.

+	3	4	1
7	10	11	8
2	5	6	3
5	8	9	6

Now we are ready to perform the trick. Have someone circle any one of the nine numerals in the box, say 10, and then cross out all the other numerals in the same row and column as 10.

+	3	4	1
7	(10)	11	8
2	5	6	3
5	8	9	6

Next circle one of the remaining numerals, say 3, and repeat the process. Circle the only remaining numeral, 9. The sum of the circled numbers is 10 + 3 + 9 = 22, as shown in the figure.

+	3	4	1
7	(10)	11	8
2	5	6	(3)
5	8	(9)	6

The interesting item here is that the sum of the three circled numbers will always be equal to 22, regardless of where you start! Furthermore, note that 22 is the sum of the six numbers outside the square. Try to use algebra to explain why this trick works, and then build a table with 16 entries.

EXERCISES

For Exercises 1 through 4 use a table of counting numbers as in Section 1-2. Assume that the grid shown in each diagram is placed on top of the number table so that each box of the grid encloses exactly one number in the table.

$n-11$	$n-10$	$n-9$
$n-1$	n	$n+1$
$n+9$	$n+10$	$n+11$

1. Select any square array of nine numbers from that table and prove that the sum of the numbers will be equal to nine times the center number. (*Hint:* Let the center number be n. Then the array of nine numbers will be as in the adjacent figure.)

2. For the arrays of numbers represented by the following diagrams, prove that the sum of all 11 numbers will be equal to 11 times the number n in the center.

(a)

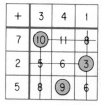

(b)

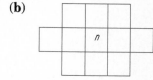

39

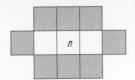

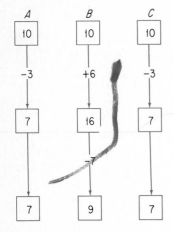

3. Prove that the sum of the numbers in the shaded regions will be equal to eight times the number n in the center for any such array in a number chart.

4. Consider the following magic trick. Have someone place an equal number of pennies (or other similar objects) in each of three positions, such as in spaces A, B, and C in the adjacent figure. For example, assume that 10 coins are placed in each position. Then provide these instructions.

   i. Take three coins from each of the positions A and C and add them to the coins in position B.

   ii. Count the number of coins in position A. Remove that number from position B.

   Start with a different number of coins and perform these same two steps. Show that the final number in position B again will be 9. Then prove that this number will always be 9. (*Hint:* Begin with n coins in each position.)

Use a calendar for any month of the year as a table of numbers.

5. Select any square array of nine numbers from the table and explain why the sums of the numbers in opposite corners must be equal.

6. Repeat Exercise 5 for a rectangular array of three rows and four columns.

7. Select any square array of nine numbers from the table, explain why each positive difference of the products of the numbers in opposite corners is the same constant, and identify that constant.

8. Repeat Exercise 7 for a rectangular array of three rows and four columns.

State the outcome of each of the following tricks. Use algebraic phrases to explain why each trick works as it does.

9. Think of a number; add 7 to this number; multiply by 3; subtract 21; divide by 3; subtract the number with which you started.

10. Think of a number; multiply this number by 6; add 12; divide by 3; subtract 4; divide by the number with which you started.

*11. Select any four-digit number, with each of its digits different, such as 3274. Form all possible *cyclic numbers* by keeping the digits in the same sequence. For example, consider the digits as on a clock and begin with a different digit each time as shown.

Possible cyclic numbers

3274

2743

7432

4327

Find the sum of the four numbers and divide by the sum of the four digits. For this example, the sum is 17,776, the sum of the four digits is 16, and the quotient $17{,}776 \div 16 = 1111$. Repeat this procedure for a different four-digit number. Then prove that the final result will always be 1111. (*Hint:* Use $1000a + 100b + 10c + d$ to represent the given number.)

*12. Assume that you give a friend a book of 20 matches and that, without telling you any of the numbers, the friend follows these instructions:

Tear out and discard any number from 1 to 9 of matches.

Count the number N of matches that remain in the book.

Add the decimal digits of N. Let the sum be S.

Tear out and discard S more matches from the book.

Finally, burn, one at a time, any number from 1 to 9 of the matches that remain in the book.

You count the number of matches that are burned and state the number that are then left in the book of matches. Explain how you can tell the number of matches that are left in the book.

*13. Try the following card trick several times. Then explain mathematically why the trick always works.

First place a predetermined card, such as the ace of spades, as the 21st card from the top of a deck of cards. Then ask a friend to select any number n from 1 through 10 and to remove that number of cards from the top of the deck. Next you spread out the next 20 cards in a row face down. Finally, you ask your friend to count cards backward to n, the number originally selected by that person. You then announce, without seeing the card, that this card is the ace of spades, the predetermined card.

*14. Verify that each of the following statements is correct.

$$(1 + 2)^2 = \qquad 1^3 + 2^3 = \qquad 9$$
$$(1 + 2 + 3)^2 = \qquad 1^3 + 2^3 + 3^3 = \qquad 36$$
$$(1 + 2 + 3 + 4)^2 = 1^3 + 2^3 + 3^3 + 4^3 = 100$$

The general pattern (formula) is

$$(1 + 2 + 3 + \cdots + n)^2 = 1^3 + 2^3 + 3^3 + \cdots + n^3 = \frac{n^2 (n + 1)^2}{4}$$

Show that this formula holds for $n = 5$.

Explorations

In this set of explorations we shall examine the ways in which any counting number (1, 2, 3, 4, 5, . . .) can be represented as the difference of squares of whole numbers (0, 1, 2, 3, . . .). Note, for example, the following.

$$5 = 3^2 - 2^2 \qquad (5 = 9 - 4)$$
$$7 = 4^2 - 3^2 \qquad (7 = 16 - 9)$$

Several interesting questions can be raised about such relationships.

(a) Can every counting number be represented as the difference of the squares of two whole numbers?

(b) Is there any pattern to those representations that exist?

The answers to these questions will be considered in the explorations that follow.

1. Represent each of the odd numbers from 1 through 15 as a difference of two squares. (Note that the answers for 5 and 7 are given above.) Study the pattern of answers and conjecture a manner in which any odd counting number can be written as a difference of squares.

2. An odd counting number can be represented by the form $2n + 1$ where n is a whole number. For example: when $n = 0$, $2n + 1 = 1$; when $n = 1$, $2n + 1 = 3$; when $n = 3$, $2n + 1 = 7$; and so on. Now consider the adjacent partial table of odd numbers $2n + 1$ and their representations as differences of squares. It seems that for every whole number n, the odd number $2n + 1$ can be written as the difference $(n + 1)^2 - n^2$. Prove that this will be true in all such cases by simplifying the expression $(n + 1)^2 - n^2$.

n	$2n + 1$	$(n+1)^2 - n^2$
0	1	$1^2 - 0^2$
1	3	$2^2 - 1^2$
2	5	$3^2 - 2^2$
3	7	$4^2 - 3^2$

n	$4n$	$(n+1)^2 - (n-1)^2$
1	4	$2^2 - 0^2$
2	8	
3	12	
4	16	

3. Any counting number that is a multiple of 4 can be represented as $4n$ where n is a counting number. Copy and complete the adjacent table for the multiples of 4, 8, 12, and 16. Use several additional examples to test this conjecture: For every counting number n, the number $4n$ can be written as the difference $(n + 1)^2 - (n - 1)^2$.

4. Prove the conjecture expressed in Exploration 3 by simplifying the algebraic expression given there.

5. Try to represent several of the remaining even numbers (2, 6, 10, 14, . . .) as a difference of two squares. What do you conclude?

6. Find a magic card trick that you can demonstrate to the class. Explain the mathematical basis for the trick. For information on card tricks, see the many books by Martin Gardner, as well as books on tricks found in most magic shops.

Chapter 1 Review

Solutions to the following exercises may be found within the text of Chapter 1. Try to complete each exercise without referring to the text.

Section 1-1 Problem-Solving Strategies

1. Find the sum

$$\frac{1}{1 \times 2} + \frac{1}{2 \times 3} + \frac{1}{3 \times 4} + \cdot \cdot \cdot + \frac{1}{9 \times 10}$$

Describe the problem-solving strategy that you use to find the sum.

2. List all the possible sets of three counting numbers whose product is 72.

3. How many squares of all sizes are there on a standard checkerboard? Describe a problem-solving strategy that can be used to answer this question.

Section 1-2 Problem Solving with Arithmetic Patterns

4. Consider the table of counting numbers and explain why the multiples of 11 are on a line through the first entry in the second row.

5. Describe two patterns concerning the multiples of 9.

6. Show how Carl Gauss is said to have found the sum of the first 100 counting numbers.

7. Use a method similar to that of Gauss to find the sum of the first 200 counting numbers.

8. State a rule that tells how to find the sum of consecutive powers of 2 beginning with 2^0. Then use the rule to find the sum $2^0 + 2^1 + 2^2 + 2^3 + 2^4 + 2^5$.

Section 1-3 Problem Solving with Geometric Patterns

9. Copy and complete this table.

Number of sides of polygon	4	5	6
Number of diagonals from a given vertex			
Number of triangles formed			

10. State a formula for the total number of possible diagonals in an n-gon.

11. A basketball team consists of five members. Each member shakes hands with every other member of the team before the game starts. How many handshakes will there be in all?

12. Find the total number of diagonals and sides for a polygon of n sides.

13. Find the total number of diagonals and sides for a polygon with ten sides.

Section 1-4 Problem Solving with Algebraic Patterns

14. For any rectangular array of three rows and four columns from the table of counting numbers in Section 1-2, prove that the sums of numbers in opposite corners will always be equal.

15. Consider a square array of nine numbers from the table of counting numbers in Section 1-2. Show that the positive difference of the products of the numbers in opposite corners is always constant, and state what that constant is.

Chapter 1 Test

1. Find the sum:

$$\frac{1}{1 \times 2} + \frac{1}{2 \times 3} + \frac{1}{3 \times 4} + \cdots + \frac{1}{49 \times 50}$$

Explain the method you used to obtain your answer.

2. Use a method similar to that of Gauss to find the sum of the first 150 counting numbers. Use a diagram to illustrate your procedure.

3. Repeat Exercise 2 for the sum of the even counting numbers from 2 through 200, that is,

$$2 + 4 + 6 + \cdots + 198 + 200$$

4. Consider a table of counting numbers in rows of ten (as in Section 1-2) and describe five different patterns that can be observed in that table. Select one of these patterns and explain why it works.

5. Repeat Exercise 4 for a table of multiplication facts from 1×1 through 9×9.

6. Make a table of the counting numbers 1 through 60 in rows of twelve. Explain why the sums of the numbers in opposite corners of any rectangular array of four rows and five columns from the table must be equal.

7. Note the sums of these powers of 2.

$$2^0 + 2^1 + 2^2 = 1 + 2 + 4 = 7$$
$$2^0 + 2^1 + 2^2 + 2^3 = 1 + 2 + 4 + 8 = 15$$
$$2^0 + 2^1 + 2^2 + 2^3 + 2^4 = 1 + 2 + 4 + 8 + 16 = 31$$

(a) Describe a pattern for finding such sums without actual addition.

(b) Use this pattern to find the sum $1 + 2 + 4 + 8 + \cdots + 256$.

8. Find the total number of diagonals in an *octagon,* a polygon of eight sides.

9. At a committee meeting the eight members present each shake hands with every other member. Describe two different mathematical procedures for determining the total number of handshakes that take place.

10. Note the sums in this table of sums of counting numbers.

$$1 = 1$$
$$1 + 2 = 3$$
$$1 + 2 + 3 = 6$$
$$1 + 2 + 3 + 4 = 10$$

(a) Describe a pattern for the sums 1, 3, 6, 10.

(b) Use this pattern to predict the sums that would be obtained if the table were continued for three more rows.

Calculators and Computers

2

2-1
Calculators

Calculators are very useful tools for problem solving. Indeed, calculators have now become a part of the way of life in our society. They enable us to complete most routine computations very quickly and with very little effort.

To use a calculator first turn it on and then press the keys to give your instructions to the calculator. Consider the subtraction problem

$$2.1278 - 1.3544 = 0.7734$$

The answer is shown in the display of the calculator in the photograph. This calculator has an $\boxed{=}$ key and performs arithmetic operations in the same manner as in algebra. This use of algebraic procedures is indicated by specifying that the calculator uses an *algebraic logic*. Instructions may be given to such calculators in the order that we would ordinarily read the instructions.

The keys to be pressed and the resulting display as 2.1278 is entered in the calculator are shown in this array.

Press: ON $\boxed{2}$ $\boxed{\cdot}$ $\boxed{1}$ $\boxed{2}$ $\boxed{7}$ $\boxed{8}$

Display: 0. 2. 2. 2.1 2.12 2.127 2.1278

The $\boxed{\cdot}$ key is pressed to fix the decimal point in that position. We shall hereafter assume that the calculator has been turned on and indicate the instructions for entering a number simply by listing the number. Then for the subtraction problem under consideration the instructions and displays would be as in the next array.

Instructions: 2.1278 $\boxed{-}$ 1.3544 $\boxed{=}$

Displays: 2.1278 2.1278 1.3544 0.7734

Does the answer shown on the calculator in the photograph have any special significance to you? If not, turn the page upside down and look again at the answer on the display of the calculator.

In ordinary arithmetic we do multiplication and division first in the order in which they occur and then we do addition and subtraction. For example,

$$2 + 3 \times 5 = 2 + 15 = 17$$
$$6 \div 2 + 1 = 3 + 1 = 4$$
$$6 + 4 \times 5 + 2 \times 8 = 6 + (4 \times 5) + (2 \times 8) = 6 + 20 + 16 = 42$$

On most calculators the matching of numerals and inverted letters is

0	1	3	4	5	7	8
0	I	E	h	S	L	B

Make up some arithmetic problems that can be used to send messages to your friends.

Press 5 $\boxed{+}$ 2 $\boxed{\times}$ 3 $\boxed{=}$ in that order on your calculator. Compare your result with the results obtained on other calculators. What should the answer be?

Parentheses may be used to specify, or to emphasize, the order in which operations are to be performed.

$$(5 + 2) \times 3 = 7 \times 3 = 21$$
$$5 + (2 \times 3) = 5 + 6 = 11$$

Most inexpensive calculators use the last x procedure.

Since there are many different types of calculators, we must study the instruction booklet for each calculator to determine the precise sequence of instructions that should be used for that calculator.

A calculator that does multiplication and division first has an **algebraic operating system** and would produce these displays for the problem $6 + 4 \times 5 + 2 \times 8$.

Instructions:	6	$\boxed{+}$	4	$\boxed{\times}$	5	$\boxed{+}$	2	$\boxed{\times}$	8	$\boxed{=}$
Displays:	6	6	4	4	5	26	2	2	8	42

Calculators that do not have the algebraic operating system perform operations on the last number in the display when the next operating instruction is given. Such a calculator uses the **last x procedure** and would produce these displays.

Instructions:	6	$\boxed{+}$	4	$\boxed{\times}$	5	$\boxed{+}$	2	$\boxed{\times}$	8	$\boxed{=}$
Displays:	6	6	4	10	5	50	2	52	8	416

Many calculators have a key $\boxed{\text{M+}}$ that is used to add the number in the display to the number in the memory and a key $\boxed{\text{MR}}$ for bringing the number in the memory to the display. The preceding problem can be done correctly on a calculator with a memory by doing the multiplications first and using instructions such as these.

Instructions:	4	$\boxed{\times}$	5	$\boxed{=}$	$\boxed{\text{M+}}$	2	$\boxed{\times}$	8	$\boxed{=}$	$\boxed{\text{M+}}$	$\boxed{\text{MR}}$	$\boxed{+}$	6	$\boxed{=}$
Displays:	4	4	5	20	20	2	2	8	16	16	36	36	6	42

If a calculator does not have a memory, then for problems such as this we must make our own record of numbers that are to be remembered.

Example 1 Restate each expression so that the new expression can be evaluated using the last x procedure to obtain the correct result.
(a) $10 + 2 \times 7$ (b) $29 + 3 \times 5 + 6$ (c) $5 - 7 + 12 \div 4$

Solution (a) $(2 \times 7) + 10$ (b) $(3 \times 5) + 29 + 6$ (c) $(12 \div 4) + 5 - 7$

In each case the parentheses are desirable to emphasize the order of the operations.

Calculators often have special keys for engineers, accountants, people doing advanced mathematics, or people with other special needs. We'll consider a few additional keys.

$\boxed{^{+}\!/\!_{-}}$ is used to change the sign of a number x. (This key may be marked $\boxed{-x}$.)

$\boxed{^{1}\!/\!_{x}}$ is used to find the **reciprocal** of any number x that is different from zero.

$\boxed{x^2}$ is used to find $x \cdot x$, that is, the **square** of any number x.

$$\sqrt{x} \cdot \sqrt{x} = x$$
$$\sqrt{9} = 3 \quad \text{since} \quad 3 \times 3 = 9$$
$$\sqrt{144} = 12 \quad \text{since} \quad 12 \times 12 = 144$$

$\boxed{\sqrt{x}}$ is used for positive numbers x to find the **square root** of x.

An expression such as 17^3 means $17 \times 17 \times 17$ and can be evaluated as $(17 \times 17) \times 17 = 4913$ using a calculator. In general, for any number m from the set 2, 3, 4, 5, . . . an expression of the form b^m is called a **power** of the **base** b. The number m is the **exponent** and tells how many times the base b is used as a factor in the product. For example,

$$5^2 = 5 \times 5 \qquad 3^4 = 3 \times 3 \times 3 \times 3 \qquad 11^3 = 11 \times 11 \times 11$$

The second power of a number is the *square* of the number. The third power of a number is the **cube** of the number. If a calculator with algebraic logic has a key for the operation $\boxed{y^x}$, the following instructions may be used to evaluate the cube of 11, that is, 11^3.

The use of calculators is encouraged as computations are encountered throughout this book.

Instructions: 11 $\boxed{y^x}$ 3 $\boxed{=}$

Displays: 11   11   3   1331

Example 2   Restate each expression so that the new expression can be evaluated using the last x procedure.
(a) $25 - 3 \times 2$ (b) $19 + 17^3$ (c) $3 + \sqrt{16}$

Solution   (a) $(3 \times 2) \times (-1) + 25$
(b) $(17 \times 17) \times 17 + 19$ or $17^3 + 19$ for a calculator with $\boxed{y^x}$
(c) $\sqrt{16} + 3$

Calculators are particularly useful for checking estimations of the results of computations. The symbol $\approx$ is used for "**is approximately equal to.**"

Example 3   In each of the following estimate the cost to the nearest dollar. Then use a calculator to find the exact cost and check your estimate.

Estimation is an important aspect of problem solving. Even in our ordinary daily activities estimations of results are often useful.

(a) Eleven booklets at 90¢ each.
(b) Twenty-three pictures at $1.05 each.
(c) One-fourth of $79.72.
(d) Two-thirds of $119.40.

Solution   (a) $11 \times 0.90 \approx 10$; $9.90
(b) $23 \times 1.05 \approx 24$; $24.15
(c) $79.72 \div 4 \approx 20$; $19.93
(d) $2 \times (119.40 \div 3) \approx 80$; $79.60

2-1 Calculators

49

The estimations in Example 3 may be considered in cents as "to the nearest hundred." To the nearest hundred 990 is 10 hundred, 2415 is 24 hundred, 1993 is 20 hundred, and 7960 is 80 hundred. These are examples of *rounding off* to the nearest hundred. Similarly, 567 to the nearest hundred is 600 since 567 is between 550 and 650. In the case of 6250 there does not exist a "nearest hundred" since 6250 is halfway between 6200 and 6300. To the nearest thousand, 6250 ≈ 6000.

EXERCISES *Restate each expression so that the new expression can be evaluated using the last x procedure. Then list in order a set of instructions for evaluating the given expression on a calculator with* $\boxed{\sqrt{x}}$ *and* $\boxed{+/-}$ *using the last x procedure. Finally, list the corresponding displays on any such calculator that is available to you.*

1. $7 + 2 \times 21$
2. $25 + 3 \times 4$
3. $12 + 18 \div 3$
4. $6 + 20 \div 5$
5. $20 - 3 \times 5$
6. $14 - 8 \div 2$
7. $5 + 11^2$
8. $3 + 2^3$
9. $36 + \sqrt{64}$
10. $20 - 2\sqrt{25}$
11. $7 + \sqrt{121} - 15$
12. $63 - 2\sqrt{49} - 27$

Estimate to the nearest hundred and use a calculator to check your estimate.

13. $157 + 335$
14. $785 - 591$
15. $1280 - 395$
16. $8268 + 1540$
17. 618×5
18. 1723×4
19. 19×21
20. 4×11^2
21. 7^3
22. $5280 \div 10$
23. $5280 \div 53$
24. $5280 \div 25$

Use a calculator to evaluate each expression. If necessary use pencil and paper to remember numbers that your calculator cannot hold in a memory. A calculator with a $\boxed{\sqrt{x}}$ *key should be used for Exercises 31 through 36.*

25. $13 \times 5 - 8 \times 7$
26. $10 \times 5^2 \div 125$
27. $9 \times 11 + 121 \div 11$
28. $5 \times 7 - 4 \times 8 + 3 \times 9$
29. $8^2 + 12^2 - 13^2$
30. $(29 - 4 \times 7)^2 + 6 \times 8$
31. $\sqrt{3^2 + 4^2}$
32. $\sqrt{15^2 + 20^2}$
33. $\sqrt{20^2 + 40^2 + 40^2}$
34. $\sqrt{31^2 + 31 + 32}$
35. $\sqrt{90^2 + 90 + 91}$
36. $\sqrt{70^2 - 70 - 69}$

Explorations

1. The letters that can be represented by inverting the numerals 0, 1, 3, 4, 5, 7, and 8 may be separated into two words by a decimal point.
    (a) What words are represented by 57108.34?
    (b) Give several arithmetic problems with 57108.34 as the answer.
    (c) Make up a story problem with either the words represented by 57108.34 or the words represented by 57108.345 as the answer to the question.

2. Make up a story problem with the word represented by the numeral 57738 as the answer.

3. List at least three common names of people that can be represented by numerals and make up problems involving these names.

4. Repeat Exploration 3 for at least ten other words and a problem involving at least two of them.

5. Repeat Exploration 4 for two-word sentences.

2-2
Problem Solving with Calculators

Calculators can be particularly useful in problem solving whenever arithmetic computations have a role in a problem-solving strategy.

Example 1   At Juan's college those students who pay a $10.00 fee ride the shuttle bus on campus for a nickel; other students pay a quarter. At least how many rides should be anticipated to make it worthwhile to pay the fee?

Solution
STRATEGY: Make a table.

Many problems can be solved in a variety of ways. Consider the strategy of making a table. Use a calculator as needed.

Number of rides:	1	2	$\cdots$	10	20	40	50	60
Cost with fee:	$10.05	10.10	$\cdots$	10.50	11.00	12.00	12.50	13.00
Cost without fee:	$ 0.25	0.50	$\cdots$	2.50	5.00	10.00	12.50	15.00

Juan should pay the fee only if he expects to ride the bus more than 50 times.

Often in problem solving the work is dramatically reduced if one observes a particular relationship among the quantities involved. For instance, in Example 1 how much does a student who has paid the $10 fee save on each bus ride? How many such savings are needed to make it worthwhile to pay the fee?

STRATEGY: Look for a relationship.

$0.25 - 0.05 = \$0.20$ saving on each ride
$\$10.00 \div \$0.20 = 50$

One must ride the bus 50 times to save the payment of the fee, more than 50 times to make it worthwhile to pay the fee.

Example 2   Sue wishes to determine the miles per gallon of her car on a trip of 1100 miles for which 40 gallons of gasoline were used.

Solution   Do you understand the problem? You might restate the problem as follows.

Always be certain that you understand the problem before you try to find a solution. Often it is helpful to restate the problem in your own words.

Sue used 40 gallons of gasoline to drive 1100 miles. She wants to know the average number of miles for 1 gallon of gasoline.

The pattern for the number of miles per gallon is

STRATEGY: Find the pattern.

$$\frac{\text{number of miles}}{\text{number of gallons}} = \text{number of miles per gallon}$$

$$\frac{1100}{40} = 27.5$$

The number of miles per gallon for Sue's car on the trip was 27.5.

At 27.5 miles per gallon Sue would drive 275 miles on 10 gallons and 1100 miles on 40 gallons; $40 \times 27.5 = 1100$.

Every problem that you solve provides a method for solving many other problems. These similar problems may involve very different situations.

Example 3   Identify one or more relationships and solve each problem. Use a calculator as needed.

STRATEGY: Recognize a related problem.

(a) John ran 5 miles in 40 minutes. Find his rate in miles per hour.

(b) Dot bought $10.00 worth of gasoline at $1.35 a gallon. Her car averages 28 miles per gallon. Can she expect to complete a 195-mile trip without needing to buy more gasoline?

Solution   (a) The rate in miles per hour is similar to the performance of a car in miles per gallon (Example 2). Thus

Since the rate is given in miles per hour, the time must be stated in hours.

$$\frac{\text{number of miles}}{\text{number of hours}} = \text{rate in miles per hour}$$

The time of 40 minutes is 2/3 hour.

$$\frac{5}{2/3} = 5 \times \frac{3}{2} = 7.5 \text{ miles per hour}$$

(b) One strategy would be to determine the number of gallons of gasoline needed (195 ÷ 28), the cost of that gasoline [1.35 × (195 ÷ 28)], and to compare the cost of $9.40 with the $10 paid. Since $9.40 is less than $10, it should not be necessary to buy more gasoline.

Another strategy would be to determine the number of gallons purchased, the distance that could be traveled on these gallons, and to compare this distance with the 195 miles of the trip. For this second strategy the relations are

$$\frac{\text{total cost}}{\text{cost per gallon}} = \text{number of gallons}; \quad \frac{10}{1.35} \approx 7.4$$

(number of gallons) × (miles per gallon) = distance; 28 × 7.4 ≈ 207

Since 207 is greater than 195, it should not be necessary to buy more gasoline.

The practice of observing the patterns of two or more relationships in a single problem, as we did in Example 3(b), is a valuable problem-solving strategy.

Example 4 The ABC Corporation paid $900,000 for 30,000 acres of marshland. The corporation gave 2000 acres to a tax-exempt wildlife organization. What was the cost of the 2000 acres of marshland?

Solution The cost per acre is

$$\frac{\text{total cost}}{\text{number of acres}} = \frac{900,000}{30,000} = \$30$$

Then the cost of 2000 acres is 2000 × 30, that is, $60,000.

Note that you could also think of this problem in terms of thousands of acres. Then the cost per thousand acres is

$$\frac{\text{total cost}}{\text{number of thousands of acres}} = \frac{900,000}{30} = \$30,000$$

and the cost of 2000 acres would be $60,000.

EXERCISES *Identify one or more useful relationships and solve each problem. Use a calculator as needed.*

1. Art ran 5 miles in half an hour. Find his rate in miles per hour.

2. Betty ran 3 miles in 15 minutes. Find her rate in miles per hour.

3. Elena walked 1/2 mile in one-eighth of an hour. Find her rate in miles per hour.

4. Bryan Allen pedaled his aircraft across the English Channel on June 12, 1979. The 23-mile trip took 2 hours and 49 minutes. Find his rate to the nearest tenth of a mile per hour.

5. The 6000-mile Australian Dingo Fence was built to protect sheep from wild dogs (dingos). If the Dingo Fence is 4500 miles longer than the Great Wall of China, how long is the Great Wall?

6. Olaf bought four books for $5. The books all had the same price. What was the price of each book?

7. Inez bought a box of paperback books for $10. She allows her friends to have books for whatever they cost her. If the box contained 100 books, (a) what should Anita pay Inez for seven books? (b) what should Rafael pay for two books?

8. If in Exercise 7 the box contained 75 books, (a) what should Anita pay Inez for six books? (b) what should Rafael pay for 15 books?

9. Gwen bought a $65,000 house, used $5,000 of her savings, had a $51,000 mortgage from a bank, and gave a friend a second mortgage for the balance. What was the amount of the second mortgage?

10. Suppose that a person eats an apple a day. How many apples would the person eat in a week? The month of September? The year 1984?

11. If the population of China is approximately 1,100,000,000 and there are approximately 3.1 people living in other parts of the world for each person living in China, what is the approximate population of the world?

12. At the rate of $5 an hour how much would a person earn for an 8-hour day? For a 40-hour week? For a 50-week year?

13. At the rate of $1 per second could you spend $1,000,000 in two weeks? Justify your answer.

*14. At the rate of $1 per minute could you spend $1,000,000 in 22 months? In two years?

15. NASA's Landsat spacecraft are in polar orbits, that is, they circle the Earth from pole to pole. These satellites are designed to survey the Earth's resources. Computers are needed to organize the data received. Each Landsat cir-

FIRST MAN ON THE MOON   UNITED STATES

cles the Earth about 14 times a day and requires 251 revolutions to scan the entire globe. How many days are required for a Landsat to scan the entire globe?

16. An astronaut on the moon weighs approximately one-sixth as much as on Earth. Find the weight on Earth of a person who weighs 28 pounds on the moon.

17. See Exercise 16 and find the weight on the moon of a person who weighs 192 pounds on Earth.

18. Apollo 7 made 163 revolutions of the Earth in about 260 hours. What was the average number of minutes for one revolution?

19. If the moon circles the Earth in 29.5 days, how many orbits does the moon make in one year?

20. If a humpback whale has 5000 herrings in its stomach, each herring has 7000 shrimp in its belly, and each shrimp has eaten 130,000 single-celled phytoplankton, how many million phytoplankton provided the basis for this whale's meal?

*21. Excluding our sun the nearest star is approximately 26,000,000,000,000 miles away. How many years are required for light traveling 11,160,000 miles per minute to reach Earth from this star?

*22. Bob found that the diesel model of his favorite car cost $1000 more than the gasoline model. Suppose that gasoline costs $1.32 a gallon and diesel fuel costs $1.25 a gallon. The car averages 30 miles per gallon on gasoline and 50 miles per gallon on diesel fuel. Bob expects to drive the car for four years and to be able to obtain $300 extra for the diesel when he trades the car in. If other costs are about the same for the two models, at least how many miles per year must Bob expect to drive if the diesel is to be less expensive to operate than the gasoline model?

*23. Anita is a teenager. Two of her great grandparents are alive and in their eighties. When an uncle asked Anita her age she told him that the square of her age was equal to the sum of the ages of her two living great-grandparents. How old is Anita?

Explorations

1. The first number and the second number of a sequence of numbers are given. The third number is obtained by dividing the first into the sum of the second number and 1. The fourth number is obtained by dividing the second into

the sum of the third and 1. Each number is obtained from its two predecessors by this same procedure. We seek the pattern, if any, in the sequence of numbers obtained. A calculator can be helpful in performing the necessary operations.

If the first number is 2 and the second number is 3, we make the computations

$$2, \quad 3, \quad \frac{3+1}{2} = 2, \quad \frac{2+1}{3} = 1, \quad \frac{1+1}{2} = 1, \quad \frac{1+1}{1} = 2, \quad \frac{2+1}{1} = 3, \quad \ldots$$

and obtain the sequence

$$2, \quad 3, \quad 2, \quad 1, \quad 1, \quad 2, \quad 3, \quad \ldots$$

Continue this sequence until you recognize a pattern in the sequence of numbers obtained.

2. **(a)** Repeat the procedure in Exploration 1 for other sequences such as

$$2, \quad 5, \quad \ldots$$
$$4, \quad 5, \quad \ldots$$
$$5, \quad 2, \quad \ldots$$

until you can identify a common pattern for the sequences obtained.

*(b) Repeat the procedure in Exploration 1 for the sequence

$$a, \quad b, \quad \ldots$$

to show that there must be a common pattern for all sequences obtained using this procedure.

3. The first number and the second number of a sequence are given. The third number is obtained by multiplying the second by 5 and then dividing by the first. The fourth number is obtained by multiplying the third by 5 and dividing by the second. Each number is obtained from its two predecessors by the same procedure. For example, we have the sequence

$$2, \quad 3, \quad \frac{15}{2}, \quad \frac{25}{2}, \quad \frac{25}{3}, \quad \frac{10}{3}, \quad \ldots$$

As in Explorations 1 and 2 find the common patterns, if any, of sequences of numbers obtained using this procedure.

4. In Exploration 3 would the replacement of "multiply by 5"

by "multiply by 7" or any other counting number affect the pattern obtained?

5. Read the introduction, try several of the activities, and report on *How to Develop Problem Solving Using a Calculator* by Janet Morris, National Council of Teachers of Mathematics, 1979.

6. Read and report on a major part of George Polya's *How to Solve It,* Doubleday and Co., 1957.

2-3 Flowcharts

At one time or another each of us has to give detailed instructions of some kind. We may give someone directions for getting to a specific store; we may give a recipe for making a certain kind of cake. Our instructions may be either oral or written. Now that people have to give instructions not only to people but also to calculators and computers, which lack the human ability to bridge even minor gaps, flowcharts are often useful as a means of organizing and conveying precise instructions.

Each flowchart indicates a step-by-step procedure, a sequence of steps. Arrows are used to show the flow of the steps. Different geometric figures are used to identify different types of steps. The following symbols are included in the standards published by the American National Standards Institute.

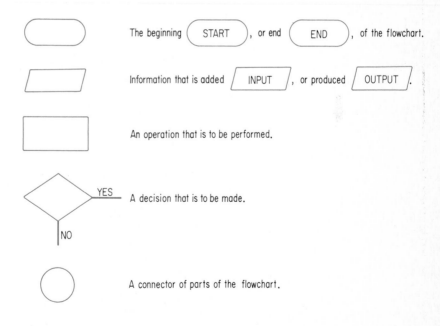

The beginning (START), or end (END), of the flowchart.

Information that is added / INPUT /, or produced / OUTPUT /.

An operation that is to be performed.

A decision that is to be made.

A connector of parts of the flowchart.

Example 1 Find the output for each flowchart.

Other symbols are sometimes used. Flowcharts may be drawn with horizontal flow lines, with vertical flow lines, or with both horizontal and vertical flow lines.

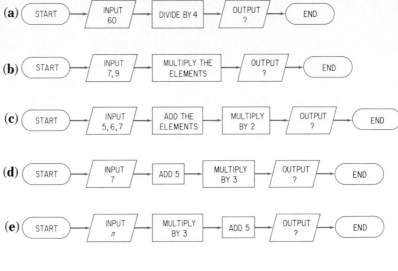

(a) START → INPUT 60 → DIVIDE BY 4 → OUTPUT ? → END

(b) START → INPUT 7, 9 → MULTIPLY THE ELEMENTS → OUTPUT ? → END

(c) START → INPUT 5, 6, 7 → ADD THE ELEMENTS → MULTIPLY BY 2 → OUTPUT ? → END

(d) START → INPUT 7 → ADD 5 → MULTIPLY BY 3 → OUTPUT ? → END

(e) START → INPUT n → MULTIPLY BY 3 → ADD 5 → OUTPUT ? → END

Solution   (a) 15   (b) 63   (c) 44   (d) 36   (e) $3n + 5$

Flowcharts provide a sequence of instructions and often are used over and over again. For example, the flowchart in Example 1(e) may be used for many values of n. If $n = 2$, the output is 11; if $n = 7$, the output is 26; and so on. Many flowcharts also have *loops* (parts) that are used again and again until a specified condition is satisfied. Some flowcharts need flow lines in both horizontal and vertical directions with arrowheads at each right-angle turn.

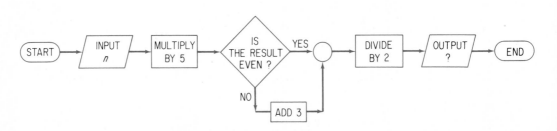

START → INPUT n → MULTIPLY BY 5 → IS THE RESULT EVEN? —YES→ ○ → DIVIDE BY 2 → OUTPUT ? → END
                                                          │NO
                                                          ADD 3

Example 2 Find the output for the given flowchart if
(a) $n = 7$ (b) $n = 12$.
The *even numbers* are 2, 4, 6, 8, 10, 12, ...

Solution   (a) 19; for $n = 7$, $5n = 35$, which is not even, $35 + 3 = 38$,
$38 \div 2 = 19$.
(b) 30; for $n = 12$, $5n = 60$, which is even, $60 \div 2 = 30$.

Example 3   Find the input for the given flowchart.

Solution   The number to which 50 is added to obtain 67 is 17, that is,
$67 - 50 = 17$ and $17 + 50 = 67$.

Example 4   Draw a flowchart to represent the given expression.
    (a) $2n + 1$ (b) $2(n + 1)$

Solution

(a) START → INPUT n → MULTIPLY BY 2 → ADD 1 → OUTPUT $2n+1$ → END

(b) START → INPUT n → ADD 1 → MULTIPLY BY 2 → OUTPUT $2(n+1)$ → END

EXERCISES   *Find the output for each flowchart* **(a)** *if* $n = 87$ **(b)** *if* $n = 215$
(c) *in terms of* n. *Compare the outputs for Exercises 1 and 2;*
compare the outputs for Exercises 3 and 4:

1. START → INPUT n → ADD 500 → SUBTRACT 17 → OUTPUT ? → END

2. START → INPUT n → ADD 483 → OUTPUT ? → END

3. START → INPUT n → MULTIPLY BY 100 → DIVIDE BY 4 → OUTPUT ? → END

4. START → INPUT n → MULTIPLY BY 25 → OUTPUT ? → END

Use the given flowchart and find the output for the given values of n.

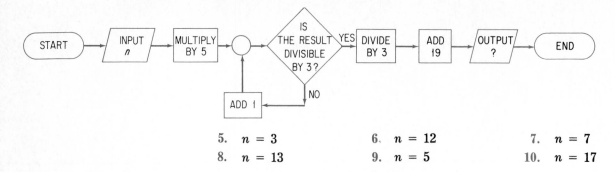

5. $n = 3$	6. $n = 12$	7. $n = 7$
8. $n = 13$	9. $n = 5$	10. $n = 17$

Find the input for the given flowchart.

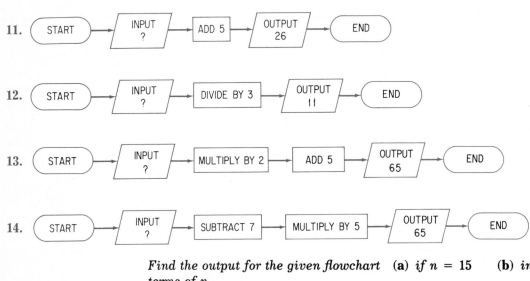

11. START → INPUT ? → ADD 5 → OUTPUT 26 → END

12. START → INPUT ? → DIVIDE BY 3 → OUTPUT 11 → END

13. START → INPUT ? → MULTIPLY BY 2 → ADD 5 → OUTPUT 65 → END

14. START → INPUT ? → SUBTRACT 7 → MULTIPLY BY 5 → OUTPUT 65 → END

Find the output for the given flowchart **(a)** *if* $n = 15$ **(b)** *in terms of n.*

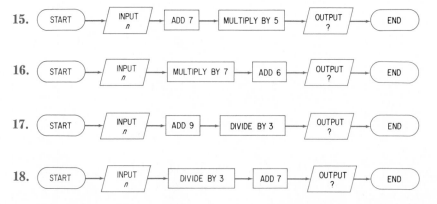

15. START → INPUT n → ADD 7 → MULTIPLY BY 5 → OUTPUT ? → END

16. START → INPUT n → MULTIPLY BY 7 → ADD 6 → OUTPUT ? → END

17. START → INPUT n → ADD 9 → DIVIDE BY 3 → OUTPUT ? → END

18. START → INPUT n → DIVIDE BY 3 → ADD 7 → OUTPUT ? → END

2 CALCULATORS AND COMPUTERS

Draw a flowchart to represent each expression.

19. $3(n - 2)$

20. $3n - 2$

21. $\frac{1}{2}(n + 6)$

22. $\frac{1}{2}n + 6$

23. $3 + n/2$

24. $(3 + n)/2$

Flowcharts can be used to generate a sequence of terms. Find the output of the given flowchart.

25.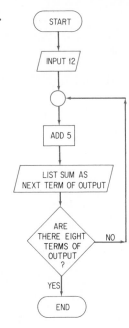

26.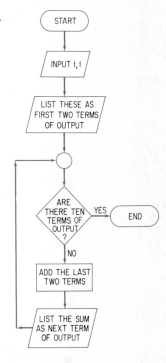

27. The sequence generated in Exercise 26 may be continued indefinitely and is the traditional **Fibonacci sequence.** Other Fibonacci sequences may be obtained using other numbers as the first two terms. Repeat Exercise 26 with input 1, 3.

28. Repeat Exercise 26 with input 3, 4.

*29. Repeat Exercise 26 with input a, b.

30. Find the sum of the ten terms of the sequence obtained in Exercise 27. Compare this sum with 11 times the seventh term.

*31. Repeat Exercise 30 for the *general Fibonacci sequence* in Exercise 29.

Explorations

1. Make a flowchart for solving any equation that is of the form $Ax = B$ where $A \neq 0$.

2. Words such as "bob" and numbers such as "2002" that read the same forward and backward are called **palindromes**. Numbers such as "22" and "55" are **two-digit palindromes**. The following steps may be used to form a palindrome from any given two-digit number.

   Write any two-digit number.

   If the number is not a palindrome, reverse its digits and add the new number to the original number.

   If the sum is not a palindrome, reverse its digits and add the new number to the sum.

   If this sum is not a palindrome, repeat the previous step.

   Make a flowchart for finding a palindrome from any given two-digit number. Check the procedure represented by your flowchart for several two-digit numbers that are not themselves palindromes.

*3. Select a common activity, such as starting an automobile engine or buying an ice cream cone of your favorite flavor, and make a flowchart of the related activites.

*4. Make a flow chart of the essential steps for you to complete this course to your satisfaction.

2-4 Computers

If you use an inexpensive calculator to solve a problem, you must provide each instruction as it is needed. If a computer is used to solve a problem it is often possible to provide all the instructions at the beginning. Under the modern concept of a **computer** it is this ability to perform both arithmetic and logical operations without human interaction that distinguishes computers from calculators.

A computer usually consists of a number of different devices that are interconnected to form a **computer system**. The power of a computer system is based on the speed and accuracy with which it can perform a few elementary operations. These operations include the following.

Input operations such as accepting data and instructions for processing.

Arithmetic operations such as addition, subtraction, multiplication, and division.

Logical operations such as determining whether one number is greater than another.

Output operations such as producing displayed or printed information.

The ENIAC occupied a space 30 feet by 50 feet and weighed 30 tons. It was programmed by connecting various wires and setting up to 6000 switches. These wires and switches had to be changed each time a program was changed.

The most striking recent advances have been with **electronic digital computers,** that is, computers using sets of electronic impulses to represent digits in a system of numeration for numbers. The first large-scale electronic digital computer was called the Electronic Numerical Integrator And Computer (ENIAC). It was completed in 1946, contained 18,000 vacuum tubes, and could perform about 350 multiplications of two numbers in one second. Computers using vacuum tubes are called *first-generation computers.*

The *second-generation computers* used transistors instead of tubes. These computers were faster, smaller, and less expensive than their predecessors. Transistors were invented at Bell Laboratories and the first transistorized computer (TRADIC) was built there in 1954. It has been estimated that there were 244 computers in use in the United States in 1955 and that altogether these 244 computers could do about 250,000 additions per second, that is, produce about the same output as

GROWTH OF APPROXIMATE NUMBER OF COMPUTERS IN THE UNITED STATES

1955	244
1958	2,550
1964	18,200
1965	26,000
1970	100,000
1980	500,000

The most powerful computer system today is the Cray-1 at the National Center for Atmospheric Research in Boulder, Colorado. It is used for analyzing atmospheric and weather information. It can process 80 million instructions per second and can store more than 1 million characters of data in its main storage unit.

Three generations of computer components

one small modern computer. By 1959 the move to the faster and less expensive transistorized computers was well established.

In 1964 IBM introduced its System/360 computers with their controlling circuitry stored on small chips instead of using transistors. These new components could be mass produced at low cost and rarely failed. The System/360 computers could perform 375,000 computations per second at a cost of $3\frac{1}{2}$ cents for each 100,000 computations. A *third generation of computers* arose using this solid logic technology.

The early concept of computers as number processors does not do justice to the wide variety of applications of computers today. Modern computers can be programmed to accept words, even spoken words, and other nonnumerical data. The most rapid growth in the use of computers is in the area of information processing. Wherever large masses of data need to be routinely processed, computers are being increasingly relied upon for doing that processing. Here are a few examples:

airline (railroad, hotel, concert, car rental) reservations
processing over 25 billion checks a year at banks
self-service automated teller terminals at banks
electronic funds transfer systems
computerized grocery checkout stands
credit card identification and records
"electronic mail" between business offices
computerized control of manufacturing processes
computer-assisted instruction in schools

Computers are controlled by human beings and are being used to meet human needs. For example, many people are now typing reports at a computer terminal. The typed material can be read on a video screen. Insertions, deletions, and other corrections can be made. The computer adjusts the spacing and prints all lines the same length unless otherwise instructed. Some computers have a program that will check the spelling of each word, that is, the computer will match words against a file of acceptable words and display words that do not match. This method of checking spelling will not recognize the substitution of one recognizable word for another, such as "too" for "two" or "advice" for "advise."

Instructions and data were communicated to early computers in machine languages in which the activities of the computer were itemized step by step. The use of these languages was very difficult and tedious. Substantial improvements have been made. FORTRAN was developed at IBM and released in 1957. Among the other high-level programming languages COBOL was released in 1960, BASIC in 1965, PL/1 in 1966, and

PASCAL in 1968. More than 200 languages have been developed during the past 30 years. The problem-solving orientation of the developers of these languages is indicated by some of the names selected for the languages,

BASIC was developed in 1965 by John Kemeny and Thomas Kurtz at Dartmouth College for the use of college students.

ALGOL   **ALG**orithmic-**O**riented **L**anguage

BASIC   **B**eginner's **A**ll-purpose **S**ymbolic **I**nstruction **C**ode

COBOL   **CO**mmon **B**usiness-**O**riented **L**anguage

FORTRAN   **FOR**mula **TRAN**slation

The high-level programming languages include many statements that are very similar to those in ordinary arithemetic. Letters are used as in algebra.

Ordinary arithmetic: $a + b$ $a - b$ $a \times b$ $a \div b$ $a < b$
BASIC: **A+B**   **A−B**   **A*B**   **A/B**   **A<B**

Capital letters are used for ease of reading. The symbols for multiplication and division are replaced to avoid confusion with the letter x and the symbol + for addition. The symbol < is read "is less than" in both representations, for example, 3 < 5. Also, as in the next display, parentheses may be used. The symbol ∅ is used for the numeral 0 to avoid confusion with the capital letter O. Notations for powers and roots are replaced so that all symbols may be written on the same horizontal line.

Ordinary arithmetic: $2(3 + 5)$ 50 a^2 b^3 $\sqrt{a}$
BASIC: **2*(3+5)**   **50**   **A↑2**   **B↑3**   **SQR(A)**

Example 1   Write in BASIC.
(a) $5(3 - 17)$ (b) $3^2 + 10$
(c) $54 \div (2^3 - 2)$ (d) $\sqrt{29} \div 3$

Solution   (a) **5*(3−17)**   (b) **3↑2+10**   (c) **54/(2↑3−2)**   (d) **SQR(29)/3**

All computers and some calculators have a memory. The **memory** (storage unit) of a computer can be used for both data (information) and instructions. The units for doing arithmetic operations (arithmetic unit), making decisions *(logic unit)*, and controlling the flow of information and instructions *(control unit)* form a **central processing unit** which is the main part of the computer system.

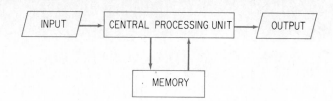

The memory makes it possible for the entire input to be scanned and the operations performed according to established rules. The following sequence of operations is followed in evaluating arithmetic expressions.

Expressions, if any, in parentheses are evaluated first working outward from the inner parentheses.

Then powers and roots, such as squares and square roots, are found.

Then multiplications and divisions are performed in order from left to right.

Finally, additions and subtractions are performed in order from left to right.

This is the same sequence of operations that you have always used in arithmetic. Practice in using these ordinary rules is provided in the exercises.

Example 2 List, in order, the operations that you would perform and evaluate each expression.

(a) $2 + 6 \times 4$ (b) $3 + 5^2$ (c) $54 \div (2^3 - 17)$

Solution (a) $6 \times 4 = 24,\ 2 + 24 = 26$
(b) $5^2 = 25,\ 3 + 25 = 28$
(c) $2^3 = 8,\ 2^3 - 17 = -9,\ 54 \div (-9) = -6$

Example 3 Evaluate.   (a) **(2∗3)↑2**   (b) **5+(12/4)↑3**

Solution (a) $(2 \times 3)^2 = 6^2 = 36$
(b) $5 + (12/4)^3 = 5 + 3^3 = 5 + 27 = 32$

EXERCISES *Use the rules for sequences of operations and evaluate.*

CHALLENGE: Do most of these exercises as mental exercises.

1. $(12 + 6) \div 2$	2. $12 + 6 \div 2$
3. $12 \div 6 \div 2$	4. $50 \div 8 \div 4$
5. $12 \div (6 \div 2)$	6. $50 \div (8 \div 4)$
7. $18 \div (3 \times 2)$	8. $18 \div 3 \times 2$

2 CALCULATORS AND COMPUTERS

9. $54 \div 3^2 - 7$

10. $2 \times 3^2 - 4^2$

11. $54 \div (3^2 - 7)$

12. $2 \times (3^2 - 4)^2$

13. $2 - 2 + 2 - 2 + 2$

14. $2 \div 2 \times 2 \div 2 \times 2$

15. $30 \div 3 \div 3 \times 3 \div 3$

16. $30 - 3 \times 3 - 3 \div 3$

17. $60 \times 4 \div 2 \times 5 \div 3$

18. $60 \div 4 \times 2 \div 5 \times 3$

Write in BASIC and evalute.

19. $5 \times 6 \div 3$

20. $2 + 3^2$

21. $6\sqrt{25 + 24}$

22. $6 - (8 \div 4)^2$

23. $8000 \div 2^5$

24. $75 - 2^6$

List in order the operations that you would perform and evaluate each expression.

25. `6+7*2`

26. `25-3↑2`

27. `6*7/3`

28. `4*5↑2`

29. `12↑2-100`

30. `25+3*2↑3`

31. `5+7*3↑2`

32. `11+3*2↑3`

33. `11+SQR(169)`

34. `SQR(3↑2+4↑2)`

35. `17+SQR(12↑2+5↑2)`

36. `SQR(3600)-1↑17`

Explorations

1. Computers can sort, arrange in a specified manner, and provide printed reports based on data that have been received from a wide variety of sources. Describe briefly one such use of a computer. For example, consider the preparation of student grade reports, salary checks, or periodic inventory reports for a large store, the recording of airline reservations or reservations for a large chain of hotels, or the assignment of college classes to available classrooms.

2. Computers have the ability to read coded instructions and to perform operations very rapidly. Describe briefly one such use of a computer. For example, consider sorting checks for distribution to the banks on which they are drawn, sorting letters according to their zip codes, providing automatic pilots for the control of airplane landings, or providing weather predictions based on reports from a worldwide network of observation stations.

3. Report on the nature of computers. For example, see Chapter 1, "Computers and Computer Programming," of *Introduction to BASIC Programming* by Gary B. Shelly and Thomas J. Cashman, Anaheim Publishing Company, 1982.

4. Report on the historical development of computer systems. For example, see Chapter 2, "The Evolution of the Electronic Computer Industry," of *Introduction to Computers and Data Processing* by Gary B. Shelly and Thomas J. Cashman, Anaheim Publishing Company, 1980.

2-5 BASIC

The selection of line numbers is arbitrary. Numbers 1, 2, 3, . . . could be used but would not allow later insertions. Numbers 10, 20, 30, . . . are frequently used.

The language BASIC is widely used to state problems so that computers may be used in solving the problems. A set of instructions in BASIC consists of a set of lines called **statements**. Each statement starts with a *line number* and a word that indicates the type of statement. If a statement is too long to fit on a single line, it is split into two statements and a second line number is assigned to the part on the second line. A similar procedure is followed for additional lines. The line number serves both as a serial number for ordering the operations of the computer and as a label for the statement. Usually, consecutive numbers are avoided so that additional data (or instructions) may be inserted. New lines, with appropriate numbers, may be added at the end of the program. Lines that contain errors may be replaced simply by adding new lines with the same number as the lines to be replaced. The computer orders and uses the statements according to their serial numbers. The instruction **LIST** may be used by the operator to obtain a list of the lines of the program in their proper order. Then, if the program is satisfactory, the instruction **RUN** may be used to implement the program. The only additional procedure needed to operate the computer would be to satisfy the machine that you were a recognized user.

Many problems involve numerous repetitions of a simple task. A computer can be programmed to perform repetitive tasks any specified number of times or until a specified objective has been accomplished. For example, the steps indicated in the following flowchart may be used to obtain a table of the cubes of the numbers 1 through 100.

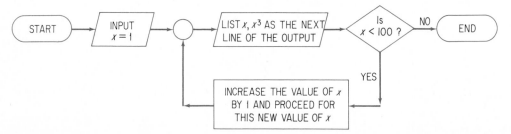

Instructions for a computer are often referred to as *commands*.

These steps may be modified slightly to obtain the following four instructions, called a **program**, which can be used to instruct a computer to produce a table of the cubes of the numbers 1 through 100.

```
10   FOR X=1 TO 100
20   PRINT X, X*X*X
30   NEXT X
40   END
```

The **FOR TO** statement instructs the computer to start with $x = 1$ and to consider in turn the values 1, 2, 3, . . . , 100 for the *variable x*. The **PRINT** statement instructs the computer to start a new line of output, to print the value of x, to leave a space at the position of the comma, and then to print the value of x^3. Either **X*X*X** or **X↑3** may be used for x^3. The printed output will be in two columns. The **NEXT X** statement instructs the computer to return for more data, that is, in this program to return to statement 10 and replace x by $x + 1$. Under this program the computer considers in order the instructions

$$10, \quad 20, \quad 30, \quad 10, \quad 20, \quad 30, \quad . . . , \quad 10, \quad 20, \quad 30, \quad . . .$$

until all the data in statement 10 have been used, that is, until lines 20 and 30 have been processed for $x = 100$. Then the computer proceeds to the next instruction, in this case line 40.

Example 1   State in ordinary notation the printed output for this program.

```
10   FOR X=1 TO 5
20   PRINT X, 2+X
30   NEXT X
40   END
```

Solution

1	3
2	4
3	5
4	6
5	7

The statements used in a BASIC program must be of specific types. These types are identified by the key words used. For example,

FOR X=1 TO 5

is a **FOR TO** statement and

PRINT X, 2+X

is a **PRINT** statement. The types of statements that may be used in BASIC programs also include the following.

DATA	The statement includes a list of numbers or other representations of facts, concepts, or instructions ready for processing.
READ	The computer is to read the first element, or specified number of elements, from the data.
LET	A variable is assigned, or reassigned, as a name for a number or expression.
GO TO	The computer is to go to the indicated line of the program.
IF THEN	A condition is stated and the computer is to go to a specified line of the program if the condition is satisfied. If the condition is not satisfied, the computer goes to the next line of the program.
LIST	The computer is to list the statements of the program in the order of their line numbers.
RUN	The computer, for an acceptable program, is to provide the output of the program.

A **DATA** statement may be useful when the number of elements is not large or is not easily described to the computer. Here is a program for a table of the cubes of the first 20 counting numbers.

```
10   READ X
20   DATA 1, 2, 3, 4, 5, 6, 7, 8, 9, 10
30   DATA 11, 12, 13, 14, 15, 16, 17, 18, 19, 20
40   LET Y=X ↑ 3
50   PRINT X, Y
60   GO TO 10
70   END
```

The printout for this table of values will end with the statement

```
10   OUT OF DATA
```

If the **DATA** statements had consisted of the numbers 1 through 100, a table of the cubes of the numbers 1 through 100 would have been obtained. Thus a table of the cubes of the numbers 1 through 100 may be produced either using **FOR TO** and **NEXT** statements as in the program at the beginning of this section or using **READ, DATA,** and **GO TO** statements as in the preceding program.

A table of cubes may also be produced using **LET** and **IF THEN** statements. In the next program the first **LET** statement assigns the initial value 0 to the variable **X**; the second **LET** statement reassigns the variable **X** to the next counting number. In other words, the computer increases the value of **X** by 1 and renames the sum as **X**, that is, reassigns the variable **X** to the new value. We instruct the computer to do this by writing

    LET X = X + 1

Such a statement cannot be true in ordinary algebra and should not be considered as an algebraic equation. Rather it is an instruction to the computer to rename the quantity **X + 1** as **X**, that is, assign a new value to **X**. The following program instructs the computer to follow the steps that are described at the beginning of this section in the flowchart for obtaining a table of cubes of the numbers 1 through 100.

```
10   LET X = 0
20   LET X = X + 1
30   PRINT X, X ↑ 3
40   IF X<100 THEN 20
50   END
```

We have considered three programs for tables of cubes in order to emphasize that often several different programs may be written to solve the same problem.

The pattern formed by the differences of the squares of the first eight successive whole numbers from the squares of their predecessors is shown in the output of the next program.

PRINTED OUTPUT

`10   READ D`	
`20   DATA 1, 2, 3, 4, 5, 6, 7, 8`	1
`30   LET P = D ↑ 2`	3
`40   LET Q = (D − 1) ↑ 2`	5
`50   LET R = P − Q`	7
`60   PRINT R`	9
`70   GO TO 10`	11
`80   END`	13
`RUN`	15

To test a conjecture regarding the pattern formed by the output of the previous program for some additional numbers, such as the first ten integers greater than one thousand, only the **DATA** statement needs to be changed and the new set of numbers inserted. Computers are frequently used to test conjectures of arithmetic patterns. A practical example of the use of computers by classroom teachers is considered in the following example.

Example 2   Suppose that you have three grades for each of your students and wish to average these grades. Prepare a BASIC program that may be used to do this.

Solution

```
10   READ G1, G2, G3
20   DATA       (List the grades for each student, being sure
                 to list together three grades for each student;
                 use extra lines 21, 22, 23, . . . , as needed.)
60   LET A=(G1+G2+G3)/3
70   PRINT G1, G2, G3, A
80   GO TO 10
90   END
```

The computer may be set so that the three grades and the average of these grades are printed with a separate line for each student.

Languages, such as BASIC, may be used on many different types of computers. A computer of any given type usually has special features that are intended to enhance its usefulness.

Programs for more difficult problems can be introduced after the problems are understood. Numerous problems in algebra, geometry, and especially advanced mathematics are routinely solved on computers throughout the industrialized countries of the world. The breadth of the problems for which computers are used is expanding daily. Most banks and big stores of all sorts use computers extensively. In some cities ordinary citizens use the touch-tone system of their telephones to instruct computers at their banks to pay their bills and complete other financial transactions.

EXERCISES   *State in ordinary notation the printed output for each program.*

1.
```
10   READ X
20   DATA 2, 4, 6, 8
30   PRINT X, X/2
40   GO TO 10
50   END
```

2.
```
10   READ X
20   DATA 3, 5, 7, 9
30   PRINT X, (X−1)/2
40   GO TO 10
50   END
```

3.
```
10   READ X
20   DATA 1, 2, 3, 4
30   PRINT X, X↑3
40   GO TO 10
50   END
```

4.
```
10   READ X
20   DATA 10, 11, 12, 13
30   PRINT X, X↑2
40   GO TO 10
50   END
```

5.
```
10   READ X, Y
20   DATA 1, 2, 3, 4, 5, 6
30   PRINT X, Y, X+Y
40   GO TO 10
50   END
```

6.
```
10   READ X, Y
20   DATA 1, 2, 3, 4, 5, 6
30   PRINT X, Y, X*Y
40   GO TO 10
50   END
```

7.
```
10   READ X, Y, Z
20   DATA 1, 2, 3, 4, 5, 6
21   DATA 7, 8, 9, 10, 11, 12
30   PRINT X, Y, Z, X+Y+Z
40   GO TO 10
50   END
```

8.
```
10   READ X, Y, Z
20   DATA 11, 12, 13, 14, 15, 16
30   PRINT X+Y, X+Z, Y+Z
40   GO TO 10
50   END
```

9.
```
10   FOR X = 10 TO 15
20   PRINT X, X↑2
30   NEXT X
40   END
```

10.
```
10   FOR X = 1 TO 5
20   PRINT X, X*3
30   NEXT X
40   END
```

<table>
<tr><td>11.</td><td>

```
10  FOR X = 1 TO 6
20  LET Y = X + 1
30  PRINT X, Y, X*Y
40  NEXT X
50  END
```
</td><td>12.</td><td>

```
10  FOR X = 1 TO 6
20  LET Y = X + 1
30  LET Z = Y↑2 − X↑2
40  PRINT X, Y, Z
50  NEXT X
60  END
```
</td></tr>
</table>

Assume that the computer has been preset to print answers with five places after the decimal point and write a BASIC program to print the specified data.

If a computer is available, modify your program as necessary and run it on the computer.

13. A table of the counting numbers 1 through 20 with their fifth powers.

14. A table of the counting numbers 1 through 50 with their square roots.

15. A table of the counting numbers 1 through 1000 with their squares and their cubes.

16. A table of the counting numbers 1 through 25 with their squares and their square roots.

17. A table of the counting numbers 1 through 10 with their squares, cubes, and square roots.

18. At 300 miles per hour a plane travels 5 miles each minute, that is, $5n$ miles in n minutes. Write a BASIC program that can be used to print a table of distances traveled at 300 miles per hour in n minutes for counting numbers 1 through 120.

*19. An investment of $1000 at simple interest of 15% for n years amounts to $1000(1 + n \times 0.15)$ dollars. For any counting number n of years 1 through 20 write a BASIC program that could be used to obtain the amounts to which $1000 would accumulate at 15% simple interest.

*20. An investment of $1000 at 12% compounded monthly for n years amounts to $1000(1 + 0.01)^{12n}$ dollars. Assume that n may be any counting number from 1 to 20 and write a BASIC program that could be used to obtain a table of the amounts to which $1000 would accumulate in n years at 12% compounded monthly.

Write a BASIC program to print the specified data.

*21. The semester average for students who take three 1-hour examinations and a 3-hour final examination. The hour examinations are equally important. The final examination and the average for the three 1-hour examinations each provide the basis for one-half of the student's final grade.

*22. Repeat Exercise 21 with the hour examinations providing the basis for two-thirds of the student's final grade and the final examination providing the basis for one-third.

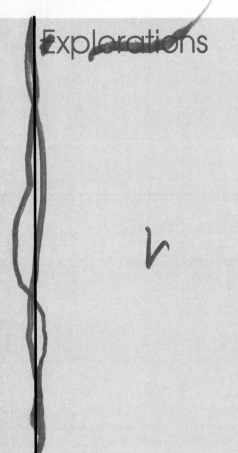

Explorations

1. The numbers of dots in square arrays are called **square numbers**.

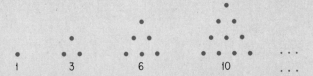

Write a BASIC program to list the square numbers less than or equal to one million.

2. The numbers of dots in triangular arrays of the form shown below are called **triangular numbers**. Write a BASIC program to list the first 100 triangular numbers.

Write a BASIC program to print the specified data.

3. A table of the counting numbers 1 through 100 and their factorials. For any counting number n the **factorial** of n, written $n!$, is equal to the product of the counting numbers that are less than or equal to n.

4. The numbers that are less than 100 in the Fibonacci sequence described in Exercise 26 of Section 2-3.

Chapter 2 Review

Solutions to the following exercises may be found within the text of Chapter 2. Try to complete each exercise without referring to the text.

Section 2-1   Calculators

1. Restate each expression so that the new expression can be evaluated using the last x procedure to obtain the correct result.

(a) $10 + 2 \times 7$ (b) $29 + 3 \times 5 + 6$ (c) $5 - 7 + 12 \div 4$
(d) $25 - 3 \times 2$ (e) $19 + 17^2$ (f) $3 + \sqrt{16}$

2. In each of the following estimate the cost to the nearest dollar. Then use a calculator to find the exact cost and check your estimate.
(a) Eleven booklets at 90¢ each.
(b) Twenty-three pictures at $1.05 each.
(c) One-fourth of $79.72.
(d) Two-thirds of $119.40.

Section 2-2   Problem Solving with Calculators

3. At Juan's college those students who pay a $10.00 fee ride the shuttle bus on campus for a nickel; other students pay a quarter. At least how many rides should be anticipated to make it worthwhile to pay the fee?

4. Identify one or more relationships and solve each problem. Use a calculator as needed.
(a) John ran 5 miles in 40 minutes. Find his rate in miles per hour.
(b) Dot bought $10.00 worth of gasoline at $1.35 a gallon. Her car averages 28 miles per gallon. Can she expect to complete a 195-mile trip without needing to buy more gasoline?

5. The ABC Corporation paid $900,000 for 30,000 acres of marshland. The corporation gave 2000 acres to a tax-exempt wildlife organization. What was the cost of the 2000 acres of marshland?

Section 2-3   Flowcharts

6. Find the output for each flowchart.

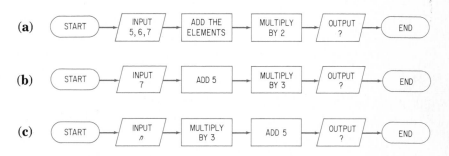

7. Find the input for the given flowchart.

8. Draw a flowchart to represent the given expression.
   (a) $2n + 1$ (b) $2(n + 1)$

Section 2-4 Computers

9. Write in BASIC.
   (a) $5(3 - 17)$ (b) $3^2 + 10$ (c) $54 \div (2^3 - 2)$
   (d) $\sqrt{29} \div 3$

10. List, in order, the operations that you would perform and evaluate each expression.
    (a) $2 + 6 \times 4$ (b) $3 + 5^2$ (c) $54 \div (2^3 - 17)$

11. Evaluate.   (a) **(2*3)↑2**   (b) **5+(12/4)↑3**

Section 2-5 BASIC

12. State in ordinary notation the printed output for this program.

    ```
    10  FOR X = 1 TO 5
    20  PRINT X, 2 + X
    30  NEXT X
    40  END
    ```

13. Suppose that you have three grades for each of your students and wish to average these grades. Prepare a BASIC program that may be used to do this.

Chapter 2 Test

Restate each expression so that the new expression can be evaluated using the last x procedure.

1. $17 + 2 \times 5$ 2. $25 - 5 \times 4$ 3. $26 - 12 \div 3$
4. $31 + 26 \div 13$ 5. $17 - 2\sqrt{11}$ 6. $36 + \sqrt{31} - 29$
7. Find the word that may be obtained by reading upside down the value of the expression $2(15,246 + 13,611)$.

Use a calculator as needed and solve each problem.

8. Abe jogged 10 miles in an hour and 20 minutes. Find his rate in miles per hour.

9. Betty needed 12 gallons of gasoline for a 348-mile drive. Find the miles per gallon for her trip.

10. On August 17, 1978 Ben Abruzzo, Maxie Anderson, and Larry Newson landed their Double Eagle II balloon in France after a 3100-mile flight from Presque Isle, Maine, in 137 hours and 6 minutes. Find to the nearest tenth of a unit the rate for their flight in miles per hour.

11. In May 1927 Charles A. Lindbergh flew from New York to Paris in $33\frac{1}{2}$ hours. To the nearest tenth of a day, how

2 CALCULATORS AND COMPUTERS

much longer was the 1978 balloon trip (see Exercise 10) than Lindbergh's 1927 flight in a single-engine plane?

12. How long does the light from the sun travel to reach Earth? The sun is about 92,960,000 miles from Earth and light travels about 11,160,000 miles per minute.

Find the output for the given flowchart for n = 20.

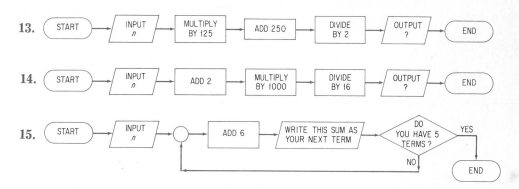

Find in terms of n the output for the flowchart given in the specified exercise.

16. Exercise 13. 17. Exercise 14. 18. Exercise 15.

Draw a flowchart to represent the given expression.

19. $5(n - 3)$ 20. $5n - 3$

21. Write in BASIC and evaluate: $6 + 2\sqrt{9} - 11^2$.

22. Evaluate: **SQR(49) – 15*8/10**.

23. State the printed output for the BASIC program:

```
10  FOR X = 1 TO 6
20  PRINT X, X ↑ 2 − 5
30  NEXT X
40  END
```

24. Write a BASIC program to print each of the even numbers from 2 through 50 with its cube.

25. Write a BASIC program to print a table of numbers and their square roots for the numbers 1 through 10,000.

Mathematics for the Consumer

3

3-1

Percent

What is the sale price of an $80 jacket that has been reduced 25%? What is the amount of a 5% tax on a $4.20 purchase? How much interest can be expected the first year on $500 in a savings account at 6% compounded quarterly? Many of the everyday applications of mathematics that affect the consumer involve the use of percent. However, as many national tests have shown, this is one topic in mathematics that is not well understood. Therefore, we shall begin this chapter with an overview of the basic concepts and skills that involve percents.

Many consumers now carry their calculators on their wrists for quick and easy reference.

The word *percent* comes from the Latin *per centum* for per hundred and can be interpreted to mean *hundredths*. Thus we may write a percent as a fraction with denominator 100, or as a decimal.

$$75\% = \frac{75}{100} = 0.75 \qquad 0.5\% = \frac{0.5}{100} = 0.005$$

$$9\% = \frac{9}{100} = 0.09 \qquad 100\% = \frac{100}{100} = 1$$

$$125\% = \frac{125}{100} = 1.25$$

In computations with percents, it is often helpful to express the percent as a fraction in simplest form. A fraction is said to be in *simplest form* when it is not possible to divide both the numerator and the denominator by the same counting number k, $k \neq 1$. Here are some of the most commonly used equivalents.

These equivalents are used often and are worth memorizing for future use.

$$25\% = \frac{25}{100} = \frac{1}{4} \qquad 20\% = \frac{20}{100} = \frac{1}{5} \qquad 33\frac{1}{3}\% = \frac{33\frac{1}{3}}{100} = \frac{1}{3}$$

$$50\% = \frac{50}{100} = \frac{1}{2} \qquad 40\% = \frac{40}{100} = \frac{2}{5}$$

$$75\% = \frac{75}{100} = \frac{3}{4} \qquad 60\% = \frac{60}{100} = \frac{3}{5} \qquad 66\frac{2}{3}\% = \frac{66\frac{2}{3}}{100} = \frac{2}{3}$$

$$100\% = \frac{100}{100} = 1 \qquad 80\% = \frac{80}{100} = \frac{4}{5}$$

The preceding illustrations indicate the procedure for expressing a percent as a fraction. Write the percent as a number of hundredths and then reduce the fraction, if possible.

Example 1   Write as a fraction in simplest form   (a) 70%   (b) 45%.

Solution   (a) $70\% = \dfrac{70}{100} = \dfrac{7}{10}$ (b) $45\% = \dfrac{45}{100} = \dfrac{9}{20}$

Example 2   Write as a percent   (a) 0.32   (b) $\dfrac{7}{10}$.

Solution   (a) $0.32 = \dfrac{32}{100} = 32\%$ (b) $\dfrac{7}{10} = \dfrac{70}{100} = 70\%$

One method of expressing a fraction as a percent is shown in Example 2. This method causes no difficulty if the given fraction can easily be rewritten as an equivalent fraction with denominator 100. Otherwise we may make use of a **proportion**, a statement of the equality of two fractions. For example, each of the following is a proportion.

A proportion may also be considered as the equivalence of two *ratios*. The ratio of *a* to *b* is *a/b*.

$$\frac{2}{3} = \frac{12}{18} \qquad \frac{3}{4} = \frac{75}{100} \qquad \frac{50}{100} = \frac{1}{2}$$

The *proportion property* states that the *cross products* of any proportion are equal. Thus in the examples shown, we have

$$2 \times 18 = 3 \times 12 \qquad 3 \times 100 = 4 \times 75 \qquad 50 \times 2 = 100 \times 1$$

The **proportion property** is an application of the definition of two *equivalent fractions* a/b and c/d.

$$\frac{a}{b} = \frac{c}{d} \quad \text{if and only if} \quad ad = bc$$

See Section 4-5 for a further discussion of the use of the proportion property with fractions.

The problem of expressing a fraction as a percent may be considered as solving (finding a missing part of) a proportion.

Example 3   Write $\dfrac{5}{8}$ as a percent.

Solution

$$\frac{5}{8} = \frac{n}{100}$$
$$5 \times 100 = 8 \times n$$
$$8n = 500$$
$$n = 62\frac{1}{2}$$
$$\frac{5}{8} = 62\frac{1}{2}\%$$

80

3   MATHEMATICS FOR THE CONSUMER

Another way to express $\frac{5}{8}$ as a percent is to recognize the fraction as the indicated quotient $5 \div 8$. Then divide, and recall that percent means hundredths.

$$\frac{5}{8} = 0.625 = \frac{62.5}{100} = 62.5\%, \quad \text{that is,} \quad 62\frac{1}{2}\%$$

$$\begin{array}{r} 0.625 \\ 8\overline{)5.000} \\ \underline{4\,8} \\ 20 \\ \underline{16} \\ 40 \\ \underline{40} \end{array}$$

We are now ready to solve problems that involve percents. Most such problems can be solved by remembering that a percent is the ratio of some number to 100. In the following three examples, we shall solve three different types of problems by using a *percent scale* to set up and solve a proportion.

Example 4   Find 25% of 80.

Solution   Since 25% means 25/100 we wish to find a number n that compares to 80 in the same way that 25 compares to 100. This comparison may be observed using scales on a line segment to represent 80, which we consider to be 100%, or all, of the given number. The adjacent line segment has been divided into 20 parts. Each part represents 1/20 of 100%, that is, 5%; five parts represent 25%. The parts are labeled above with numbers based on 80 and labeled below with percents based on 100. We read from these scales.

$$\frac{n}{80} = \frac{25}{100}$$

(The ratio of n to 80 on the scale of numbers is the same as the ratio of 25 to 100 on the scale of percents.)

$$100 \times n = 80 \times 25$$
$$100n = 2000$$
$$n = 20 \qquad (2000 \div 100 = 20)$$

Therefore 25% of 80 is 20.

Example 5   20 is what percent of 80?

Solution   Consider this percent scale, with the known parts shown.

$$\frac{20}{80} = \frac{n}{100}$$

(The ratio of 20 to 80 is the same as the ratio of n to 100 on the percent scale.)

$$20 \times 100 = 80 \times n$$
$$80n = 2000$$
$$n = 25 \qquad (2000 \div 80 = 25)$$

Therefore 20 is 25% of 80.

Example 6 20 is 25% of what number?

Solution We first construct a percent scale.

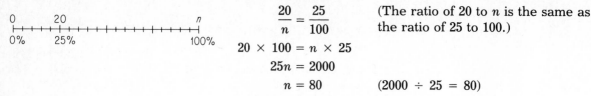

$$\frac{20}{n} = \frac{25}{100}$$

(The ratio of 20 to n is the same as the ratio of 25 to 100.)

$$20 \times 100 = n \times 25$$
$$25n = 2000$$
$$n = 80 \qquad (2000 \div 25 = 80)$$

Therefore 20 is 25% of 80.

In each of the examples shown, other means of solution are possible. Because the numbers selected in these examples caused no arithmetical difficulties, other methods may be easier. For example, consider again the problem of finding 25% of 80. We can also obtain the answer by either of the following methods.

USING FRACTIONS

$$25\% = \frac{25}{100} = \frac{1}{4}$$

$$\frac{1}{4} \text{ of } 80 = \frac{1}{4} \times 80 = 20$$

USING DECIMALS

$$25\% = 0.25$$

$$25\% \text{ of } 80 = 0.25 \times 80 = 20$$

Although these approaches may seem somewhat easier here, the procedures described in Examples 4, 5, and 6 are applicable to all problems of these types, regardless of the numerical values in the problems. Consider, for example, one additional problem solved by the proportion method and where the answer is not readily seen in advance.

Example 7 There are 22 students who arrive by school bus. These students represent 55% of the students in the class. How many students are there in the class?

Solution Here we consider the ratio of students bused to the total number in the class as 22/n. This is equal to the ratio of 55% to the total, which is always 100%. Thus we have the following.

$$\frac{22}{n} = \frac{55}{100}$$
$$22 \times 100 = 55 \times n$$
$$55n = 2200$$
$$n = 40 \qquad (2200 \div 55 = 40)$$

There are 40 students in the class.

EXERCISES *Write each percent as a decimal.*

1. 57% 2. 89% 3. 3% 4. 225% 5. 0.95%
6. 1% 7. 250% 8. 125% 9. 100% 10. 0.5%

Write each percent as a fraction in simplest form.

11. 50% 12. 40% 13. 95% 14. 65% 15. 99%
16. 130% 17. 150% 18. 125% 19. 8% 20. 0.8%

Write each decimal as a percent.

21. 0.35 22. 0.01 23. 0.45 24. 1.35 25. 0.9
26. 0.08 27. 1.45 28. 0.006 29. 0.001 30. 1.00

Write each fraction as a percent.

31. $\dfrac{9}{100}$ 32. $\dfrac{36}{100}$ 33. $\dfrac{92}{100}$ 34. $\dfrac{125}{100}$ 35. $\dfrac{150}{100}$

36. $\dfrac{16}{20}$ 37. $\dfrac{38}{50}$ 38. $\dfrac{9}{10}$ 39. $\dfrac{17}{10}$ 40. $\dfrac{42}{20}$

Copy and complete each table.

	Fraction	Decimal	Percent
41.	$\frac{3}{10}$	0.3	30
42.		0.89	
43.	$\frac{55}{100}$	0.55	55%
44.		0.7	
45.	$\frac{17}{20}$		
46.			175%

	Fraction	Decimal	Percent
47.			5%
48.		0.17	
49.	$\frac{11}{10}$		
50.		0.01	
51.	$\frac{143}{200}$		
52.			120%

Use an equivalent fraction to find each percent.

53. Find 75% of 120. 54. Find 40% of 240.

55. Find 125% of 48. 56. Find $33\frac{1}{3}$% of 210.

57. Find $66\frac{2}{3}$% of 72. 58. Find 80% of 320.

Write and solve a proportion.

59. Find 40% of 60. 60. Find 80% of 140.
61. Find 35% of 80. 62. Find 120% of 75.

63. 20 is what percent of 160?
64. 15 is what percent of 120?
65. 120 is what percent of 160?
66. 160 is what percent of 120?
67. 25 is 20% of what number?
68. 40 is 75% of what number?
69. 80 is 125% of what number?
70. 80 is 20% of what number?

Explorations

The following figure illustrates a **percent chart**. Here is how this chart may be used to find 50% of 60. First a line is drawn from 0 on the percent scale to 60 on the base scale. Next 50 is located on the percent scale. Then the answer is found by reading across to the line and down to 30 on the base scale. Thus 50% of 60 is 30.

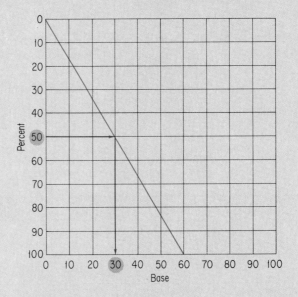

Construct a percent chart to illustrate each of the following problems.

1. Find 25% of 80.
2. What percent of 50 is 40?
3. 20 is 40% of what number?

3-2
Applications of Percent

Insisting that business was bad, an employer ordered a 20% cut in salary for all of his workers. Several months later, after extended negotiations, the employer agreed to give everyone a 20% increase in wages. Should the employees be satisfied that their original salaries had been reinstated?

Let us consider the case of an employee who earns $100 a week and see what happens after these two transactions:

Recall that $20\% = \dfrac{20}{100} = \dfrac{1}{5}$.

$$20\% \text{ of } \$100 = \frac{1}{5} \times \$100 = \$20$$

Salary after cut in salary: $\$100 - \$20 = \$80$

$$20\% \text{ of } \$80 = \frac{1}{5} \times \$80 = \$16$$

Salary after raise in salary: $\$80 + \$16 = \$96$

Thus the employee is earning $4 less per week than was the case prior to these two actions. In general, a 20% decrease followed by a 20% increase results in a number ($96) that is 96% of the original amount; that is, it is equivalent to a 4% decrease in salary.

Although we used $100 as a base figure to simplify computations, the outcome would be the same for other initial amounts. Show that a $250 salary subject to a 20% cut, followed by a 20% increase, will give a final amount that is 96% of the original figure.

Suppose that the employer had granted a 20% increase first, and then had followed this by a 20% cut in salary. Would the final salary be better, worse, or the same as the original? First guess; then try to justify your answer by using a specific initial salary.

One of the most frequently encountered uses of percent involves discounts on sale items. Thus we frequently see advertisements such as the following in the newspapers.

SAVE — 25% OFF

Of course we cannot tell what the *selling price* of the item is from the information above unless we also know the *regular price*. Often a store will only give the sale price and state that this represents a certain *discount*, such as in this case.

SALE PRICE — $90
SAVE 25%

In this case the consumer must assume that the sale price does indeed represent a saving of 25% of the regular price.

Assume that a coat normally sells for $120 and is offered on sale at a discount of 25%. How can we find the sale price? First we find the amount that can be saved by buying the coat on sale; that is, we find 25% of $120.

USING FRACTIONS	USING DECIMALS

$$25\% = \frac{25}{100} = \frac{1}{4} \qquad\qquad 25\% = 0.25$$

$$\frac{1}{4} \times \$120 = \$30$$

$$\begin{array}{r} \$1\ 20 \\ \times \quad 0.25 \\ \hline 6\ 00 \\ 24\ 0 \\ \hline \$30.00 \end{array}$$

The amount saved by buying on sale is $30. To find the sale price, we must subtract the amount saved from the regular price.

$$\begin{array}{rl} \$120 & \text{(regular price)} \\ - \quad 30 & \text{(amount saved)} \\ \hline \$\ 90 & \text{(sale price)} \end{array}$$

There is an alternate method that may be used to solve this last example. Since the discount is 25%, the amount that you pay for the coat must be 100% minus 25%, that is, 75%. Therefore, the sale price is 75% of the regular price as in this diagram.

We can find 75% of 120 using a percent scale and a proportion.

Thus the ratio of n to 120 is the same as the ratio of 75% to 100%:

$$\frac{n}{120} = \frac{75}{100}$$

$$100n = 9000$$

$$n = 90 \qquad 75\% \text{ of } \$120 \text{ is } \$90.$$

Note that we think of 75% *of* $120 as meaning 75% × $120.

We may also find 75% of $120 by either of the following ways.

USING FRACTIONS	USING DECIMALS

$$75\% = \frac{75}{100} = \frac{3}{4} \qquad\qquad 75\% = 0.75$$

$$\frac{3}{4} \times \$120 = \$90$$

$$\begin{array}{r} \$\ 120 \\ \times \quad 0.75 \\ \hline 6\ 00 \\ 84\ 0 \\ \hline \$90.00 \end{array}$$

Example 1   A television set costs $245. For one week only it is offered on sale at 22% off. How much can be saved by buying the set during the sale week?

Solution   We need to find 22% of $245. In this case the easiest approach is to use decimals.

$$22\% = 0.22$$

$$
\begin{array}{r}
\$2\,45 \\
\times\ \ 0.22 \\
\hline
4\,90 \\
49\,0 \\
\hline
\$53.90
\end{array}
$$

The saving during the sale is $53.90.

Very often the consumer is not necessarily interested in an exact answer, but rather needs to make a quick estimate to determine whether or not a sale is worthwhile. Thus, in Example 1, you might think of the television set as costing *about* $250, and the discount as *approximately* 20%. Since 20% = 1/5, the amount saved is approximately 1/5 of $250. This computation can be done mentally to obtain an estimate of $50 for the amount saved by buying on sale.

Example 2   The regular price of a radio is $80. The sale price is $64. What percent discount is being offered during this sale?

Solution   The amount saved is $80 − $64, that is, $16. Here is the question that needs to be considered.

16 is what percent of 80?

Using a proportion, the percent is found as follows.

$$\frac{16}{80} = \frac{n}{100}$$

$$16 \times 100 = 80 \times n$$

$$80n = 1600$$

$$n = 20$$

The radio is being offered at a 20% discount.

In Example 2 it is important to note that we compared the saving with the original, or regular, price. If you compare the saving with the sale price, you will find that 16 is 25% of $64. However, it is *incorrect* to claim that this sale offers a 25% discount.

The solution for Example 2 can be checked mentally. Since 20% = 1/5, we need merely to find 1/5 of $80.

$$\frac{1}{5} \times 80 = 16$$

This checks since the discount was found to be $16.

There are many times in daily life when it is helpful to be able to compute a percent mentally.

A tip in a restaurant is almost universally computed at 15% of the cost of the food consumed. (That is, you need not pay 15% of any tax that may be added to the bill.) The tip can often be computed mentally by thinking of 15% as the sum of 10% and 5%. We can find 10% of a number because 10% = 1/10. To find 1/10 of a number, just divide by 10; that is, move the decimal point one decimal place to the left. Here are some examples.

$$10\% \text{ of } \$240 = \frac{1}{10} \times \$240 = \$24$$

$$10\% \text{ of } \$8.20 = \frac{1}{10} \times \$8.20 = \$0.82$$

Now suppose that the total for a bill is $8.20 and you wish to compute a tip of 15%. First find 10% of that amount. Then find 5% of the amount, which will be one-half of 10%, and add.

$$\begin{array}{ll} 10\% \text{ of } \$8.20 = \$0.82 & \\ + \ \underline{5\% \text{ of } \$8.20 = \$0.41} & (0.82 \div 2 = 0.41) \\ 15\% \text{ of } \$8.20 = \$1.23 & \end{array}$$

EXERCISES   *Select the best estimate mentally.*

1. 25% of $198   (a) $25   (b) $50   (c) $100

The ability to estimate is an important skill used often in daily life. Use Exercises 1 through 8 to practice this skill. Thereafter try to estimate your answers before you actually compute.

2.	34% of $241	(a) $8	(b) $60	(c) $80
3.	19% of $352	(a) $35	(b) $70	(c) $90
4.	49% of $81	(a) $20	(b) $42	(c) $50
5.	1% of $120	(a) $1	(b) $12	(c) $120
6.	9% of $230	(a) $2.30	(b) $23	(c) $2300
7.	26% of $81	(a) $2	(b) $20	(c) $25
8.	5% of $320	(a) $1.60	(b) $16	(c) $32

Find the selling price of an item for each given regular price and discount.

9. $240, less 25%   10. $320, less 10%   11. $180, less 20%
12. $48, less $33\frac{1}{3}$%   13. $90, less 40%   14. $60, less 15%
15. $68, less 35%   16. $76, less 18%   17. $46, less 22%
18. $7.80, less 15%   19. $8.50, less 12%   20. $6.80, less 25%

Find the percent discount being offered. (Round answers to the nearest percent.)

	REGULAR PRICE	SALE PRICE		REGULAR PRICE	SALE PRICE
21.	$80	$60	22.	$48	$32
23.	$8.50	$6.50	24.	$12.50	$9.80
25.	$120	$100	26.	$150	$130
27.	$32.50	$24.75	28.	$45.50	$37.50

29. A storekeeper advertises a sale offering discount of $33\frac{1}{3}$% on all merchandise. A coat on display is marked with regular price $120 and sale price $90. Is this the correct selling price? If not, what should it be and how can you account for the merchant's error?

30. Wendy bought a coat on sale for $35.55. The regular price of the coat was $45.00.
(a) What percent discount did she receive?
(b) The sale price is what percent of the regular price of the coat?

31. The school population in a certain elementary school rose from 700 to 800 in a recent year. What was the percent of increase? That is, the increase was what percent of the original enrollment?

32. Roberto went on a diet and reduced his weight from 150 pounds to 135 pounds. What was the percent of decrease in his weight? That is, his loss of weight was what percent of his original weight?

*33. A television set that regularly costs $250 is advertised on sale at a 15% discount. A week later it is further reduced by 10% of the same price. Find a single discount that is equivalent to these two successive discounts.

*34. A storekeeper pays $80 for a coat and sells it at a markup of 15%. Later it is marked up an additional 10% of the selling price. Find a single markup equivalent to these two successive markups.

*35. What single action on a given initial salary would be equivalent to a 20% cut followed by a 25% increase? Answer the same question for a 25% increase followed by a 20% cut.

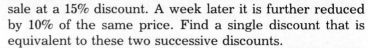

Explorations

1. You purchase $50 worth of books in a bookstore that offers a 20% discount. However, there is also a 5% state sales tax on all such purchases. Is it to your advantage to have the store give you the discount first and then add the sales tax, or add the sales tax first and then give you the discount? First guess which you think would be better for you, and then compute the cost using each approach.

2. Repeat Exploration 1 to determine which approach would be best for the storekeeper and which would be best for the state.

3. Investigate the procedure used by your bookstore in the event that they offer a discount and also must add a state sales tax to the purchase.

4. Begin a collection of as many applications of percent as you can find in newspapers and other periodicals.

3-3
Simple and Compound Interest

Almost every consumer is concerned about *interest*. We wish to receive maximum interest on our savings. And we often need to pay interest on loans taken for a college education or an automobile bought "on time." In order to understand the usual procedures for determining interest, we first need to understand *simple interest*.

Suppose that you borrow $2000 from a friend at the rate of 12% per year simple interest. The interest on the loan each year is

$$0.12 \times 2000, \quad \text{that is,} \quad \$240$$

The interest rates used here are for illustrative purposes only. In actual practice they vary greatly, especially in periods of rapid inflation.

The original amount of the loan is called the **principal** (*P*). The interest (*I*) is a percent (*r*) of the principal each year. The duration (time, *t*) of the loan is stated in years. The amount needed to pay off a loan depends on the duration of the loan.

3 MATHEMATICS FOR THE CONSUMER

The amount needed in any one of the first five years to pay off a loan of $2000 at 12% per year simple interest is shown in the following table.

AT THE END OF	PRINCIPAL	ACCUMULATED INTEREST	AMOUNT DUE
1 year	$2000	$240	$2240
2 years	2000	480	2480
3 years	2000	720	2720
4 years	2000	960	2960
5 years	2000	1200	3200

> The **simple interest** (I) on a principal (P) at a rate (r) per year for t years is
>
> $$I = Prt$$

Example 1 Find the simple interest on a loan of $2500 at 6% for 3 years.

Solution $P = \$2500, \quad r = 6\%, \quad t = 3$:

$$I = Prt = \$2500 \times 0.06 \times 3 = \$450$$

The interest is $450; at the end of 3 years the amount due is $2500 + $450, that is $2950.

In general practice, the interest charged on loans and paid on savings accounts is **compound interest,** that is, interest is charged (or paid) on the previously due (or credited) interest. On credit card balances, the interest is compounded and added monthly. On bank deposits the interest is usually added quarterly, but often is compounded daily. The effect of compounding may be seen in the following development where we shall consider a sum of money that is deposited in a bank where interest is compounded. To simplify the discussion, we begin by considering interest compounded yearly. The computations can be extended to interest compounded more frequently.

Suppose that $2500 is deposited at 6% interest compounded annually. Then, as in the next table, the interest is added to the account at the end of each year and the interest for the following year is computed on this new amount.

	PRINCIPAL (at the start of the year)	INTEREST	AMOUNT (at the end of the year)
First year	$2500	$2500 \times 0.06 \times 1 = \$150$	$2500 + \$150 = \$2650$
Second year	2650	$2650 \times 0.06 \times 1 = 159$	$2650 + 159 = 2809$
Third year	2809	$2809 \times 0.06 \times 1 = 168.54$	$2809 + 168.54 = 2977.54$

Each year the procedure is the same. The **amount** *(A)* at the end of the year is equal to the sum of the principal *(P)* and the interest *(Pr)* for that year.

$$A = P + Pr = P(1 + r)$$

First year	$A = 2500(1 + 0.06) = 2650$
Second year	$A = 2650(1 + 0.06) = [2500(1 + 0.06)](1 + 0.0$
	$= 2500(1 + 0.06)^2 = 2809$
Third year	$A = 2809(1 + 0.06) = 2500(1 + 0.06)^3 = 2977$

The pattern continues.

For ten years	$A = 2500(1 + 0.06)^{10}$
For n years	$A = 2500(1 + 0.06)^n$

A calculator is especially helpful for computations of interest. For example, if you have an exponential key $\boxed{y^x}$, one possible set of steps to compute $2500(1.06)^3$ is the following.

Instructions: 1.06 $\boxed{y^x}$ 3 $\boxed{\times}$ 2500 $\boxed{=}$

Displays: 1.06 1.06 3 1.191016 2500 2977.54

We make use of the *last x procedure* here, using the form $(1.06)^3 \times 2500$.

Without an exponential key, the preceding computation can be completed by continued multiplications as follows.

Instructions: 2500 $\boxed{\times}$ 1.06 $\boxed{\times}$ 1.06 $\boxed{\times}$ 1.06 $\boxed{\times}$

Displays: 2500 2500 1.06 2650 1.06 2809 1.06 2977.54

Note that this procedure can be continued indefinitely. Each time you press 1.06 followed by $\boxed{\times}$ (or by $\boxed{=}$) the amount for an additional year is found.

Since interest is paid (charged) on interest, compound interest is always more than simple interest. For example, at 6% the simple interest on $2500 is $150 per year and $450 for 3 years, as in Example 1. Thus after 3 years

$2977.54	*(amount at compound interest)*
− 2950.00	*(amount at simple interest)*
$ 27.54	*(difference)*

Although this difference over a 3-year period may seem inconsequential, the accumulation for larger sums over a longer period of time is quite substantial. Furthermore, interest is usually compounded more frequently than once a year. Thus interest may be compounded twice a year (semiannually), or four times a year (quarterly), or more often. Indeed, as noted earlier, some funds compound money on a daily basis. In general, the more often money is compounded the more interest is earned.

3 MATHEMATICS FOR THE CONSUMER

Example 2 Find the amount on deposit at the end of 1 year if $2000 is deposited at 5% compounded semiannually.

Solution A rate of 5% per year compounded semiannually is equivalent to a rate of $2\frac{1}{2}$% per half year.

	PRINCIPAL	INTEREST	AMOUNT
First half year	$2000	$2000 \times 0.05 \times \dfrac{1}{2} = \$50$	$\$2000 + \$50 = \$2050$
Second half year	$2050	$2050 \times 0.05 \times \dfrac{1}{2} = \$51.25$	$\$2050 + \$51.25 = \$2101.25$

The amount on deposit at the end of 1 year is $2101.25.

In Example 2 the total interest earned in one year on $2000 at 5% compounded semiannually is $101.25, which is 5.0625% of the original investment. This equivalent rate of 5.0625% simple interest is called the **effective annual rate** of interest.

> The amount A, when interest is compounded annually for n years, is
> $$A = P(1 + r)^n$$

We can modify this general formula to account for cases when interest is compounded more frequently than once a year. Thus in Example 2 the **period** (interval for compounding) is 6 months. There are two periods per year, the rate is $r/2$ per period, and the number of periods in n years is $2n$. Thus, when interest is compounded semiannually for n years, the amount A is

$$A = P\left(1 + \frac{r}{2}\right)^{2n}$$

Similarly, if interest is compounded quarterly, the rate per period is $r/4$, there are $4n$ periods in n years, and

$$A = P\left(1 + \frac{r}{4}\right)^{4n}$$

This pattern may be extended for compounding k times per year to obtain

$$\boxed{A = P\left(1 + \frac{r}{k}\right)^{kn}}$$

Example 3 Give an expression for the amount on deposit in each case.

(a) $1000 is deposited at 5% interest compounded annually for 15 years.

(b) $2000 is deposited at 6% interest compounded semiannually for 20 years.

(c) $5000 is deposited at 8% interest compounded quarterly for 6 years.

Solution

Computations of these amounts can be very tedious. Fortunately, it is a relatively simple matter with an appropriate calculator.

(a) $A = 1000(1 + 0.05)^{15}$, that is, $\$1000(1.05)^{15}$

(b) $A = 2000\left(1 + \dfrac{0.06}{2}\right)^{2 \times 20}$, that is, $\$2000(1.03)^{40}$

(c) $A = 5000\left(1 + \dfrac{0.08}{4}\right)^{4 \times 6}$, that is, $\$5000(1.02)^{24}$

Example 4 Show the instructions and displays for using a calculator to find the amount at the end of 2 years for a deposit of $2000 at 8% interest compounded semiannually.

Solution The amount may be expressed as $2000(1.04)^4$. Then one possible way to compute this amount on a calculator is as follows.

Instructions: 2000 $\boxed{\times}$ 1.04 $\boxed{\times}$ 1.04 $\boxed{\times}$ 1.04 $\boxed{\times}$ 1.04 $\boxed{\times}$

Display: 2000 2000 1.04 2080 1.04 2163.20 1.04 2249.73 1.04 2339.72

A compound interest table may also be used to find amounts for different rates and periods. Thus Table I is a table of values of $(1 + i)^n$ where i is the interest rate per period and n is the number of periods. Let us use the table to evaluate the amount in Example 3(a), $\$1000(1.05)^{15}$.

Because of rounding procedures used, the answer obtained by use of the table may differ slightly from that obtained with a calculator.

Use $i = 5\%$ and $n = 15$

From Table I, $(1 + i)^n = 2.0789$

$A = \$1000(1.05)^{15} = \$1000(2.0789) = \$2078.90$

Note that your money has more than doubled. Compare this with the amount that you would have had at 5% simple interest for 15 years.

Example 5 Use Table I to find the amount for a deposit of $400 **(a)** at 6% compounded semiannually for 4 years **(b)** at 12% compounded quarterly for 5 years.

Solution (a) $P = \$400$, $i = 3\%$, $n = 8$:

$A = 400(1.03)^8 = 400(1.2668) = 506.72$

The amount is $506.72.

(b) $P = \$400, \quad i = 3\%, \quad n = 20$:

$$A = 400(1.03)^{20} = 400(1.8061) = 722.44$$

The amount is $722.44.

TABLE I COMPOUND INTEREST $(1 + i)^n$

n	1%	$1\frac{1}{2}\%$	2%	$2\frac{1}{2}\%$	3%	4%	5%	6%	7%	8%
1	1.0100	1.0150	1.0200	1.0250	1.0300	1.0400	1.0500	1.0600	1.0700	1.0800
2	1.0201	1.0302	1.0404	1.0506	1.0609	1.0816	1.1025	1.1236	1.1449	1.1664
3	1.0303	1.0457	1.0612	1.0769	1.0927	1.1249	1.1576	1.1910	1.2250	1.2597
4	1.0406	1.0614	1.0824	1.1038	1.1255	1.1699	1.2155	1.2625	1.3108	1.3605
5	1.0510	1.0773	1.1041	1.1314	1.1593	1.2167	1.2763	1.3382	1.4026	1.4693
6	1.0615	1.0934	1.1262	1.1597	1.1941	1.2653	1.3401	1.4185	1.5007	1.5869
7	1.0721	1.1098	1.1487	1.1887	1.2299	1.3159	1.4071	1.5036	1.6058	1.7138
8	1.0829	1.1265	1.1717	1.2184	1.2668	1.3686	1.4775	1.5938	1.7182	1.8509
9	1.0937	1.1434	1.1951	1.2489	1.3048	1.4233	1.5513	1.6895	1.8385	1.9990
10	1.1046	1.1605	1.2190	1.2801	1.3439	1.4802	1.6289	1.7908	1.9672	2.1589
11	1.1157	1.1779	1.2434	1.3121	1.3842	1.5395	1.7103	1.8983	2.1049	2.3316
12	1.1268	1.1956	1.2682	1.3449	1.4258	1.6010	1.7959	2.0122	2.2522	2.5182
13	1.1381	1.2136	1.2936	1.3785	1.4685	1.6651	1.8856	2.1329	2.4098	2.7196
14	1.1495	1.2318	1.3195	1.4130	1.5126	1.7317	1.9799	2.2609	2.5785	2.9372
15	1.1610	1.2502	1.3459	1.4483	1.5580	1.8009	2.0789	2.3966	2.7590	3.1722
16	1.1726	1.2690	1.3728	1.4845	1.6047	1.8730	2.1829	2.5404	2.9522	3.4259
17	1.1843	1.2880	1.4002	1.5216	1.6528	1.9479	2.2920	2.6928	3.1588	3.7000
18	1.1961	1.3073	1.4282	1.5597	1.7024	2.0258	2.4066	2.8543	3.3799	3.9660
19	1.2081	1.3270	1.4568	1.5987	1.7535	2.1068	2.5270	3.0256	3.6165	4.3157
20	1.2202	1.3469	1.4859	1.6386	1.8061	2.1911	2.6533	3.2071	3.8697	4.6610
21	1.2324	1.3671	1.5157	1.6796	1.8603	2.2788	2.7860	3.3996	4.1406	5.0338
22	1.2447	1.3876	1.5460	1.7216	1.9161	2.3699	2.9253	3.6035	4.4304	5.4365
23	1.2572	1.4084	1.5769	1.7646	1.9736	2.4647	3.0715	3.8197	4.7405	5.8715
24	1.2697	1.4295	1.6084	1.8087	2.0328	2.5633	3.2251	4.0489	5.0724	6.3412
25	1.2824	1.4509	1.6406	1.8539	2.0938	2.6658	3.3864	4.2919	5.4274	6.8485
26	1.2953	1.4727	1.6734	1.9003	2.1566	2.7725	3.5557	4.5494	5.8074	7.3964
27	1.3082	1.4948	1.7069	1.9478	2.2213	2.8834	3.7335	4.8223	6.2139	7.9881
28	1.3213	1.5172	1.7410	1.9965	2.2879	2.9987	3.9201	5.1117	6.6488	8.6271
29	1.3345	1.5400	1.7758	2.0464	2.3566	3.1187	4.1161	5.4184	7.1143	9.3173
30	1.3478	1.5631	1.8114	2.0976	2.4273	3.2434	4.3219	5.7435	7.6123	10.0627
31	1.3613	1.5865	1.8476	2.1500	2.5001	3.3731	4.5380	6.0881	8.1451	10.8677
32	1.3749	1.6103	1.8845	2.2038	2.5751	3.5081	4.7649	6.4534	8.7153	11.7371
33	1.3887	1.6345	1.9222	2.2589	2.6523	3.6484	5.0032	6.8406	9.3253	12.6760
34	1.4026	1.6590	1.9607	2.3153	2.7319	3.7943	5.2533	7.2510	9.9781	13.6901
35	1.4166	1.6839	1.9999	2.3732	2.8139	3.9461	5.5160	7.6861	10.6766	14.7853
36	1.4308	1.7091	2.0399	2.4325	2.8983	4.1039	5.7918	8.1473	11.4239	15.9682
37	1.4451	1.7348	2.0807	2.4933	2.9852	4.2681	6.0814	8.6361	12.2236	17.2456
38	1.4595	1.7608	2.1223	2.5557	3.0748	4.4388	6.3855	9.1543	13.0793	18.6253
39	1.4741	1.7872	2.1647	2.6196	3.1670	4.6164	6.7048	9.7035	13.9948	20.1153
40	1.4889	1.8140	2.2080	2.6851	3.2620	4.8010	7.0400	10.2857	14.9745	21.7245
41	1.5038	1.8412	2.2522	2.7522	3.3599	4.9931	7.3920	10.9029	16.0227	23.4625
42	1.5188	1.8688	2.2972	2.8210	3.4607	5.1928	7.7616	11.5570	17.1443	25.3395
43	1.5340	1.8969	2.3432	2.8915	3.5645	5.4005	8.1497	12.2505	18.3444	27.3666
44	1.5493	1.9253	2.3901	2.9638	3.6715	5.6165	8.5572	12.9855	19.6285	29.5560
45	1.5648	1.9542	2.4379	3.0379	3.7816	5.8412	8.9850	13.7646	21.0025	31.9204

Find the amount for each of these deposits at simple interest.

1. $450, 5%, 8 years
2. $600, 6%, 5 years
3. $1250, 4%, 3 years
4. $2050, 5%, 6 years
5. $3500, $4\frac{1}{2}$%, 4 years
6. $4000, $5\frac{1}{4}$%, 2 years
7. $2000, 9%, $3\frac{1}{2}$ years
8. $2750, 8%, $4\frac{1}{2}$ years

Find the amount for each of these deposits at compound interest.

For Exercises 9 through 16 it is suggested that you find the answer using Table I and check your results using a calculator. The results by the two methods may differ slightly due to approximations arising from the number of places in the table or the capacity of the calculator.

9. $1500, 6% annually, 4 years
10. $3200, 5% annually, 5 years
11. $750, 6% semiannually, 5 years
12. $850, 5% semiannually, 3 years
13. $150, 8% quarterly, 8 years
14. $800, 8% quarterly, 10 years
15. $3250, 12% quarterly, 6 years
16. $7500, 10% quarterly, 4 years

Use Table I to answer each of the following.

17. Approximately how many years does it take for money to double in amount if the money is left on deposit at 6% interest compounded annually?

18. Repeat Exercise 17 for (a) 5% (b) 8%.

19. Find the effective annual rate of interest if $100 is deposited for one year at 8% interest compounded (a) semiannually (b) quarterly.

20. Repeat Exercise 19 for an initial rate of 10%.

21. A deposit of $2000 earns 12% interest compounded annually. How much additional interest would be earned the first year if the interest were compounded (a) semiannually (b) quarterly?

22. Repeat Exercise 21 for interest compounded monthly. Use a calculator to compute the interest since Table I does not provide the necessary information.

23. Assume that you earn $12,000 per year and that the rate of inflation is 10% per year. How much must you earn 5 years from now just to keep up with inflation? (*Hint:* Compute $12,000(1.10)^5$.)

24. Repeat Exercise 23 for each of these annual inflation rates that have been experienced in other countries.
    (a) 30% (b) 50% (c) 100% (d) 150%

The answer to Exercise 25 is called the **present value** of $15,000 under the specified conditions.

*25. Let us explore one further use of the compound interest table. Suppose you determine that you must have $15,000 in 10 years in order to finance a child's education. Furthermore, suppose that your bank offers 6% interest compounded quarterly. To the nearest dollar, how much should you deposit today to have $15,000 available when it is needed? In this problem we know the value for A; and we need to find the value of P in the compound interest formula $A = P(1 + i)^n$ where

$$i = \frac{0.06}{4} = 1.5\% \qquad n = 4 \times 10 = 40$$

and from the table

$$(1.015)^{40} = 1.8140$$

Therefore,

$$15{,}000 = P(1.8140)$$

Solve for P by completing this division.

$$P = \frac{15{,}000}{1.8140}$$

*26. Repeat Exercise 25 assuming that your money can earn 12% compounded quarterly.

*27. You wish to have $10,000 in 10 years and are making a single deposit for that purpose.
   (a) How much money would you need if you deposit your money at 6% interest compounded quarterly?
   (b) How much would you need if you deposit your money at 8% interest compounded semiannually?
   (c) What is the difference between the amounts needed in parts (a) and (b)?

Explorations

1. Visit a local bank and obtain information concerning the different types of accounts offered and the interest rates used. Try to find out what method is used for compounding interest, that is, the frequency with which interest is compounded. Then compare your findings with that for other banks in your community.

2. At one time the United States sold Treasury bills in denominations of $10,000 for 3 months at a discount rate of 12.35% per year. This means that you are credited with the interest when you purchase the bill and receive the face value of $10,000 when it matures in 3 months. How

much interest can you earn on such a purchase? How much must you invest to purchase a $10,000 Treasury bill? What is the equivalent annual rate of simple interest on your investment?

3. Obtain details on Christmas savings clubs at one or more banks and compare this approach with other methods of saving.

4. Find out what is meant by a bank's **prime rate**.

5. In recent years interest rates have fluctuated widely from month to month. Watch the financial pages of a newspaper and keep a record of interest rates offered by one of your local banks for six-month **Certificates of Deposit**.

3-4
Using Tables

Most consumers need to use many different types of tables in their daily activities. Although the ability to read tables and interpret tabular data is often taken for granted in schools, many consumers do not feel confident of their skills. For example, many consumers pay experts to prepare their income tax returns rather than rely upon their own abilities to follow the directions and use a tax table.

INCOME TAX TABLE

This table is one that was used in a recent year. However, tables and tax laws change almost yearly. Visit a local IRS office and obtain their latest tax booklet.

SCHEDULE X
SINGLE TAXPAYERS

Use this schedule if you checked Filing Status Box 1 on Form 1040—

If the amount on Form 1040, line 34 is:		Enter on line 2 of the worksheet on this page:	
Over—	But not Over—		of the amount over—
$0	$2,300	—0—	
2,300	3,400	.... 14%	$2,300
3,400	4,400	$154 + 16%	3,400
4,400	6,500	314 + 18%	4,400
6,500	8,500	692 + 19%	6,500
8,500	10,800	1,072 + 21%	8,500
10,800	12,900	1,555 + 24%	10,800
12,900	15,000	2,059 + 26%	12,900
15,000	18,200	2,605 + 30%	15,000
18,200	23,500	3,565 + 34%	18,200
23,500	28,800	5,367 + 39%	23,500
28,800	34,100	7,434 + 44%	28,800
34,100	41,500	9,766 + 49%	34,100
41,500	55,300	13,392 + 55%	41,500
55,300	81,800	20,982 + 63%	55,300
81,800	108,300	37,677 + 68%	81,800
108,300	....	55,697 + 70%	108,300

The tax table shown was taken from material distributed by the Internal Revenue Service in a recent year and is one of many that appear in their tax booklet. You must read instructions carefully to determine which table to use for your particular tax return. Let us assume that you must use Schedule X and that your taxable income is $13,500. In order to determine your tax, you must find the interval in which this amount occurs.

Read down the table until you come to an interval that includes $13,500, that is, the row that gives the tax for amounts between $12,900 and $15,000.

| $12,900 | $15,000 | $2059 + 26% | $12,900 |

This means that your tax will be $2059 plus 26% of the amount of your income *in excess of* $12,900. To determine your tax, you must find the amount of your taxable income that exceeds $12,900, find 26% of that amount, and add this to $2059. Since your assumed taxable income is $13,500, the amount that is over $12,900 is the difference $13,500 − $12,900.

$$
\begin{array}{rl}
 & \$13,500 \\
- & \underline{12,900} \\
 & \$\ \ \ 600 \quad \text{(excess)}
\end{array}
\qquad
\begin{array}{rl}
 & \$600 \\
\times & \underline{0.26} \\
 & 36\ 00 \\
 & \underline{120\ 0\ \ } \\
 & \$156.00 \quad \text{(26\% of excess)}
\end{array}
$$

Your tax is the sum of the amount ($2059) given in the table and $156, which is 26% of the excess of $600.

$$
\begin{array}{rl}
 & \$2059 \\
+ & \underline{\ \ \ 156} \\
 & \$2215
\end{array}
$$

Your tax is $2215. It is often of interest to determine what percent your tax is of your taxable income. That is, $2215 is what percent of $13,500?

To solve for n, use a calculator and complete the division $221{,}500 \div 13{,}500$.

$$\frac{2215}{13,500} = \frac{n}{100}$$

$$13,500n = 221,500$$

$$n \approx 16.41$$

To the nearest percent, the tax rate for a taxable income of $13,500 is 16%.

Example 1 Use the tax table to determine the tax on a taxable income of $18,750.

Solution

First find the appropriate line in the tax table.

$18,200	$23,500	$3565 + 34%	$18,200

Find 34% of the excess of $18,750 over $18,200.

Be careful to locate the correct line in the table. Note that $18,750 is *between* $18,200 and $23,500.

$$
\begin{array}{r}
\$18{,}750 \\
-\ \ 18{,}200 \\
\hline
\$\ \ \ \ 550
\end{array}
\qquad
\begin{array}{r}
\$5\ 50 \\
\times\ \ 0.34 \\
\hline
22\ 00 \\
165\ 0 \\
\hline
\$187.00
\end{array}
$$

Add to find the total tax.

$$
\begin{array}{r}
\$3565 \\
+\ \ \ \ 187 \\
\hline
\$3752
\end{array}
$$

Next we shall look at a table that is helpful when one wishes to purchase an item on credit and determine the interest being paid. Consumers often are unaware of the actual rates of interest that they pay when purchasing items on an installment plan. Therefore Congress passed a *Truth-in-Lending Act* in 1969 that requires that a company inform the consumer of both the total interest (finance) charge and the *true annual interest rate,* also called the *effective annual rate,* that is being charged.

As an example, consider the purchase of a television set for $350 which can be financed by making monthly payments of $32.10 for a year. Thus the total amount paid to the merchant is found by this product:

$$12 \times \$32.10 = \$385.20$$

The total interest charge is found by subtracting the original cost from the actual amount paid under the installment plan.

$$\$385.20 - \$350.00 = \$35.20$$

Now if one computes what percent this interest charge is of the purchase price, the rate obtained is 10.06%.

35.20 is what percent of 350? This can be solved using a proportion.

$$\frac{35.20}{350} = \frac{n}{100}$$
$$350n = 3520$$
$$n = \frac{3520}{350}$$
$$n \approx 10.06$$

$$\frac{35.20}{350} \approx 0.1006 = 10.06\% \qquad (\$10.06 \text{ per } \$100)$$

Actually the true annual interest rate being charged is considerably higher than 10.06% because you do not have the full use of the $350 being financed for the entire year. Since you are obligated to repay the loan with monthly payments, during the final month you only owe $32.10 but are nevertheless paying interest on the full amount of $350.

To find the true annual interest rate for monthly payments, the Federal Reserve Bank has prepared the next table.

Number of Payments	10%	12%	14%	14½%	15%	15½%	16%	16½%	17%	18%	20%	25%	30%
6	$ 2.94	3.53	4.12	4.27	4.42	4.57	4.72	4.87	5.02	5.32	5.91	7.42	8.93
12	5.50	6.62	7.74	8.03	8.31	8.59	8.88	9.16	9.45	10.02	11.16	14.05	16.98
18	8.10	9.77	11.45	11.87	12.29	12.72	13.14	13.57	13.99	14.85	16.52	20.95	25.41
24	10.75	12.98	15.23	15.80	16.37	16.94	17.51	18.09	18.66	19.82	22.15	28.09	34.19
30	13.43	16.24	19.10	19.81	20.54	21.26	21.99	22.72	23.45	24.92	27.89	35.49	43.33
36	16.16	19.57	23.04	23.92	24.80	25.68	26.57	27.46	28.35	30.15	33.79	43.14	52.83
42	18.93	22.96	27.06	28.10	29.15	30.19	31.25	32.31	33.37	35.51	39.85	51.03	62.66
48	21.74	26.40	31.17	32.37	33.59	34.81	36.03	37.27	38.50	41.00	46.07	59.15	72.83

To use the table complete these two steps:

1. Divide the amount of interest paid by the amount financed.
2. Multiply by 100 to find the finance (interest) charge for each $100 financed.

In the example under discussion we have the following:

$$\frac{\text{Amount of interest}}{\text{Amount financed}} = \frac{35.20}{350} \approx 0.1006$$

$$0.1006 \times 100 = 10.06$$

Thus the finance charge is $10.06 for each $100 financed. Now enter the table in the row showing 12 payments and read across to the entry that is closest to 10.06. Note that there is an entry of 10.02; read up at this point to find 18% in the top row of the table. Thus the purchase of the television set under the conditions given actually costs the consumer a true annual interest rate of approximately 18%.

Example 2 A tape recorder costs $275. It can be bought for $50 down and payments of $14.50 per month for 18 months. What is the approximate true annual interest rate?

Solution The amount to be financed is the cash price less the down payment; that is, $275 − $50, or $225. Then

$$\text{Payments:} \quad 18 \times \$14.50 = \$261$$

$$\text{Interest:} \quad \$261 - \$225 = \$36$$

$$\frac{\text{Amount of interest}}{\text{Amount financed}} \times \$100 = \frac{36}{225} \times \$100 = \$16$$

Now go to the table and read across the row showing 18 payments. The closest entry to $16 is $16.52; read up to find the true annual interest rate of 20%. Actually, since $16 is somewhat less than $16.52, the true rate will be a little less than 20%.

Let us turn our attention to one more table that affects consumers although it is one that typical consumers will never encounter. Insurance companies compile data on deaths in the form of a mortality table, and use this to make predictions of life expectancy and to determine insurance rates.

The mortality table is based on past experiences. Starting with 10,000,000 newborn infants, 70,800 can be expected to die during their first year. Thus the death rate during the first year is 7.08 per thousand and 9,929,200 are expected to reach their first birthday. Similarly, it is assumed that 9,664,994 reach their 20th birthday, and 17,300 can be expected to die during their 20th year. The death rate for people of age 20 is 1.79 per thousand. The mortality table, Table II on page 103, can be used in several ways as illustrated in the following examples.

The tables used by insurance companies vary according to the use that is to be made of them. Also tables are changed as life expectancies increase or catastrophies, such as war, cause life expectancies to decrease. However, since it is based on very large numbers, the table provides a useful approximation of actual occurrence.

Example 3 According to the mortality table, what percent of the individuals who reach their 20th birthday can be expected to die before reaching the age of 40?

Solution Read the following information from the table.

$$
\begin{array}{rl}
9,664,994 & \text{(reach their 20th birthday)} \\
- \; 9,241,359 & \text{(reach their 40th birthday)} \\
\hline
423,635 & \text{(by subtraction, die before reaching their 40th birthday)}
\end{array}
$$

Next we need to determine what percent this difference is of the number who reached their 20th birthday. That is, 423,635 is what percent of 9,664,994?

$$\frac{423,635}{9,664,994} = \frac{n}{100}$$
$$9,664,994n = 42,363,500$$
$$n = 4.38 \quad \text{(to the nearest hundredth)}$$

To the nearest percent, we conclude that 4% of those alive at age 20 die before reaching age 40. A more optimistic point of view is that approximately 96%, that is, 100% minus 4%, of those who reach age 20 can expect to reach the age of 40.

Example 4 Compute the death rate per 1000 for people of age 25.

Solution We can compute this by means of the following proportion.

$$\frac{x}{1000} = \frac{\text{number dying during year}}{\text{number living at start of year}}$$

3 MATHEMATICS FOR THE CONSUMER

TABLE II COMMISSIONER'S 1958 STANDARD ORDINARY TABLE OF MORTALITY

Age	Number Living	Deaths Each Year	Deaths Per 1000	Age	Number Living	Deaths Each Year	Deaths Per 1000
0	10,000,000	70,800	7.08	50	8,762,306	72,902	8.32
1	9,929,200	17,475	1.76	51	8,689,404	79,160	9.11
2	9,911,725	15,066	1.52	52	8,610,244	85,758	9.96
3	9,896,659	14,449	1.46	53	8,524,486	92,832	10.89
4	9,882,210	13,835	1.40	54	8,431,654	100,337	11.90
5	9,868,375	13,322	1.35	55	8,331,317	108,307	13.00
6	9,855,053	12,812	1.30	56	8,223,010	116,849	14.21
7	9,842,241	12,401	1.26	57	8,106,161	125,970	15.54
8	9,829,840	12,091	1.23	58	7,980,191	135,663	17.00
9	9,817,749	11,879	1.21	59	7,844,528	145,830	18.59
10	9,805,870	11,865	1.21	60	7,698,698	156,592	20.34
11	9,794,005	12,047	1.23	61	7,542,106	167,736	22.24
12	9,871,958	12,325	1.26	62	7,374,370	179,271	24.31
13	9,769,633	12,896	1.32	63	7,195,099	191,174	26.57
14	9,756,737	13,562	1.39	64	7,003,925	203,394	29.04
15	9,743,175	14,225	1.46	65	6,800,531	215,917	31.75
16	9,728,950	14,983	1.54	66	6,584,614	228,749	34.74
17	9,713,967	15,737	1.62	67	6,355,865	241,777	38.04
18	9,698,230	16,390	1.69	68	6,114,088	254,835	41.68
19	9,681,840	16,846	1.74	69	5,859,253	267,241	45.61
20	9,664,994	17,300	1.79	70	5,592,012	278,426	49.79
21	9,647,694	17,655	1.83	71	5,313,586	287,731	54.15
22	9,630,039	17,912	1.86	72	5,025,855	294,766	58.65
23	9,612,127	18,167	1.89	73	4,731,089	299,289	63.26
24	9,593,960	18,324	1.91	74	4,431,800	301,894	68.12
25	9,575,636	18,481	1.93	75	4,129,906	303,011	73.37
26	9,557,155	18,732	1.96	76	3,826,895	303,014	79.18
27	9,538,423	18,981	1.99	77	3,523,881	301,997	85.70
28	9,519,442	19,324	2.03	78	3,221,884	299,829	93.06
29	9,500,118	19,760	2.08	79	2,922,055	295,683	101.19
30	9,480,358	20,193	2.13	80	2,626,372	288,848	109.98
31	9,460,165	20,718	2.19	81	2,337,524	278,983	119.35
32	9,439,447	21,239	2.25	82	2,058,541	265,902	129.17
33	9,418,208	21,850	2.32	83	1,792,639	249,858	139.38
34	9,396,358	22,551	2.40	84	1,542,781	231,433	150.01
35	9,373,807	23,528	2.51	85	1,311,348	211,311	161.14
36	9,350,279	24,685	2.64	86	1,100,037	190,108	172.82
37	9,325,594	26,112	2.80	87	909,929	168,455	185.13
38	9,299,482	27,991	3.01	88	741,474	146,997	198.25
39	9,271,491	30,132	3.25	89	594,477	126,303	212.46
40	9,241,359	32,622	3.53	90	468,174	106,809	228.14
41	9,208,737	35,362	3.84	91	361,365	88,813	245.77
42	9,173,375	38,253	4.17	92	272,552	72,480	265.93
43	9,135,122	41,382	4.53	93	200,072	57,881	289.30
44	9,093,740	44,741	4.92	94	142,191	45,026	316.66
45	9,048,999	48,412	5.35	95	97,165	34,128	351.24
46	9,000,587	52,473	5.83	96	63,037	25,250	400.56
47	8,948,114	56,910	6.36	97	37,787	18,456	488.42
48	8,891,204	61,794	6.95	98	19,331	12,916	668.15
49	8,829,410	67,104	7.60	99	6,415	6,415	1000.00

For people of age 25 we have

$$\frac{x}{1000} = \frac{18,481}{9,575,636}$$

$$x = \frac{18,481,000}{9,575,636} \approx 1.93$$

That is, for people of age 25, we can expect approximately two deaths per 1000 persons. Note that the mortality table includes a column with deaths per 1000 already computed.

The data in the mortality table can also be presented in graphical form. The following graphs present a general picture of these data, although specific numerical facts can be only approximated from these graphs.

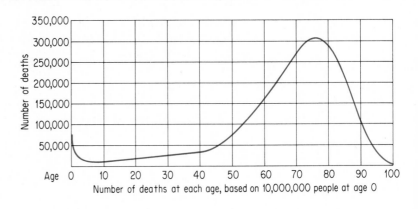

Number of deaths at each age, based on 10,000,000 people at age 0

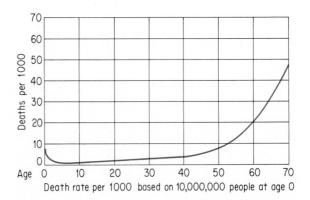

Death rate per 1000 based on 10,000,000 people at age 0

Consider a group of 100 20-year-olds each of whom wishes to purchase a $1000 twenty-year **term policy**. Under this plan,

3 MATHEMATICS FOR THE CONSUMER

Term insurance provides for protection only, and involves no savings feature. As such, it enables a person to purchase for a given period of time (term) the maximum amount of insurance for any fixed sum of money.

the company will pay $1000 to the person designated as a *beneficiary* upon the death of the insured. If the insured lives for 20 years, that is, until the person is age 40, then the policy terminates.

To help determine the premium to charge for such a policy, the company turns to the mortality table and notes that about 4% of all 20-year-olds will die before reaching the age of 40 (see Example 2). Therefore, 4% of the 100 to be insured, or 4 persons, can be expected to die within the next 20 years according to past experience. Thus the company will have to pay their beneficiaries 4 × $1000, or $4000. This fact, together with prevailing interest rates and company expenses, will help an *actuary* employed by the insurance company determine the annual charge for each of the 100 persons who will, in a sense, share the risk of death.

If the 100 persons had decided to purchase a 20-year **endowment policy**, the charges would be quite different. Under this plan, the company agrees to pay $1000 to any named beneficiary upon the death of the insured. Therefore, the company can still expect to pay approximately $4000 in death benefits during this time. However, the company also agrees to pay $1000 to anyone of the original group who is still alive at age 40! In order to accumulate sufficient funds to pay these benefits, the company obviously must charge a larger premium than they would for a term policy.

There are many other types of life insurance policies and all have premiums that are determined on the basis of experience, usually as represented by mortality tables such as the one given in Table II.

EXERCISES *Use the tax table given in this section and find the tax for each of the given incomes.*

1. $4000 2. $6500 3. $10,750 4. $18,800 5. $20,000
6. $10,200 7. $22,100 8. $28,200 9. $35,000 10. $25,750

11. Find the amount of tax, and find what percent this is of the income for a taxable income of $10,500.

12. Repeat Exercise 11 for a taxable income of $19,500.

*13. Craig computes his tax and finds it to be $5000, using the tax table of this section. What was his taxable income to the nearest dollar?

*14. Repeat Exercise 13 for a tax of $2500.

15. Find the tax on an income of $10,100. Then find the tax on an income of $10,200. Compare your answers and determine how much of the extra $100 you actually keep if your income increases from the first to the second amount.

Use the True Annual Interest Rate table on page 101 to answer Exercises 16 through 19.

16. A bicycle can be bought for $125 cash. On an installment plan it costs $20 down and $10 per month for 12 months. What is the approximate true annual interest rate under this plan?

17. Repeat Exercise 16 for a plan that calls for a down payment of $45 and payments of $14 per month for 6 months.

18. A used automobile is selling for $3000. It can also be bought for a down payment of $300 and payments of $90 per month for 3 years. Find the approximate true annual interest rate under this installment plan.

19. Repeat Exercise 18 for a plan that calls for a down payment of $750 and payments of $80 per month for 3 years.

Use Table II to answer Exercises 20 through 22.

20. Find the percent of the people who reach age 20 that can be expected to die before reaching the specified age.
    (a) 50   (b) 60   (c) 70   (d) 80

21. Find the percent of the people who reach age 20 that can be expected to reach the specified age.
    (a) 30   (b) 45   (c) 65   (d) 75

22. Use the mortality table to find the age at which the number of people still living is approximately one-half of the number that reached age 20. The difference between this age and 20 is called the *median future lifetime* at age 20. That is, it represents the number of years it will take before one-half of the 20-year-olds will have died. Find the median future lifetime at age 20; at age 30.

Use the graphs based on the mortality table to answer Exercises 23 through 25.

23. The graph of the death rate per 1000 stops at age 70. What happens to this curve after age 70?

24. Explain why both graphs show an initial decrease before both start to rise again.

25. Both graphs have the same shape at first. The latter portions, however, are quite different. Explain why this is so.

Explorations

1. Collect as many examples as you can find of different types of tables from newspapers and other periodicals.
2. Speak to a life insurance agent and attempt to find the difference between a term policy and a whole-life policy. Determine the advantages and disadvantages of each and report on this to your class.
3. Store clerks often use a table to determine sales taxes. Attempt to obtain a copy of such a table for your state and explain its use.
4. Select a specific type of insurance, such as fire insurance on a house, automobile insurance with specified coverages, or theft insurance. Describe some of the circumstances that affect the cost of the selected type of insurance.

3-5 Measurement

Most people expect mathematics to serve two basic needs: to compute and to measure. The typical consumer often needs to measure to find the dimensions of a room in order to buy furniture, the size of a lawn in order to buy seed, the area of a wall in order to buy wallpaper, and so on. We also need to understand measurement when we note the weight or capacity of a container or read road signs concerning distances. In fact, throughout the ages people have had a need to measure. As the Nile River overflowed its banks each year, the ancient Egyptians needed to survey the flooded area to determine boundary marks, and they are believed to have used ropes stretched to form right angles in doing so. Because of the process used, these surveyors were actually referred to as "rope-stretchers." From the Pythagorean theorem we know that if a rope is knotted to form 3, 4, and 5 units, respectively, a right triangle and therefore a right angle can be formed, as shown in the figure.

Palm

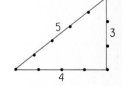

Span

Various parts of the human body were used as early units of measure. Records of early Egyptian and Babylonian civilizations indicate that lengths were first measured by such units as a **palm** and a **span,** the distance covered by an outstretched hand.

Another widely used early unit of measurement was a **cubit,** the distance from one's elbow to the tip of the middle finger when the hand is held straight. For most people, this is a distance of approximately 18 inches.

Mention of the cubit has been found in excavations that date back to 3000 B.C., and it is a unit of measure frequently mentioned in the Bible. Thus in the sixth chapter of Genesis we find a description of Noah's ark that states "the length of the ark shall be 300 cubits, the breadth of it 50 cubits, and the height of it 30 cubits."

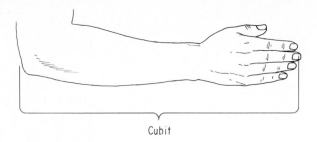

Cubit

There are many interesting stories about the origins of units of measure. It is said that King Henry I of England decreed that the distance from the tip of his nose to the end of his thumb should be the official equivalent of one yard. Then in the thirteenth century, King Edward I declared that one-third of a yard should be called the foot. Later, in the sixteenth century, Queen Elizabeth I declared that the traditional *Roman mile* of 5000 feet would be replaced by one of 5280 feet. However, it was the Romans who used 12 in their measurements and provided a basis for our use of 12 inches to a foot.

Although the use of a unit of measure such as the cubit was more meaningful than such units as "a bundle" or "a day's journey," it certainly was not a standardized unit and obviously varied from person to person. There was a great need throughout the world for the establishment of a uniform standard for all weights and measures.

About 300 years ago Gabriel Mouton, Vicar of St. Paul in Lyons, France, recognized the need for a standardized system of measurement that would be accepted throughout the world. In 1670 he proposed a decimal system of measurement that was based on a standard unit equal to the length of one minute of arc of a great circle of the earth. In 1790, in the midst of the French Revolution, the National Assembly of France asked the French Academy of Sciences to create a uniform standard for all weights and measures.

In establishing a uniform standard there are a number of fundamental principles that need to be kept in mind.

1. The standard unit used should be based on some invariant factor in the physical universe.

2. Basic units of length, capacity (volume), and weight (mass) should be interrelated.
3. The multiples and subdivisions of the basic unit should be in terms of the decimal system, that is, in terms of powers of ten.

With these principles in mind, the Academy proposed that the basic unit of length be one ten-millionth of an arc drawn from the North Pole to the equator. They called this basic unit the **metre (m)**, as it is commonly known in most parts of the world and in scientific work. The word *metre* is derived from the Greek *metron,* "a measure," and is usually spelled as **meter** in the United States. One meter is a little longer than a yard. One-tenth of a meter is 1 **decimeter;** one-tenth of a decimeter is 1 **centimeter (cm)**. We can also think of these units of measure in terms of groups of ten in this way.

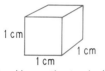

1 centimeter (cm)

$$1 \text{ meter} = 10 \text{ decimeters}$$
$$1 \text{ decimeter} = 10 \text{ centimeters}$$

We conclude that 1 meter = 100 centimeters, or that 1 centimeter is one-hundredth of a meter.

The name **gram (g)** was assigned to a metric unit of mass (weight). The gram was defined as the weight of 1 **cubic centimeter** of water. A cubic centimeter is a cube with each edge of length 1 centimeter.

1 cm
1 cm
1 cm
1 cubic centimeter (cc)

As a basic measure of capacity, the Academy chose the **liter** and defined this as a **cubic decimeter,** that is, the volume of a cube with each edge of length 1 decimeter.

1 decimeter

In 1840 France made it compulsory to use the metric system, and soon thereafter many other nations followed suit. In 1875 the Metric Convention met and 17 nations (including the United States) signed the "Treaty of the Meter," which established permanent metric standards for length and weight. By 1900 most of the nations of Europe and South America, a total of 35 in all, had officially adopted the metric system of measurement. There is now an international General Conference on Weights and Measures, to which the United States belongs.

About 1890 the International Bureau of Weights and Measures defined the meter as the distance between two marks on a platinum bar that was kept at a constant temperature of 32° Fahrenheit. This distance was approximately equal to one ten-millionth of the distance from the North Pole to the equator. However, in 1960 the Eleventh General Conference on Weights and Measures abandoned the meter bar as the international standard of length, and redefined the meter in terms of 1 650 763.73 wavelengths of the orange-red line in the spectrum of krypton 86. With this definition, lengths of one meter could be reproduced in scientific laboratories anywhere in the world. The 1960 Conference adopted a revision and simplification of the metric system, *Le Systéme International d' Unités* (International System of Units), now known as SI.

Commas are replaced by spaces in the metric system of notation for numbers.

The SI system of measure is really quite simple. First a basic unit, such as the meter, gram, or liter, is adopted. Thereafter, multiples and subdivisions of this unit are always given in terms of powers of ten. Greek prefixes are used to denote multiples, and Latin prefixes to denote subdivisions. Some of the basic prefixes used are shown in the following table.

tera-	one trillion	1 000 000 000 000	10^{12}
giga-	one billion	1 000 000 000	10^{9}
mega-	one million	1 000 000	10^{6}
kilo-	one thousand	1 000	10^{3}
hecto-	one hundred	100	10^{2}
deka-	ten	10	10^{1}
BASE UNIT	one	1	10^{0}
deci-	one-tenth	0.1	10^{-1}
centi-	one-hundredth	0.01	10^{-2}
milli-	one-thousandth	0.001	10^{-3}
micro-	one-millionth	0.000 001	10^{-6}
nano-	one-billionth	0.000 000 001	10^{-9}
pico-	one-trillionth	0.000 000 000 001	10^{-12}

There are several observations that should be made concerning the table. Note the increasing (or decreasing) powers of ten as one reads up (or down) the table. In the SI system, spaces are used instead of commas to separate groups of three digits starting from the decimal point in either or both directions. The reason for this is the use, in certain countries, of commas to represent decimal points.

Although these units will be discussed in detail in the following section, let us briefly explore the use of these prefixes with the basic unit of 1 meter. Using the table, you should be able to see such relationships as the following.

1 *kilo*meter = 1000 meters   1 meter = 0.001 kilometer
1 *centi*meter = 0.01 meter   1 meter = 100 centimeters

In the metric system any measurement in terms of one unit may be expressed in terms of a larger unit very easily.

Simply divide by an appropriate power of ten; that is, move the decimal point an appropriate number of places to the left, as in Example 1.

Example 1   (a) Change 358 centimeters to meters.
                 (b) Change 7495 meters to kilometers.

Solution   (a) 100 centimeters = 1 meter;   358 centimeters = 3.58 meters
              (b) 1000 meters = 1 kilometer;
                  7495 meters = 7.495 kilometers

Any measurement in terms of one unit may be expressed in terms of a smaller unit. Simply multiply by an appropriate power of ten; that is, move the decimal point an appropriate number of places to the right, as in Example 2.

Example 2   (a) Change 8.35 meters to centimeters.
                 (b) Change 15.755 kilometers to meters.

Solution   (a) 1 meter = 100 centimeters;   8.35 meters = 835 centimeters
              (b) 1 kilometer = 1000 meters;
                  15.755 kilometers = 15 755 meters

EXERCISES   *Classify each statement as true or false.*

1.   100 meters = 1 kilometer
2.   1 meter = 100 centimeters
3.   1 centimeter = 100 millimeters
4.   10 millimeters = 1 centimeter
5.   1 decimeter = 10 centimeters
6.   1 meter = 1000 millimeters
7.   1 millimeter = 0.01 centimeter
8.   10 meters = 1 decimeter
9.   1 centimeter = 0.01 meter
10.   1 centimeter = 0.01 decimeter

Complete each statement.

11.   5 meters = _____ centimeters
12.   3 kilometers = _____ meters
13.   4 centimeters = _____ millimeters
14.   475 centimeters = _____ meters
15.   8350 meters = _____ kilometers
16.   75 millimeters = _____ centimeters

Use decimal notation to change as indicated.

17. 358 centimeters to meters
18. 482 millimeters to centimeters
19. 3785 meters to kilometers
20. 7500 millimeters to meters
21. 85 kilometers to meters
22. 9.82 meters to centimeters
23. 5.8 centimeters to millimeters
24. 3.5 meters to millimeters

Explorations

1. Make a collection of illustrations of popular uses of measurements that are not standardized. For example, consider "a pinch of salt."
2. Estimate the length of your car in terms of cubits. Then use your arm to find its length using this measure.
3. Repeat Exploration 2 for the distance across the front of your classroom.
4. Since it takes 100 pennies to equal one dollar, we may write such equivalents as 249¢ = $2.49 and $3.75 = 375¢. Discuss how this idea may be used to help illustrate translations from a number of centimeters to a number of meters and also from a number of meters to a number of centimeters.

3-6
The Metric System

The United States has initiated steps toward officially adopting the metric system during the decade of the 1980s. Almost all of the scientific laboratories in this country use the metric system and have done so for many years.

Today most of the civilized world uses the metric system of measurement. We make use of a number of metric units in our daily lives. For example, many people own or at least have heard of a 35-millimeter camera, and contents of vitamins are usually given in terms of multiples of a gram. Furthermore, some road signs give distances in both miles and kilometers. The general public needs to understand such measurements.

The metric system is based on powers of ten. Since our system of numeration is also based on powers of ten, computations using metric units are relatively simple. For practical everyday use, there are surprisingly few units of measure to learn. The basic unit for measuring lengths is the meter (m). A meter is approximately 39 inches in length. The following scale drawing shows the comparison between 1 meter and 1 yard.

Normally, the yard is divided into 36 parts, each 1 inch in length. The meter, as we have seen, is divided into 100 parts,

Centimeters
0 100
0 36
Inches

each 1 centimeter in length. It is well to have an intuitive feeling for the size of 1 centimeter. The centimeter is about 0.4 of an inch. A nickel is about 2 centimeters in diameter (wide). An inch is *defined* to be 2.54 centimeters and we often use the approximation

$$1 \text{ inch} \approx 2\tfrac{1}{2} \text{ centimeters}$$

The figures that follow show the *actual size* of 1 centimeter, the *actual size* of 1 inch, and a comparison of the lengths of centimeters and inches on a part of a ruler that is marked in both centimeters and inches.

1 centimeter: |_____|

1 inch: |_____|

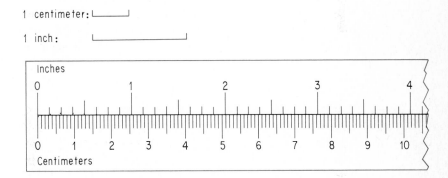

As in the previous figure each centimeter is divided into ten parts, each 1 millimeter (mm) in length.

$$1 \text{ centimeter} = 10 \text{ millimeters} \qquad (1 \text{ cm} = 10 \text{ mm})$$
$$1 \text{ millimeter} = 0.1 \text{ centimeter} \qquad (1 \text{ mm} = 0.1 \text{ cm})$$

Also as in the previous figures, 10 centimeters ≈ 4 inches.

Note that the symbol cm has no period. The same convention (no periods) is followed for all symbols for metric units.

To help you visualize the size of a millimeter note that a nickel is about 2 millimeters thick; 35 millimeter film is 35 millimeters in width.

For measuring greater distances we use the kilometer (km), which is equal to 1000 meters. A kilometer is approximately equal to 0.6 of a mile, so that 10 kilometers is equal to about 6 miles.

For normal, everyday usage in a country that has adopted the metric system, the average citizen uses the centimeter for small measurements, the meter for somewhat larger ones, and the kilometer for greater distances. Thus, in countries that have adopted the metric system, distances between cities are given in terms of kilometers. However, the meter is the basic unit of length. Several common multiples of this unit are shown in the following table. Note that the prefixes used for multiples of a meter are those for the basic units in the metric (SI) system as shown in Section 3-5.

1 *kilo*meter	=	1000 meters
1 *hecto*meter	=	100 meters
1 *deka*meter	=	10 meters
1 meter	=	1 meter
1 *deci*meter	=	0.1 meter
1 *centi*meter	=	0.01 meter
1 *milli*meter	=	0.001 meter

Example 1 The distance between the cities of New York and San Francisco is 4120 kilometers. About how many miles is this?

Solution Recall that 1 kilometer is approximately 0.6 mile. Since 0.6 × 4120 = 2472, the distance is about 2472 miles. Actually, a motorist would probably want a rough approximation only and would therefore round off the distance to 4000 kilometers. In miles, this would be approximately equal to the product 0.6 × 4000 = 2400.

Example 2 Normally, metric rulers are marked in centimeters and subdivided into millimeters. Read each indicated point on this metric ruler in millimeters and in centimeters.

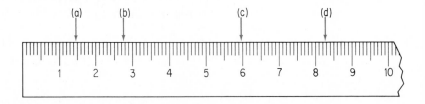

Solution Recall that 1 centimeter = 10 millimeters.
(**a**) 15 mm = 1.5 cm   (**b**) 28 mm = 2.8 cm
(**c**) 60 mm = 6.0 cm   (**d**) 83 mm = 8.3 cm

Example 3 Complete (**a**) 8 cm = _____ mm   (**b**) 35 mm = _____ cm
(**c**) 3 km = _____ m   (**d**) 2500 m = _____ km.

Solution (a) 1 cm = 10 mm;   8 cm = 80 mm
(b) 10 mm = 1 cm;   35 mm = 3.5 cm
(c) 1 km = 1000 m;   3 km = 3000 m
(d) 1000 m = 1 km;   2500 m = 2.5 km

To measure volume, the basic unit used is the **liter.** One liter is just a little greater than a quart and is the metric measure of capacity that one is most likely to encounter in daily activities. However, various multiples of 1 liter are obtained by using the set of prefixes previously listed. Again, note the use of decimal notations, with each unit being ten times as great as the one listed beneath it in this table.

INTERNATIONAL SYSTEM OF
WEIGHTS AND MEASURES

GRAM   LITRE   METRE

WEIGHT   CAPACITY   LENGTH

PAKISTAN   20P
پاکستان   POSTAGE

1 *kilo*liter	=	1000 liters
1 *hecto*liter	=	100 liters
1 *deka*liter	=	10 liters
1 liter	=	1 liter
1 *deci*liter	=	0.1 liter
1 *centi*liter	=	0.01 liter
1 *milli*liter	=	0.001 liter

One might encounter the **milliliter (ml)** in small measurements. For example, in cooking and baking, it would be important to know that 5 milliliters is approximately equivalent to 1 teaspoonful.

The *gram* is a unit of metric measure for *mass.* The *weight* of an object is determined by its mass and the force of gravity to which it is subjected. For example, an astronaut's body has the same mass on earth as when he or she is in a weightless condition orbiting the earth. However, most of us stay relatively close to the surface of the earth. Therefore, we are concerned with objects that are subjected to an approximately constant force of gravity. Thus we shall use the term *weight* instead of using the technical term and concept of *mass.* Then the *gram* is also a unit for weight; a nickel weighs about 5 grams. As before, other multiples of a gram are obtained as in the following table.

1 *kilo*gram	=	1000 grams
1 *hecto*gram	=	100 grams
1 *deka*gram	=	10 grams
1 gram	=	1 gram
1 *deci*gram	=	0.1 gram
1 *centi*gram	=	0.01 gram
1 *milli*gram	=	0.001 gram

Of these various multiples of the gram, the one that is most likely to be encountered when the United States finally "goes metric" is the **kilogram** (**kg**). A kilogram is a little over 2 pounds, about 2.2 pounds, and is generally referred to as a **kilo**. Thus a shopper ordering 1 kilo of apples can expect to receive a little more than 2 pounds of apples; by ordering one-half of a kilo, one would receive about 1 pound.

The **milligram** (**mg**) is a frequently used unit of measure for various medicines and vitamins. A typical multipurpose vitamin pill might contain 250 milligrams of vitamin C. Inasmuch as 1 milligram is equal to one-thousandth of a gram, 250 milligrams is equal to one-fourth of a gram in weight.

Example 4 An athlete weighs 100 kilograms. About how many pounds is this?

Solution One kilogram is approximately equal to 2.2 pounds. Therefore, 100 kilograms $\approx 2.2 \times 100 = 220$ pounds.

Example 5 Complete (**a**) 3 kg = _____ g (**b**) 2500 g = _____ kg (**c**) 5 g = _____ mg (**d**) 7000 mg = _____ g.

Solution (**a**) 1 kg = 1000 g; 3 kg = 3000 g
(**b**) 1000 g = 1 kg; 2500 g = 2.5 kg
(**c**) 1 g = 1000 mg; 5 g = 5000 mg
(**d**) 1000 mg = 1 g; 7000 mg = 7 g

EXERCISES *Each of the following sentences uses a metric measure. State whether each seems likely, or unlikely.*

1. A college football player weighs 200 kilograms.
2. A college basketball player is 2 meters tall.
3. An empty tank in a compact automobile will hold 40 liters of gasoline.
4. A quarter weighs 25 grams.
5. The average student drinks 5 liters of water daily.
6. The diameter of a nickel is about 5 centimeters.
7. A new pencil is about 20 centimeters long.
8. When airborne a small plane was flying at the rate of 800 kilometers per hour.
9. A typical man's wristwatch has a diameter of about 25 millimeters.
10. A student ate a steak that weighed 2 kilograms.

Select the most likely answer for each situation.

11. The weight of a quarter is approximately equal to
  (a) 1 gram   (b) 5 grams
  (c) 15 grams   (d) 30 grams

12. The weight of this book is approximately equal to
  (a) 10 grams   (b) 50 grams
  (c) 1 kilogram   (d) 10 kilograms

13. Ten gallons of gasoline is approximately equal to
  (a) 20 liters   (b) 30 liters
  (c) 40 liters   (d) 80 liters

14. The distance between New York City and Washington, D.C., is approximately equal to
  (a) 50 kilometers   (b) 125 kilometers
  (c) 200 kilometers   (d) 400 kilometers

15. Some doctors recommend that a person should spend an hour per day walking. In this time, one would probably walk
  (a) 1 kilometer   (b) 5 kilometers
  (c) 10 kilometers   (d) 15 kilometers

Change each of the following as indicated.

16. 5 meters to centimeters
17. 8 kilometers to meters
18. 3 centimeters to millimeters
19. 80 millimeters to centimeters
20. 350 centimeters to meters
21. 15 000 meters to kilometers
22. 3000 grams to kilograms
23. 7 kilograms to grams
24. 7500 milligrams to grams
25. 2500 milliliters to liters

Read each indicated point on this metric scale
(a) *in millimeters*   (b) *in centimeters.*

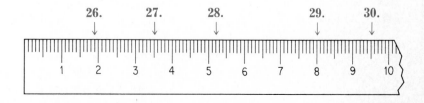

31. A weight of 200 pounds is about how many kilograms? (Use 1 kilogram $\approx$ 2.2 pounds; give your answer to the nearest kilogram.)

32. A weight of 150 kilograms is about how many pounds?

33. A length of 5 inches is approximately how many centimeters? (Use 1 inch $\approx$ 2.54 centimeters.)

34. Note that 1 inch is equal to 2.54 centimeters, and complete these statements.
    (a) 1 cm $\approx$ _____ inch   (b) 40 cm $\approx$ _____ inches

35. A distance of 400 kilometers is about how many miles? (Use 1 kilometer $\approx$ 0.6 mile.)

36. Note that 1 kilometer is approximately equal to $\frac{3}{5}$ of a mile, and complete these statements.
    (a) 1 mile $\approx$ _____ km   (b) 10 miles $\approx$ _____ km

*37. The prefix *mega* represents 10^6, that is, 1 000 000. Thus 1 megameter = 1 000 000 meters. Complete each statement.
    (a) 1 megaliter = _____ liters
    (b) 1 megagram = _____ grams
    (c) 1 megameter = _____ kilometers
    (d) 5000 kilometers = _____ megameters

*38. The prefix *micro* represents 10^{-6}, that is, 0.000 001. Complete each statement.
    (a) 1 micrometer = _____ meter
    (b) 1 meter = _____ micrometers
    (c) 5 grams = _____ micrograms
    (d) 3 000 000 micrograms = _____ grams

Explorations

1. Collect as many popular sayings as you can find that involve units of measure. For example, here are two such sayings.
    (a) "An ounce of prevention is worth a pound of cure."
    (b) "Give him an inch and he'll take a mile."
    Then translate each into metric units of measure.

2. Obtain a metric ruler and find the measure of the width of your palm, your hand span, and the distance from the tip of your elbow to the end of your outstretched middle finger, all to the nearest centimeter. Compare these measurements with those of other members of your class.

3. Prepare a collection of everyday objects that show measurements in metric units.

4. Use a metric ruler and find the dimensions of a one-dollar bill to the nearest millimeter.

5. Use a metric ruler and find the diameter of a quarter to the nearest millimeter.

6. Mark off, as carefully as possible, a 10-centimeter scale on a strip of paper about 2 centimeters wide. Then make nine more such scales and fasten them together to form a strip one meter long and subdivided into centimeters.

7. Construct an open box with each edge 10 centimeters long. Then the volume of the box will be one cubic decimeter, which is, by definition, one liter.

Chapter 3 Review

Solutions to the following exercises may be found within the text of Chapter 3. Try to complete each exercise without referring to the text.

Section 3-1 Percent

1. Write as a fraction in simplest form (a) 70%   (b) 45%.

2. Write as a percent (a) 0.32   (b) $\dfrac{7}{10}$.

3. Write $\dfrac{5}{8}$ as a percent.

4. Find 25% of 80.

5. 20 is what percent of 80?

6. 20 is 25% of what number?

Section 3-2 Applications of Percent

7. A television set costs $245. For one week only it is offered on sale at 22% off. How much can be saved by buying the set during the sale week?

8. The regular price of a radio is $80. The sale price is $64. What percent discount is being offered during this sale?

9. You wish to leave a tip of 15% for a bill of $8.20. Find the amount of the tip mentally.

Section 3-3 Simple and Compound Interest

10. Find the simple interest on a personal loan of $2500 at 6% for 3 years.

11. Find the amount on deposit at the end of 1 year if $2000 is deposited at 5% interest compounded semiannually.

12. Use Table I and find the amount for a single deposit of $400 **(a)** at 6% compounded semiannually for 4 years **(b)** at 12% compounded quarterly for 5 years.

Section 3-4   Using Tables

13. Use the tax table to determine the tax on a taxable income of $18,750.

14. A tape recorder costs $275. It can be bought for $50 down and payments of $14.50 per month for 18 months. What is the approximate true annual interest rate?

15. According to Table II, what percent of the individuals who reach their 20th birthday can be expected to die before reaching age 40?

16. Compute the death rate per 1000 for people of age 25.

Section 3-5   Measurement

17. **(a)** Change 358 centimeters to meters.
    (b) Change 7495 meters to kilometers.

18. **(a)** Change 8.35 meters to centimeters.
    (b) Change 15.755 kilometers to meters.

Section 3-6   The Metric System

19. The distance between the cities of New York and San Francisco is 4120 kilometers. About how many miles is this?

20. Complete

    (a) 8 cm = ____ mm   **(b)** 35 mm = ____ cm
    (c) 3 km = ____ m   **(d)** 2500 m = ____ km

21. Complete

    (a) 3 kg = ____ g   **(b)** 2500 g = ____ kg
    (c) 5 g = ____ mg   **(d)** 7000 mg = ____ g

Chapter 3 Test

1. Write each percent as a decimal.
   (a) 93%   **(b)** 1%   **(c)** 125%   **(d)** 0.1%   **(e)** 100%

2. Write each percent as a fraction in simplest form.
   (a) 45%   **(b)** 5%   **(c)** 80%   **(d)** 120%   **(e)** 0.6%

3. Write each decimal as a percent.
   (a) 0.27   **(b)** 0.02   **(c)** 0.93   **(d)** 1.25   **(e)** 0.008

4. Write each fraction as a percent.
   (a) $\dfrac{17}{100}$ **(b)** $\dfrac{132}{100}$ **(c)** $\dfrac{19}{20}$ **(d)** $\dfrac{27}{50}$ **(e)** $\dfrac{13}{10}$

Use an equivalent fraction to find each percent.

5. **(a)** 75% of 240 **(b)** 40% of 320
6. **(a)** 125% of 72 **(b)** 66⅔% of 63
7. **(a)** 80% of 420 **(b)** 33⅓% of 252

Write a proportion and solve the problem.

8. Find 45% of 70. 9. Find 115% of 60.
10. 45 is what percent of 75? 11. 18 is 20% of what number?

Find the selling price of an item for each given regular price and discount.

12. $120, less 15% 13. $80, less 23%

Find the percent discount being offered. (Round answers to the nearest percent.)

14. Regular price: $72 15. Regular price: $84.00
    Sale price: $60 Sale price: $71.40

16. Find the amount for a deposit of $1600 at 6% simple interest for 2 years.
17. Find the amount for a deposit of $1200 at 8% interest compounded semiannually for 5 years.
18. Use the tax table in this chapter to find the tax on an income of $16,750.
19. Use Table II to find the percent of the individuals who reach their 18th birthday that can be expected to be alive at age 60.
20. A guitar sells for $300 cash. It can also be bought for a down payment of $50 and payments of $12 per month for 2 years. Find the approximate true annual interest under this installment plan.

Change each of the following as indicated.

21. **(a)** 3 meters to centimeters
    (b) 5 kilometers to meters
    (c) 2 centimeters to millimeters
    (d) 3.75 meters to centimeters
22. **(a)** 5000 meters to kilometers
    (b) 3500 milligrams to grams
    (c) 12 grams to milligrams
    (d) 1500 milliliters to liters

Select the best answer.

23. A distance of 500 kilometers is approximately how many miles?
    (a) 100   (b) 200   (c) 300   (d) 400   (e) 500

24. A weight of 90 kilograms is approximately how many pounds?
    (a) 50   (b) 100   (c) 150   (d) 200   (e) 250

25. A capacity of 20 liters is approximately how many gallons?
    (a) 5   (b) 10   (c) 20   (d) 40   (e) 80

Sets
and
Logic

4

4-1
Set Notation

The authors welcome you to the *set* of readers of this book. You may be a *member* of a college class for which this book is the assigned text. If so, the members of your class who read this page before it was discussed in class form a *subset* of the set of members of the class. Relative to the *universal set* of members of your class, the set of members who did not read this page before it was discussed in class is the *complement* of the set of members of the class who did. If each person in your class has exactly one copy of the textbook of his or her own, then there is a *one-to-one correspondence* between the set of members of your class and the set of their textbooks; that is, the set of members of your class and the set of their textbooks are *equivalent sets.*

The terms set, member (or element), subset, universal set, complement, one-to-one correspondence, and equivalent sets, as used in the preceding paragraph, are part of the vocabulary of the language of mathematics. Each term has a very precise meaning. We shall use examples, counterexamples, exercises, and explorations to reinforce and "sharpen" your understanding of the terms considered.

There is a classic problem of which came first, the chicken or the egg. We have a similar problem in the language of mathematics — we cannot define everything. Thus we must assume that you already understand some terms from your previous experiences. For example, we assume that you know what is meant by a *set* (collection) of elements and also what is meant by a particular element *being a member of (belonging to)* a specified set.

Consider these two sets.

The set of letters of the English alphabet.
The set of states of the United States of America on January 1 of the bicentennial year 1976.

You can tell whether or not any specified element belongs to either of these given sets; that is, each set is a **well-defined set.**

Whenever there can be doubt as to the membership of an element in a set, the set is not a well-defined set. For example, the following sets are not well-defined sets.

The set of good tennis players.
The set of successful country music singers.

Mathematics is primarily concerned with well-defined sets. The elements of these sets may be numbers, points, geo-

The study of mathematics may be considered as the study of a new language. Mathematics has its own vocabulary and its own grammar (rules for combining its terms and symbols).

These sets are not well-defined sets because there can be differences of opinion, possibly based on different standards, whether or not a particular person is a good tennis player or a successful country music singer.

4 SETS AND LOGIC

metric figures, blocks of various shapes (sizes, colors, materials, and so on), statements (equations, inequalities, and so on), people, or other identifiable entities.

Consider these sets of numbers.

$$C = \{1, 2, 3, 4, \ldots\} \qquad \text{the set of \textbf{counting numbers}}$$
$$W = \{0, 1, 2, 3, \ldots\} \qquad \text{the set of \textbf{whole numbers}}$$
$$D = \{0, 1, 2, 3, \ldots, 9\} \qquad \text{the set of \textbf{decimal digits}}$$

We have used braces and indicated the sets C, W, and D in set notation. The three dots are used to indicate elements that are not explicitly listed. The list continues in the indicated pattern either indefinitely as in sets C and W or until the specified last element is reached as in set D. The sets C and W do not have a last element (they continue indefinitely) and are **infinite sets.** The set D has a last element and is a **finite set.**

In general, the order in which the elements of a set are listed is unimportant. However, when three dots are used to indicate elements that are not explicitly listed, it is necessary to state the elements in some order so that a pattern can be observed and used to identify the missing elements. Any such ordered set of elements is a **sequence.**

The *elements* of a set are the members of the set. The membership symbol is $\in$. For example, we may write

$$2 \in \{1, 2, 3, 4\}; \qquad \text{that is, 2 is an element of the set } \{1, 2, 3, 4\}$$

$$7 \notin \{1, 2, 3, 4\}; \qquad \text{that is, 7 is not an element of the set } \{1, 2, 3, 4\}$$

The symbol "$\in$" should remind you of "E" and thus *element*. We read "$\in$" as "is an element of the set."

Frequently capital letters are used to name sets as we did for the set C of counting numbers. Sometimes it is convenient to seek verbal descriptions for given sets of elements that have already been listed.

Example 1 Give a verbal description for the set
$$Y = \{1, 3, 5, 7, 9\}$$

Solution There are several correct responses. Two of these are: "The set of odd numbers 1 through 9" and "The set of odd numbers between 0 and 11." Note that the word "between" implies that the first and last numbers (0 and 11) are not included as members of the given set.

Two sets that have precisely the same elements are **equal sets.** We write

$$\{1, 2, 3\} = \{3, 2, 1\}$$

since order of listing does not affect membership. Furthermore, we shall not repeat elements within a listing since listing an element more than once does not increase the number of different members of the set. In general, we write $A = B$ to show that sets A and B have the same members; that is, A and B are *two names for the same set*. Similarly, $A \neq B$ indicates that sets A and B do not have the same members and thus are not equal sets.

Two sets $X = \{x_1, x_2, x_3, \ldots\}$ and $Y = \{y_1, y_2, y_3, \ldots\}$ are said to be in **one-to-one correspondence** if we can find a pairing of the x's and y's such that each x corresponds to one and only one y and each y corresponds to one and only one x. Any two equal sets may be placed in one-to-one correspondence since each element may be made to correspond to itself as in the adjacent illustration.

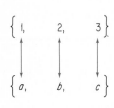

Also, it may be possible to place the members of two sets with different elements in one-to-one correspondence. Any two sets that can be placed in one-to-one correspondence are **equivalent sets** in the sense that they have the *same number of elements*. In general, any two equal sets are equivalent sets but, as in the accompanying illustration, two equivalent sets are not necessarily equal sets.

Example 2 Show that $\{c, r, a, b\}$ and $\{f, i, s, h\}$ are
(a) equivalent sets of letters   (b) not equal sets

Solution   (a) $\begin{matrix} \{c, & r, & a, & b\} \\ \updownarrow & \updownarrow & \updownarrow & \updownarrow \\ \{f, & i, & s, & h\} \end{matrix}$

(b) The element c, for example, is a member of the first set and not a member of the second set.

$A' = \mathcal{U} - A.$

The set containing all of the elements for any particular discussion is called the **universal set** $\mathcal{U}$, and may vary for each discussion: the set of readers of this book, the set of decimal digits, the set of counting numbers, and so forth. The **complement** A' (also written $\overline{A}$) of a given set A is the set of those elements of the universal set $\mathcal{U}$ that are not elements of A. This complement of A relative to $\mathcal{U}$ may also be denoted by $\mathcal{U} - A$, that is, $A' = \mathcal{U} - A$.

For any two sets A and B, the set of elements of B that are not elements of A is the **difference set** $B - A$. The complement of the universal set $\mathcal{U}$ is the **empty set (null set)** and may be denoted by either $\varnothing$ or $\{\ \}$. The empty set is the set that contains no elements.

4 SETS AND LOGIC

Example 3   Let $\mathcal{U} = \{1, 2, 3, \ldots, 10\}$, $A = \{1, 2, 3, 4, 5, 7, 9\}$, and
$B = \{2, 4, 6, 8\}$. Find **(a)** A' **(b)** B' **(c)** $A - B$
(d) $B - B$

Solution   **(a)** $A' = \{6, 8, 10\}$ **(b)** $B' = \{1, 3, 5, 7, 9, 10\}$
(c) $A - B = \{1; 3, 5, 7, 9\}$ **(d)** $B - B = \varnothing$

Consider the set B of counting numbers 1 through 9.

$$B = \{1, 2, 3, 4, 5, 6, 7, 8, 9\}$$

A set A is a **subset** of B if the set A has the property that each element of A is also an element of B. We write $A \subseteq B$ (read "A is included in B"). There are many subsets of B; a few of them are listed below.

$A_1 = \{1, 2, 3\}$ $A_2 = \{1, 5, 7, 8, 9\}$ $A_3 = \{2\}$ $A_4 = \{1, 2, 3, 4, 5, 6, 7, 8, 9\}$

Note in particular that set A_4 contains each of the elements of B and is classified as a subset of B. Any set is said to be a subset of itself. Note also that if a set A is a subset of B, then A does not contain any elements that are not elements of B. This statement can be taken as a definition of *subset*. This interpretation is needed to explain why the empty set is a subset of every set.

$A_1 \subset B$
$A_2 \subset B$
$A_3 \subset B$
$A_4 \subseteq B$

A *set A* is a **proper subset** of a set B if A is a subset of B and there is at least one element of B that is not an element of A. We write $A \subset B$ (read "A is a proper subset of B"). Intuitively, we speak of a proper subset as part of, but not all of a given set. The sets A_1, A_2, A_3, and A_4 are subsets of B; the sets A_1, A_2, and A_3 are proper subsets of B; and the set A_4 is not a proper subset of B.

Example 4   List three proper subsets of $\{1, 2, 3, 4\}$.

Solution   Among others: $\{1, 2\}$ $\{1, 3, 4\}$ $\{1\}$.

All possible subsets of $\{1, 2, 3, 4\}$ may be found by considering the elements of the set none at a time (the empty set), one at a time, two at a time, three at a time, and four at a time (the given set). This procedure may be simplified by pairing the subsets with their complements. This pairing is done vertically in the following array.

$\varnothing$	$\{1\}$	$\{2\}$	$\{3\}$	$\{4\}$	$\{1, 2\}$	$\{1, 3\}$	$\{1, 4\}$
$\{1, 2, 3, 4\}$	$\{2, 3, 4\}$	$\{1, 3, 4\}$	$\{1, 2, 4\}$	$\{1, 2, 3\}$	$\{3, 4\}$	$\{2, 4\}$	$\{2, 3\}$

In this array the selections of elements two at a time were made by considering the first element with each of the other elements. It was not necessary to continue the selections on the first row of the array since all other subsets of two or more elements had already been listed on the second row. Similar procedures may be used to find all subsets of any given set.

Example 5 List all possible subsets of $\{1, 2, 3\}$.

Solution

$$\varnothing \qquad \{1\} \qquad \{2\} \qquad \{3\}$$
$$\{1, 2, 3\} \qquad \{2, 3\} \qquad \{1, 3\} \qquad \{1, 2\}$$

$A \cap A' = \varnothing$

For any two sets A and B, the set of elements that are members of *both* A and B is the **intersection** of the sets A and B, written $A \cap B$. For example,

if $A = \{1, 2, 3, 4, 5\}$ and $B = \{4, 5, 6, 7\}$, then $A \cap B = \{4, 5\}$.

$A \cup A' = \mathcal{U}$

The set of elements that are members of *at least one* of the sets A and B is the **union** of A and B, written $A \cup B$. For the preceding example of sets A and B
$$A \cup B = \{1, 2, 3, 4, 5, 6, 7\}$$

The set of elements x such that $x \in \{1, 2, 3, 4, 5\}$ may be written in **set-builder notation** as

$$\{x \mid x \in \{1, 2, 3, 4, 5\}\}$$

and is precisely the set $\{1, 2, 3, 4, 5\}$. Note that the vertical line is read "such that." We may use set-builder notation to define both intersection and union. If, as before, $A = \{1, 2, 3, 4, 5\}$ and $B = \{4, 5, 6, 7\}$, then

$$A \cap B = \{x \mid (x \in A) \quad \text{and} \quad (x \in B)\} = \{4, 5\}$$
$$A \cup B = \{x \mid (x \in A) \quad \text{or} \quad (x \in B)\} = \{1, 2, 3, 4, 5, 6, 7\}$$

If the intersection of two sets is the empty set, that is, if the sets have no elements in common, then the sets are **disjoint sets**. If A and A' are complementary sets, then

$$A \cap A' = \varnothing \quad \text{and} \quad A \cup A' = \mathcal{U}$$

that is, A and A' are disjoint sets whose union is the universal set.

Note that *equality for sets* such as $A \cap A'$ and $\varnothing$ indicates that $A \cap A'$ and $\varnothing$ are names for the same set. This corresponds to the use of the *equality symbol* ($=$) for numbers to show that two symbols or expressions are names for the same numbers. For example, we write $2 + 3 = 5$ to show that $2 + 3$ and 5 are names for the same number; we write $2 + 2 \neq 5$ to show that $2 + 2$ and 5 are not names for the same number.

4 SETS AND LOGIC

Consider each of the sets (a) {s, e, n, d}   (b) {m, o, r, e}
(c) {m, o, n, e, y}.

1. Is *e* a member of the given set?
2. Is *m* a member of the given set?
3. Is *n* a member of the given set?
4. Is *s* a member of the given set?
5. Is it true that *x* is not a member of the given set?
6. Is it true that *y* is not a member of the given set?

State whether or not the given set is a well-defined set.

7. The set of cities that are state capitals in the United States of America.
8. The set of states with good climates in the United States of America.
9. The set of cities with good governments in the United States of America.
10. The set of states in the United States of America that do not have any seacoast.

State whether or not the given sets are (a) *equivalent sets*
(b) *equal sets.*

11. {t, o, n},   {n, o, t}
12. {c, a, r, t},   {r, a, c, k}
13. {t, a, m, p},   {m, a, p}
14. {l, a, z, y},   { f, a, s, t}
15. {d, o, n, t},   {d, o, n, e}
16. {h, o, t},   {h, e, a, t}

Give a verbal description for each of the following sets.

17. $R = \{1, 2, 3, \ldots, 99\}$
18. $S = \{51, 52, 53, \ldots\}$
19. $M = \{5, 10, 15, 20, \ldots\}$
20. $K = \{10, 20, 30, \ldots, 150\}$
21. $T = \{1, 4, 9, 16, 25, 36\}$
22. $P = \{0, 2, 6, 12, 20, \ldots, 72, 90\}$

In Exercise 22 look for a pattern. Find the differences between successive elements of the set.

For each of the following sets list the elements in (a) $A \cup B$
(b) $A \cap B.$

23. $A = \{1, 3, 4\}, \quad B = \{1, 3, 5, 7\}$
24. $A = \{3, 4, 5\}, \quad B = \{4, 5, 6, 7\}$
25. $A = \{2, 4, 6, 8\}, \quad B = \{4, 6, 7, 8\}$
26. $A = \{1, 3, 5, \ldots\}, \quad B = \{2, 4, 6, \ldots\}$
27. $A = \varnothing, \quad B = \{1, 2, 3, \ldots\}$
28. $A = \{1, 2, 3, \ldots\}, \quad B = \{1, 3, 5, \ldots\}$

For each of the given universal sets, list the elements in
(a) A' **(b)** B' **(c)** $A' \cup B'$ **(d)** $A' \cap B'$.

29. $\mathcal{U} = \{1, 2, 3, 4, 5\}$; $A = \{1, 2\}$, $B = \{1, 3, 5\}$
30. $\mathcal{U} = \{1, 2, 3, \ldots, 10\}$; $A = \{1, 3, 5, 7, 9\}$,
    $B = \{2, 4, 6, 8, 10\}$
31. $\mathcal{U} = \{1, 2, 3, \ldots\}$; $A = \{1, 3, 5, \ldots\}$,
    $B = \{2, 4, 6, \ldots\}$
32. $\mathcal{U} = \{1, 2, 3, 5, 4, 6, 7\}$; $A = \varnothing$, $B = \{1, 2, 3, 4, 5, 6, 7\}$
33. $\mathcal{U} = \{1, 2, 3\}$; $A = \{1\}$, $B = \{3\}$

List all proper subsets of the given set.

34. **(a)** $\{p\}$ **(b)** $\{p, q\}$ **(c)** $\{r, s, t, u\}$
35. **(a)** $\{a, b, c\}$ ***(b)** $\{a, b, c, d, e\}$ ***(c)** $\varnothing$

State whether you would expect each of the following statements to be always true or not always true. If the statement is not always true, then give at least one counterexample.

*36. **(a)** $(A \cap B) \subseteq A$ **(b)** $(A \cap B) \subset B$
*37. **(a)** $B \subset (A \cup B)$ **(b)** $A \subseteq (A \cup B)$
*38. **(a)** $(A - B) \subset A$ **(b)** $(A - B) \subseteq A$

Describe, if possible, the conditions on the sets A and B such that each statement will always be true.

*39. $(A - B) = A$ *40. $(A \cap B) = A$
*41. $(A \cup B) = A$ *42. $(A \cap B) = (B \cap A)$
*43. $(A \cup B) = (B \cup A)$ *44. $[A \cap (A \cup B)] = A$
*45. $[A \cap (A \cup B)] = B$ *46. $(A \cup B) = (A \cap B)$

Explorations

1. Name at least 30 words that indicate sets. For example, consider a *school* of fish, a *swarm* of bees, and a *squadron* of planes.

Warning! Exploration 3 is tricky.

2. Select a newspaper or news magazine and list several of the references to sets and subsets that are made in that paper or magazine. Don't miss the committees and sub-committees.

3. Give at least two verbal descriptions for the set {8, 5, 4, 9, 1, 7, 6, 3, 2, 0}. Include a verbal description that identifies the order in which the elements are listed but does not include a listing of them.

4. Use specific examples as necessary and complete the following table.

Number of elements	0	1	2	3	4	5	6	10	50
Number of subsets	1								
Number of proper subsets									

5. Use the results obtained in Exploration 4 to conjecture for a set consisting of n elements (a) a formula for the number N of subsets (b) a formula for the number P of proper subsets.

6. Given two sets A and B make a flowchart for finding (a) $A \cap B$ (b) $A \cup B$.

4-2
Venn Diagrams

Suppose that in a class of 32 students, 25 students did the odd-numbered problems, 20 did the even-numbered problems, and 15 did all the problems. Is it possible that everyone did either the odd- or the even-numbered problems? Try to solve this problem before reading the solution near the end of this section.

These diagrams are called Venn diagrams in honor of the English logician John Venn (1834–1923). The diagrams are also sometimes called *Euler circles* in honor the Swiss mathematician Leonhard Euler (1707–1783), who first used circular regions in the discussion of principles of logic.

Relationships among sets are often represented by sets of points. In this text, we use a rectangular region to represent the universal set. Then a particular subset A is represented by a circular, or other convenient, region. In the second figure, the dashed line indicates that the points of the circle are elements of A and are *not* elements of A'.

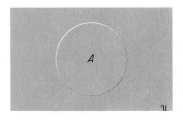

The shading is often omitted when the meaning is clear without the shading. With or without the shading, the figures are called **Venn diagrams.** We may use Venn diagrams to show the intersection of two sets.

LEONHARD EULER

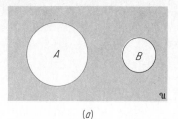

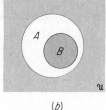

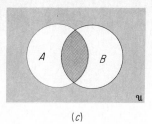

(a) (b) (c)

Note that in part (**a**) of the preceding figure, $A \cap B$ is the empty set; in part (**b**), $A \cap B = B$. We may also use Venn diagrams to show the union of two sets. Note that $A \cup B = A$ in part (**b**) of the next figure.

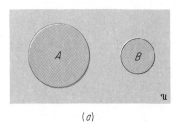

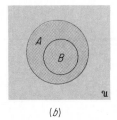

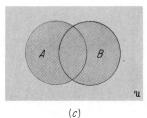

(a) (b) (c)

The figures for intersection and union may be used to illustrate the following properties of any two sets A and B.

$$(A \cap B) \subseteq A \qquad (A \cap B) \subseteq B$$
$$A \subseteq (A \cup B) \qquad B \subseteq (A \cup B)$$

We consider only well-defined sets (Section 4-1), and thus each element of the universal set $\mathcal{U}$ is a member of exactly one of the sets A and A'. Unless a given set A is specified as the universal set, we assume that the set A determines two sets A and A'. Similarly, unless other relations are specified, we assume that any two given sets A and B determine four subsets of the universal set. Then we draw the Venn diagram accordingly.

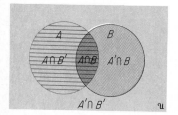

$A \cap B;$ that is, $\{x \mid (x \in A) \text{ and } (x \in B)\}$
$A \cap B';$ that is, $\{x \mid (x \in A) \text{ and } (x \notin B)\}$
$A' \cap B;$ that is, $\{x \mid (x \notin A) \text{ and } (x \in B)\}$
$A' \cap B';$ that is, $\{x \mid (x \notin A) \text{ and } (x \notin B)\}$

Since these four sets include all elements of the universal set and no element belongs to more than one of the four sets, the universal set is said to be *partitioned* into four subsets by the sets A and B. The Venn diagram is subdivided into four regions.

    4 SETS AND LOGIC

Venn diagrams may be used to show that two sets refer to the same set of points, that is, are equal.

Example 1   Show by means of a Venn diagram that $(A \cup B)' = A' \cap B'$.

Solution   We make separate Venn diagrams for $(A \cup B)'$ and $A' \cap B'$. The diagram for $(A \cup B)'$ is made by shading the region for A horizontally, shading the region for B vertically, identifying the region for $A \cup B$ as consisting of the points in regions that are shaded in any way (horizontally, vertically, or both horizontally and vertically), and identifying the region for $(A \cup B)'$ as consisting of the points in the region without horizontal or vertical shading.

The statements
$$(A \cup B)' = A' \cap B'$$
$$(A \cap B)' = A' \cup B'$$
are **De Morgan's laws for sets.**

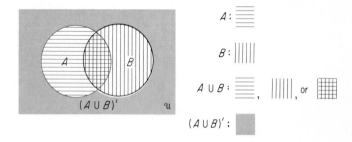

The diagram for $A' \cap B'$ is made by shading the region for A' horizontally, shading the region for B' vertically, and identifying the region for $A' \cap B'$ as consisting of all points in regions that are shaded both horizontally and vertically.

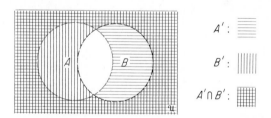

The solution is completed by observing that the region for $(A \cup B)'$ in the first Venn diagram is identical with the region for $A' \cap B'$ in the second Venn diagram; the two sets have the same elements and therefore are equal.

In the most general situation three sets A, B, and C partition the universal set into eight subsets. We assume that this situation holds unless other relations are specified. The corresponding Venn diagram with eight regions is usually drawn as follows.

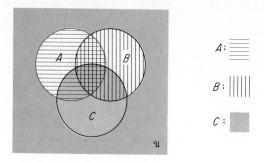

Example 2 Show that $A \cap (B \cup C) = (A \cap B) \cup (A \cap C)$.

Solution Set A is shaded with vertical lines; $B \cup C$ is shaded with horizontal lines. The intersection of these sets, $A \cap (B \cup C)$, is the region that has both vertical and horizontal shading.

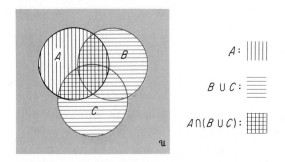

The set $A \cap B$ is shaded with horizontal lines; $A \cap C$ is shaded with vertical lines. The union of these sets is the subset of $\mathcal{U}$ that is shaded with lines in either or in both directions.

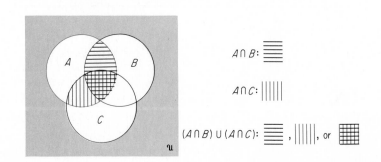

Note that the final results in the two diagrams designate the same regions, thus $A \cap (B \cup C)$ and $(A \cap B) \cup (A \cap C)$ are shown to be equal sets.

The *number of elements in a set S* is denoted by $n(S)$. For the set A in the figure for Example 3, we have

$$n(A) = 3 + 2 + 1 + 4 = 10.$$

Example 3 For the given Venn diagram find
(a) $n(A \cap B \cap C)$ (b) $n(A \cap B' \cap C)$.

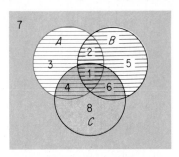

Solution (a) There is one element in the intersection of all three sets. Thus $n(A \cap B \cap C) = 1$.
(b) There are four elements that are in both sets A and C, but not in set B. Thus $n(A \cap B' \cap C) = 4$.

Example 4 In a group of 35 students, 15 are studying French, 22 are studying English, 14 are studying Spanish, 11 are studying both French and English, 8 are studying English and Spanish, 5 are studying French and Spanish, and 3 are studying all three subjects. How many are taking only English? How many of these students are not taking any of these subjects?

CHALLENGE: Try to solve Example 4 before reading the solution.

Solution This problem can easily be solved by means of a Venn diagram with three circles to represent the set of students in each of the listed subject-matter areas. It is helpful to start with the information that there are 3 students taking all three subjects. We write the number 3 in the region that is the intersection of all three circles. Then we work backward: Since 5 are taking French and Spanish, and 3 of these have already been identified as also taking English, there must be exactly 2 taking only French and Spanish. That is, there must be 2 in the region representing French and Spanish but *not* English. Continuing in this manner, we enter the given data in the figure.

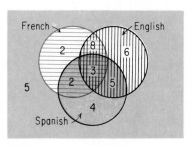

Reading directly from the figure, we find that there are 6 students taking only English. Since the total of the numbers in the various areas is 30, there must be 5 students not taking any of these specified subjects.

EXERCISES *Consider the given diagram and find each number.*

1. (a) $n(A \cap B)$
    (b) $n(A)$
    (c) $n(B \cap A')$
    (d) $n(B \cup A)$

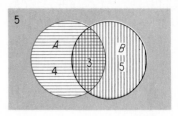

2. (a) $n(P \cup Q)$
    (b) $n(P' \cap Q')$
    (c) $n(P' \cup Q)$
    (d) $n(P \cup Q')$

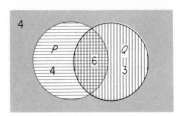

Represent each set by a Venn diagram.

3. $A' \cup B$ 4. $A' \cap B$ 5. $A \cap B'$ 6. $A \cup B'$

Show each relation by Venn diagrams.

7. $(A \cap B)' = A' \cup B'$ 8. $A \cup B' = (A' \cap B)'$

Consider the given diagram and find each number.

9. (a) $n(A \cap B \cap C)$
    (b) $n(A \cap B \cap C')$
    (c) $n(A \cap B' \cap C')$
    (d) $n(A)$
    (e) $n(A \cup B)$
    (f) $n(B \cup C)$

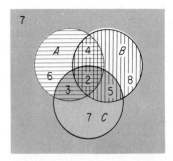

10. (a) $n(R' \cap S \cap T)$
     (b) $n(R')$
     (c) $n(R' \cup S)$
     (d) $n(S' \cup T')$
     (e) $n(R' \cup S' \cup T')$
     (f) $n(R \cup S' \cup T)$

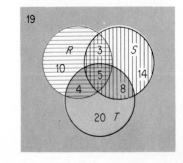

11. (a) $n(A \cup B)$
    (b) $n(B \cap C)$
    (c) $n(A \cap B')$
    (d) $n(A \cup B \cup C)$
    (e) $n(A \cup B' \cup C')$
    (f) $n(A \cap B' \cap C')$

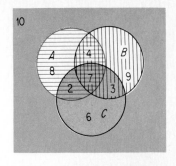

12. (a) $n(X \cup Y)$
    (b) $n(X \cap Z)$
    (c) $n(X')$
    (d) $n(X \cup Y')$
    (e) $n(X' \cap Y \cap Z)$
    (f) $n(X \cap Y' \cap Z')$

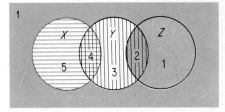

Represent each of the following by a Venn diagram.

13. (a) $A \cap B \cap C$ (b) $A \cap B \cap C'$
    (c) $A \cap B' \cap C$ (d) $A \cap B' \cap C'$

14. (a) $A' \cap B \cap C$ (b) $A' \cap B \cap C'$
    (c) $A' \cap B' \cap C$ (d) $A' \cap B' \cap C'$

Identify each gray region by set notation.

15.

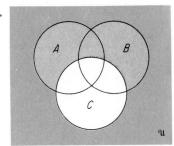

16.

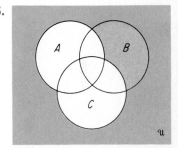

17.

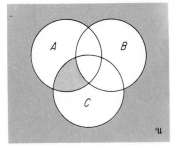

18.

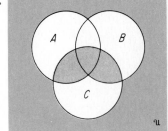

Use Venn diagrams to solve each problem.

Study Example 4 before trying Exercises 19 through 23.

19. In a survey of 50 students, the following data were collected: There were 19 taking biology, 20 taking chemistry, 19 taking physics, 7 taking physics and chemistry, 8 taking biology and chemistry, 9 taking biology and physics, 4 taking all three subjects.
    (a) How many of the group are not taking any of the three subjects?
    (b) How many are taking only chemistry?
    (c) How many are taking physics and chemistry but not biology?

For the question in the margin near the beginning of this section consider the following Venn diagram.

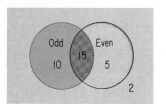

Two students failed to do either the odd- or the even-numbered problems.

20. Fifty cars belong to students in a certain dormitory. Of these cars 42 have radios, 15 have tape decks, 10 have air conditioners, 2 have all three, 6 have a radio and an air conditioner, 5 have a tape deck and an air conditioner, and 10 have a radio and a tape deck.
    (a) How many of these cars have an air conditioner but no radio?
    (b) How many have no radio, no tape deck, and no air conditioner?

21. Repeat Exercise 20 for 4 cars having all three items and all other data as before.

22. Suppose that the student who collected the data for Exercise 20 stated that 5 of the cars had all three items and gave the other data as before.
    (a) Did the student make a careful survey?
    (b) Explain your answer to part (a).

23. A survey was taken of 30 students enrolled in three different clubs, A, B, and C. Show that the following data that were collected are inconsistent: 18 in A, 10 in B, 9 in C, 5 in B and C, 6 in A and B, 9 in A and C, 3 in A, B, and C.

Explorations

1. Select any sets A and B with $n(A) = 6$, $n(B) = 4$, and such that
    (a) $n(A \cup B) = 7$ (b) $n(A \cup B) = 6$
    (c) $n(A \cup B) = 9$ (d) $n(A \cap B) = 2$

2. Select sets A and B as in Exploration 1 and find
    (a) $n(A \cap B)$ if $n(A \cup B) = 7$.
    (b) $n(A \cap B)$ if $n(A \cup B) = 6$.
    (c) $n(A \cap B)$ if $n(A \cup B) = 10$.
    (d) $n(A \cup B)$ if $n(A \cap B) = 2$.
    (e) $n(A \cup B)$ as an expression in terms of $n(A)$, $n(B)$, and $n(A \cap B)$.

4 SETS AND LOGIC

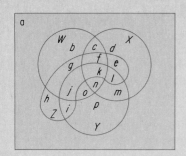

3. Venn diagrams may be drawn for four sets W, X, Y, and Z. The region Z must intersect each of the eight regions of the Venn diagram for W, X, and Y. The new Venn diagram has 16 regions as lettered in the adjacent figure. Identify each of the 16 regions by its letter and as an intersection of four of the sets W, X, Y, Z, W', X', Y', Z'.

Use letters from the figure for Exploration 3 to identify each region. For example, the region $W \cap Y$ consists of the regions j, k, n, and o.

4. W 5. Z 6. X'
7. $(W \cup X) \cap Z$ 8. $W' \cap (X \cup Z)$ 9. $(W \cup Y) \cap (X' \cup Z')$

Use a Venn diagram for four sets to determine whether each statement is true or false.

10. $A \cap (B \cup C \cup D) = (A \cap B) \cup (A \cap C) \cup (A \cap D)$
11. $A \cup (B \cap C \cap D) = (A \cup B) \cap (A \cup C) \cap (A \cup D)$

12. Lewis Carroll, the author of *Alice's Adventures in Wonderland*, was actually the mathematician C. L. Dodgson. His *Symbolic Logic* (fourth edition, 1896) and *The Game of Logic* (1886) were published as a single paperback volume in 1958 by Dover Publications, Inc.

     Symbolic Logic includes multitudes of intriguing illustrative examples, solutions, and an "Appendix addressed to teachers." The appendix includes comparisons of Euler's method of diagrams, Venn's method of diagrams, and Dodgson's method of diagrams as well as symbolic and other methods.

     Review *Symbolic Logic*, do not miss the introduction "To Learners," and report on the aspects of the book that you find most interesting. Dodgson claimed that he used all of these materials with children at most 14 years old.

4-3
Simple and Compound Statements

Commands such as "Stand up and be counted" and greetings such as "Hello" are neither true nor false and are not considered to be statements as we are using the word here.

A sentence that can be identified as true or identified as false is often called a **statement**. Each of the following sentences is an example of a **simple statement**.

I have made the last payment on my car.

I have recently purchased all of the books required for this semester.

A **compound statement** is formed by combining two or more simple statements, as in the following example.

I have made the last payment on my car *and* I have purchased all of the books required for this semester.

In this illustration the two simple statements are combined by the connective *and*. Other connectives could have been used. Consider the same simple statements used with the connective *or*.

I have made the last payment on my car *or* I have purchased all of the books required for this semester.

We shall consider such compound statements and determine the conditions under which they are true or false, assuming that the simple statements are true or false. In doing this we use symbols to represent connectives and letters or variables to represent statements. The most common **connectives** are *and* ($\wedge$), *or* ($\vee$), and *not* ($\sim$). Then compound statements may be written in symbolic form and symbolic statements may be translated into words. For example, we may use p and q to represent these simple statements.

p: I have made the last payment on my car.

q: I have purchased all of the books required for this semester.

Then we may write the following compound statements.

Connectives
$\wedge$: and
$\vee$:   or
$\sim$:   not

$p \wedge q$: I have made the last payment on my car and I have purchased all of the books required for this semester.

$p \vee q$: I have made the last payment on my car or I have purchased all of the books required for this semester.

$\sim p$: I have not made the last payment on my car.

Example 1 Use p and q as in the preceding discussion and translate.
(a) $p \wedge (\sim q)$ (b) $\sim(\sim p)$

Solution (a) I have made the last payment on my car and I have not purchased all of the books required for this semester.

The statement $\sim(\sim p)$ is equivalent to, has the same meaning as, the statement p.

(b) It is not true that I have not made the last payment on my car; that is, I have made the last payment on my car.

Example 2 Write each statement in symbolic form.
(a) I have not made the last payment on my car or I have not purchased all of the books for this semester.

4 SETS AND LOGIC

(b) I have completed neither the last payment on my car nor the purchase of all the books required for the semester.

Solution Use p and q as in the preceding discussion.

(a) $(\sim p) \vee (\sim q)$ **(b)** $(\sim p) \wedge (\sim q)$

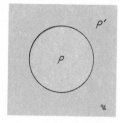

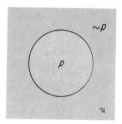

Suppose that P is the set of instances in which a given statement p is true. Then for any given universal set $\mathcal{U}$ there are two sets P and P' such that every element of the universal set belongs to exactly one of the sets P and P'. If $x \in P$, then $x \notin P'$; if $y \in P'$, then $y \notin P$. In terms of the given statement p, the points of P represent the instances in which the given statement p is true; the points of P' represent the instances in which p is false, that is, $\sim p$, the **negation** of p, is true. For each element of the universal set exactly one of the statements p, $\sim p$ is true and the other statement is false. This conclusion may be represented by a **truth table** (a table of truth values).

For any two statements p and q we have two subsets P and Q of the universal set. These two sets determine four regions which may be identified in terms of the sets P and Q, in terms of the statements p and q, or in other ways as on the lines of the next array.

p	$\sim p$
T	F
F	T

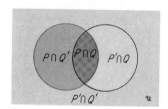

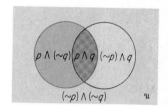

			p	q
$P \cap Q$	$p \wedge q$	p is true and q is true	T	T
$P \cap Q'$	$p \wedge (\sim q)$	p is true and q is false	T	F
$P' \cap Q$	$(\sim p) \wedge q$	p is false and q is true	F	T
$P' \cap Q'$	$(\sim p) \wedge (\sim q)$	p is false and q is false	F	F

The relationship between $\cap$ and $\wedge$ and the relationship between $\cup$ and $\vee$ may be stated in terms of set-builder notation.

$$P \cap Q = \{x \mid (x \in P) \quad \text{and} \quad (x \in Q)\} = \{x \mid (x \in P) \wedge (x \in Q)\}$$

$$P \cup Q = \{x \mid (x \in P) \quad \text{or} \quad (x \in Q)\} = \{x \mid (x \in P) \vee (x \in Q)\}$$

Venn diagrams for $p \wedge q$ and $p \vee q$ are shown in the following figures.

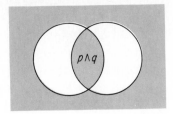 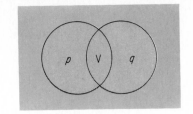

The truth values of compound statements depend on the truth values of the simple statements that are used.

~p   is true if p is false, false if p is true; that is, ~p and p are **contradictory statements**; each statement is the *negation* of the other.

$p \wedge q$ is true if *both* p and q are true, false in all other cases.

$p \vee q$ is true if *at least one of* the statements p, q is true; false if both p and q are false.

The last statement describes the *inclusive use* of *or* since if p or q or both p and q are true, then $p \vee q$ is true. The word *or* is widely used in everyday speech. Consider these statements.

At nine o'clock tomorrow morning I shall either be in a conference in New York or in my dentist's office in Boston.

Jane looks either very tired or ill.

The first of these two illustrations shows the *exclusive use* of *or*, since it is not possible for both statements to be true. The **exclusive or** (often denoted by $\veebar$, called *vel*) is true when, and only when, exactly one of the given statements is true. The second illustration shows the *inclusive use* of *or* since at least one, or possibly both, of the statements may be true. In mathematics, *or* is assumed to be the inclusive *or* unless otherwise specified.

We may use truth tables with all four possible combinations of truth values to define compound statements that involve two given statements. The arrangement used in the following definitions of $p \wedge q$ and $p \vee q$ is conventional. However, the order in which the pairs of truth values are presented is entirely arbitrary; other orders may be found in other textbooks.

The statement $p \wedge q$ is true if both p and q are true, but it is false otherwise; $p \vee q$ is true if at least one of the statements p, q is true, and false otherwise. Note that the statement $p \vee q$ is true unless both p and q are false.

p	q	$p \wedge q$
T	T	T
T	F	F
F	T	F
F	F	F

p	q	$p \vee q$
T	T	T
T	F	T
F	T	T
F	F	F

The statement $p \wedge q$ is the **conjunction** of p and q; the statement $p \vee q$ is the **disjunction** of p and q. Both conjunctions and disjunctions may be negated. A preface of "It is not the case that" is sufficient, but other forms may be used. For symbolic statements, the procedure is the same as for simple statements; also, truth tables are constructed in the same manner as for other compound statements.

p	q	$p \wedge q$	$\sim(p \wedge q)$	$p \vee q$	$\sim(p \vee q)$
T	T	T	F	T	F
T	F	F	T	T	F
F	T	F	T	T	F
F	F	F	T	F	T

Example 3   Write two forms of the negation of the given statement.
(a) The weather is hot and I need to work outside.
(b) Today is a rainy day or it is not Saturday.

Solution   (a) It is not the case that the weather is hot and I need to work outside. Note that, as in the truth table, the negation is true under three conditions.

The weather is hot and I do not need to work outside.

The weather is not hot and I need to work outside.

The weather is not hot and I do not need to work outside.

The negation of the given statement also may be stated as

Either the weather is not hot or I do not need to work outside.

(b) It is not the case that today is a rainy day or it is not Saturday. Note that, as in the truth table, the negation is true under only one condition. Thus the negation also may be stated as follows.

Today is not a rainy day and it is Saturday.

Examples 3(a) and 3(b) provide illustrations of **De Morgan's laws for statements.**

Compare with De Morgan's laws for sets (Section 4-2).

$\sim(p \wedge q)$	is equivalent to	$(\sim p) \vee (\sim q)$
$\sim(p \vee q)$	is equivalent to	$(\sim p) \wedge (\sim q)$

The British mathematician Augustus De Morgan (1806–1871) was an outstanding writer and teacher. A champion of religious and intellectual tolerance, he was one of the founders of the British Association for the Advancement of Science (1831). He is well known for his *Budget of Paradoxes*, a collection of many of his witticisms that was edited after his death by his widow, Sophia Elizabeth De Morgan.

We may use either truth tables or Venn diagrams to summarize the truth values of a wide variety of compound statements. We shall emphasize truth tables since the steps used in forming a truth table are easily identified and are also precisely the steps that must be used to form the Venn diagram. To illustrate this procedure, we shall construct a truth table for the statement $p \wedge (\sim q)$. First set up a table with the headings.

p	q	p	$\wedge$	$(\sim q)$
T	T			
T	F			
F	T			
F	F			

Now complete the column headed "p" by using the truth values that appear under p in the first column. In the column headed "$\sim q$" write the negation of each of the values given under q in the second column.

p	q	p	$\wedge$	$(\sim q)$
T	T	T		F
T	F	T		T
F	T	F		F
F	F	F		T

Finally, find the conjunctions of the values given in the third and fifth columns. The completed table appears as follows, with the order in which the columns were considered indicated by the alphabetical order of the labels (a), (b), and (c), and the final results in bold print.

4 SETS AND LOGIC

We can summarize this table by saying that the statement $p \wedge (\sim q)$ is true only in the case when p is true and q is false.

p	q	p	$\wedge$	$(\sim q)$
T	T	T	F	F
T	F	T	T	T
F	T	F	F	F
F	F	F	F	T

(a) (c) (b)

Consider, for example, these simple statements.

p: This class has 28 members.

q: Some member of this class is absent today.

Then the statement $p \wedge (\sim q)$ may be stated as follows.

This class has 28 members and no one is absent today.

This statement, as in the general case, is true only if p is true and q is false.

EXERCISES   *Think of "short" as "not tall" and use p and q as follows.*

p: *Jim is tall.*

q: *Bill is not tall.*

1. Write each of these statements in symbolic form.
   (a) Jim is short and Bill is tall.
   (b) Neither Jim nor Bill is tall.
   (c) Jim is not tall and Bill is short.
   (d) It is not the case that Jim and Bill are both tall.
   (e) Either Jim or Bill is tall.
2. Assume that Bill and Jim are both tall. Which of the statements in Exercise 1 are true?

Think of "sad" as "not happy" and use p and q as follows.

p: *Joan is happy.*

q: *Mary is sad.*

3. Write each of these statements in symbolic form.
   (a) Joan and Mary are both happy.
   (b) Either Joan is happy or Mary is happy.
   (c) Neither Joan nor Mary is happy.
   (d) It is not the case that Joan and Mary are both sad.

(e) It is not the case that neither Joan nor Mary is happy.

4. Assume that Joan and Mary are both happy. Which of the statements in Exercise 3 are true?

Give each of these statements in words. Use p and q as follows.

p: *I like this book.*

q: *I like mathematics.*

5. (a) $p \wedge q$ (b) $\sim q$
   (c) $\sim(\sim p)$ (d) $(\sim p) \wedge (\sim q)$

6. (a) $(\sim p) \wedge q$ (b) $p \vee q$
   (c) $\sim(p \wedge q)$ (d) $\sim[(\sim p) \wedge q]$

7. Assume that you like this book and that you like mathematics. Which of the statements in Exercises 5 and 6 are true for you?

8. Assume that you like this book but that you do not like mathematics. Which of the statements in Exercises 5 and 6 are true for you?

Construct a truth table for the given statement.

9. $(\sim p) \wedge q$ 10. $(\sim p) \vee q$ 11. $(\sim p) \vee (\sim q)$

12. $(\sim p) \wedge (\sim q)$ 13. $\sim(p \wedge q)$ 14. $p \vee (\sim q)$

In Exercises 15 through 18 copy and complete each truth table.

15.

p	q	$\sim$	$[p$	$\vee$	$(\sim q)]$
T	T				
T	F				
F	T				
F	F				
		(d)	(a)	(c)	(b)

16.

p	q	$\sim$	$[(\sim p)$	$\vee$	$q]$
T	T				
T	F				
F	T				
F	F				
		(d)	(a)	(c)	(b)

17.

p	q	$\sim$	$[(\sim p)$	$\wedge$	$(\sim q)]$
T	T				
T	F				
F	T				
F	F				
		(d)	(a)	(c)	(b)

18.

p	q	$\sim$	$[(\sim p)$	$\vee$	$(\sim q)]$
T	T				
T	F				
F	T				
F	F				
		(d)	(a)	(c)	(b)

4 SETS AND LOGIC

Write two forms of the negation of the given statement.

19. Today is Monday.
20. My car is a Ford.
21. These two cars are not made by the same company.
22. These two lines do not intersect.
23. I am young and I am happy.
24. I worked hard and I did not pass.
25. Pedro has $50 and he has two tickets to the game.
26. Marcia will drive her car or she will fly.
27. The textbook is expensive or it is not used.
28. A vacation should be enjoyable or it should be profitable.

Complete the indicated statement so that it has the same meaning as the given statement.

29. There is no college that is not expensive.
    Every . . . .
30. There are courses with no required textbooks.
    Not all . . . .
31. All cars are expensive.
    There are no . . . .

For Exercises 32 and 33 consider these pictures of children with hats as shown.

(i)   (ii)   (iii)

Use the numbers (i), (ii), (iii) to designate the pictures for which the given statement is true.

32. (a) Every child has a hat.
    (b) No child has a hat.
    (c) No child is without a hat.
    (d) Every child is without a hat.
33. (a) Not every child has a hat.
    (b) Not every child is without a hat.
    (c) There is a child with a hat.
    (d) There is a child without a hat.

34. Construct a truth table for $p \veebar q$, which was defined as "p or q but not both."

35. Use the given statements p and q and state the conditions under which each of the statements in Exercises 9 and 12 is true.

 p: I like this book.

 q: I like mathematics.

*36. Construct a truth table for $p \mid q$, which we define to be true when p and q are not both true and to be false otherwise.

*37. In Exercise 36, express $p \mid q$ in terms of other connectives that we have previously defined.

Explorations

Logical statements are often represented by electric circuits. For a single statement q, think of an electric light cord from a wall outlet to an electric light bulb and with a switch in the middle of the cord.

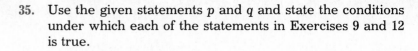

Wall outlet   Switch   Light bulb

If the switch is closed the bulb is on; if the switch is open the bulb is off.

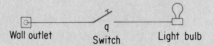

Closed Light on   Open Light off

For two statements p and q, we use two switches. The following circuits are particularly useful and common.

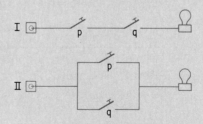

1. Assume that circuit I is in the specified condition and indicate whether the light is on or off.
 (a) p closed, q closed.   (b) p closed, q open.
 (c) p open, q closed.   (d) p open, q open.

2. Explain why it seems reasonable to call circuit I a $p \land q$ circuit.

3. Repeat Exploration 1 for circuit II.

4. Explain why it seems reasonable to call circuit II a $p \lor q$ circuit.

The statement $(p \lor q) \land r$ may be represented by the following circuit, where each circuit is left open until instructions are received for that switch.

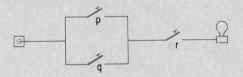

Sketch a circuit for each of these statements.

5. $(p \land q) \land r$ 6. $(p \land q) \lor r$ 7. $p \land (q \lor r)$

4-4
Conditional Statements

Many of the statements that we make in everyday conversation are based upon a condition. For example, consider the following statements.

If the telephone rings, then Jane will answer it.

If I have no homework, then I shall go bowling.

Each of these statements is expressed in the **if-then** form

If p, then q.

Any if-then statement can be expressed in symbols as $p \rightarrow q$, which is read "if p, then q." The **conditional symbol** $\rightarrow$ is another connective. It is used to form a compound statement $p \rightarrow q$, a **conditional statement**.

Note that definitions may be made to fit observed facts.

Our first task is to consider the various possibilities for p and q in order to define $p \rightarrow q$ for each of these cases. One way to do this would be to present a completed truth table and to accept this as our definition of $p \rightarrow q$. However, since a useful definition should fit the observed facts of our everyday world, let us attempt to justify the entries in such a table. Consider again the conditional statement

If the telephone rings, then Jane will answer it.

p	q	$p \rightarrow q$
T	T	T
T	F	F
F	T	T
F	F	T

If the telephone rings and Jane answers it, then the given statement is obviously true. On the other hand, the statement is false if the telephone rings and Jane does not answer it. Assume now that the telephone does not ring. Then since there is no circumstance in which Jane fails to answer the telephone, there is no circumstance in which the given statement is false. Since the given statement cannot be false, we *define* it to be true. We can summarize these assertions of the truth values of a conditional statement as in the given truth table.

Consider the statement

If it rains, then I shall give you a ride home.

Have I lied to you

1. If it rains and I give you a ride home?
2. If it rains and I do not give you a ride home?
3. If it does not rain and I give you a ride home?
4. If it does not rain and I do not give you a ride home?

The given statement $p \rightarrow q$ is true unless the *premise* p is true and the *conclusion* q is false.

According to the accepted meanings of the words used, you have a right to feel that I lied to you only if it rains and I do not give you a ride home. In any conditional statement $p \rightarrow q$, the statement p is the **premise** and the statement q is the **conclusion**. Thus the conditional statement is true unless the premise is true and the conclusion is false.

If p is false, then the statement $p \rightarrow q$ is accepted as true regardless of the truth value of q. Under this definition each of the following statements is true.

If General George Washington, born in 1732, is alive today, then he is now the President of the United States of America.

If $2 + 3 = 7$, then the moon is made of green cheese.

If Wednesday is the day after Monday, then July is the month after June.

If you have difficulty accepting any of these statements as true, then you should review the definition of the truth values of if-then statements. Remember also that there need be no relationship between p and q in an if-then statement although we tend to use related statements in this way in everyday life.

Example 1   Give the truth value of each statement.
  (a) If $5 + 7 = 12$, then $6 + 7 = 13$.
  (b) If $5 \times 7 = 35$, then $6 \times 7 = 36$.
  (c) If $5 + 7 = 35$, then $6 + 7 = 13$.
  (d) If $5 + 7 = 35$, then $6 \times 7 = 36$.

Solution   Think of each statement in the form $p \rightarrow q$.
  (a) For p true and q true, the statement $p \rightarrow q$ is *true*.
  (b) For p true and q false, the statement $p \rightarrow q$ is *false*.
  (c) For p false and q true, the statement $p \rightarrow q$ is *true*.
  (d) For p false and q false, the statement $p \rightarrow q$ is *true*.

Given any conditional statement, $p \rightarrow q$, three other related statements may be identified.

STATEMENT: $p \rightarrow q$		If p, then q.
CONVERSE: $q \rightarrow p$		If q, then p.
INVERSE: $(\sim p) \rightarrow (\sim q)$		If not p, then not q.
CONTRAPOSITIVE: $(\sim q) \rightarrow (\sim p)$		If not q, then not p.

Here are three examples of conditional statements with their converses, inverses, and contrapositives. In each case only the form of the statement is important. It is not necessary to understand "icy road," "garage," or any other terms.

1. *Statement*: If $x = 3$, then $x \neq 0$.
    Converse: If $x \neq 0$, then $x = 3$.
    Inverse: If $x \neq 3$, then $x = 0$.
    Contrapositive: If $x = 0$, then $x \neq 3$.

2. *Statement*: If the road is icy, then I leave my car in the garage.
    Converse: If I leave my car in the garage, then the road is icy.
    Inverse: If the road is not icy, then I do not leave my car in the garage.
    Contrapositive: If I do not leave my car in the garage, then the road is not icy.

3. *Statement*: $p \rightarrow (\sim q)$.
    Converse: $(\sim q) \rightarrow p$.
    Inverse: $(\sim p) \rightarrow \sim(\sim q)$, which can be simplified as $(\sim p) \rightarrow q$.
    Contrapositive: $\sim(\sim q) \rightarrow (\sim p)$, or simply $q \rightarrow (\sim p)$.

		Statement			Converse			Inverse			Contrapositive		
p	q		$p \to q$			$q \to p$			$(\sim p) \to (\sim q)$			$(\sim q) \to (\sim p)$	
T	T	T	T	T	T	T	T	F	T	F	F	T	F
T	F	T	F	F	F	T	T	F	T	T	T	F	F
F	T	F	T	T	T	F	F	T	F	F	F	T	T
F	F	F	T	F	F	T	F	T	T	T	T	T	T

In the tables the contrapositive of a conditional statement has the same truth values as the statement. Also note that the converse and the inverse of any conditional statement have the same truth values; that is, the statements are *equivalent statements*. In the next table note that $p \to q$ is equivalent to $\sim[p \wedge (\sim q)]$; the negation of $p \to q$ is equivalent to $p \wedge (\sim q)$.

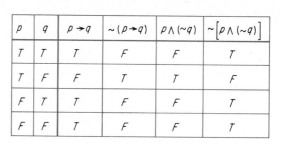

p	q	$p \to q$	$\sim(p \to q)$	$p \wedge (\sim q)$	$\sim\left[p \wedge (\sim q)\right]$
T	T	T	F	F	T
T	F	F	T	T	F
F	T	T	F	F	T
F	F	T	F	F	T

The negation of $p \to q$ is $p \wedge (\sim q)$.

The equivalence of any conditional statement $p \to q$ and $\sim[p \wedge (\sim q)]$ is often used to obtain the negation of a given conditional statement. For example the conditional statement

If Sue is wearing a coat, then the temperature is below freezing

is equivalent to

It is not the case that Sue is wearing a coat and the temperature is not below freezing

and has as its negation

Sue is wearing a coat and the temperature is not below freezing.

Example 2 Write (**a**) the negation, and (**b**) the contrapositive, of the following statement.

4 SETS AND LOGIC

If I work hard, then I shall pass the course.

Solution

The contrapositive (b) has the same truth values as the original statement. If you fail the course, then you have not worked hard, because if you had worked hard, then you would have passed the course.

(a) I work hard and I shall not pass the course.
(b) If I do not pass the course, then I have not worked hard.

Compound statements may involve three or more statements. For example, the columns (c) and (f) of the next truth table show that $p \rightarrow (q \vee r)$ and $(p \rightarrow q) \vee (p \rightarrow r)$ are equivalent statements. Note that since r may be true or false for each of the four pairs of truth values for p and q, this truth table has eight rows.

p	q	r	$p \longrightarrow (q \vee r)$			$(p \longrightarrow q)$	$\vee$	$(p \longrightarrow r)$
T	T	T	T	T	T	T	T	T
T	T	F	T	T	T	T	T	F
T	F	T	T	T	T	F	T	T
T	F	F	T	F	F	F	F	F
F	T	T	F	T	T	T	T	T
F	T	F	F	T	T	T	T	T
F	F	T	F	T	T	T	T	T
F	F	F	F	T	F	T	T	T
			(a)	(c)	(b)	(d)	(f)	(e)

EXERCISES *Consider the following statements.*

> p: *You pass the examination.*
>
> q: *You pass the course.*

Then translate each symbolic statement into an English sentence.

1. $p \rightarrow q$
2. $q \rightarrow p$
3. $(\sim p) \rightarrow (\sim q)$
4. $(\sim q) \rightarrow (\sim p)$

Repeat the specified exercise for the following statements.

> p: *John drives a red car.*
>
> q: *John lives in a red house.*

5. Exercise 1.
6. Exercise 2.
7. Exercise 3.
8. Exercise 4.

9. Give the truth value of each statement.
   (a) If $2 \times 3 = 5$, then $2 + 3 = 6$.
   (b) If $2 \times 3 = 5$, then $2 + 3 = 5$.
   (c) If $2 + 3 = 5$, then $2 \times 3 = 5$.
   (d) If $2 + 3 = 5$, then $2 \times 3 = 6$.

10. Give the truth value of each statement.
    (a) If $5 \times 6 = 56$, then $5 - 6 = 11$.
    (b) If $5 \times 6 = 42$, then $5 - 6 = 10$.
    (c) If $5 \times 6 = 30$, then $5 + 6 = 10$.
    (d) If $5 + 6 = 11$, then $5 \times 6 = 30$.

11. Assume that $2x = 6$, $x = 3$, and $x \neq 4$. Then give the truth value of each statement.
    (a) If $2x = 6$, then $x = 3$.
    (b) If $2x = 6$, then $x = 4$.
    (c) If $3 = 4$, then $x = 4$.
    (d) If $3 = 4$, then $x = 3$.

12. Assume that $a \times b = c$, $b \times c = d$, and $c \neq d$. Then give the truth value of each statement.
    (a) If $a \times b = c$, then $b \times c = d$.
    (b) If $a \times b = d$, then $b \times c = c$.
    (c) If $a \times b = d$, then $b \times c = d$.
    (d) If $a \times b = c$, then $b \times c = c$.

Write the negation, converse, inverse, and contrapositive of each statement.

13. If $x = 1$, then $x \neq 2$.
14. If $2x = 6$, then $x = 3$.
15. If we can afford it, then we shall buy a new car.
16. If we play tennis, then you will win the game.

Exercises 17 through 20 refer to the statements given in Exercises 13 and 14. Tell whether or not you accept the indicated statement as always true.

17. The given statement.
18. The converse of the given statement.
19. The inverse of the given statement.
20. The contrapositive of the given statement.

Show by means of truth tables that the statements have the same truth values.

21. $\sim(p \wedge q)$ and $(\sim p) \vee (\sim q)$.
22. $\sim(p \vee q)$ and $(\sim p) \wedge (\sim q)$.

4 SETS AND LOGIC

23. $p \rightarrow q$ and $q \vee (\sim p)$.
24. $(\sim p) \rightarrow (\sim q)$ and $q \rightarrow p$.
25. $p \wedge (q \vee r)$ and $(p \wedge q) \vee (p \wedge r)$.
26. $p \vee (q \wedge r)$ and $(p \vee q) \wedge (p \vee r)$.
27. $p \rightarrow (q \wedge r)$ and $(p \rightarrow q) \wedge (p \rightarrow r)$.
28. $(p \vee q) \rightarrow r$ and $(p \rightarrow r) \wedge (q \rightarrow r)$.

Explorations

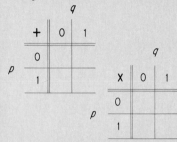

Boolean algebra is named after the English mathematician George Boole (1815–1864), who was a friend of Augustus De Morgan. Boole emphasized that logic should be associated with mathematics and that mathematics should be much broader than the science of magnitude and number. Boole's work was the start of a new era in the development of logic. The new approaches have been extensively used in the development of telephone systems and computers.

1. The $p \vee q$ circuit provides a basis for defining addition of two elements:

   1 presence of current at light bulb or through a switch.

   0 absence of current at light bulb or through a switch.

   Use sketches of a $p \vee q$ circuit as necessary and complete the adjacent *addition table*.

2. Use sketches of a $p \wedge q$ circuit as necessary and complete the adjacent *multiplication table*.

The algebra of the two elements 0 and 1 with addition and multiplication defined as in Explorations 1 and 2 is the **binary Boolean algebra.** We compare this algebra of the numbers 0 and 1 with the algebra of sets $\varnothing$ and $\mathcal{U}$, where $\varnothing$ is the empty set and $\mathcal{U}$ is the universal set.

3. Interpret $+$ as $\cup$ and complete the following addition table.

+	$\varnothing$	$\mathcal{U}$
$\varnothing$		
$\mathcal{U}$		

4. Interpret $\times$ as $\cap$ and complete the following multiplication table.

$\times$	$\varnothing$	$\mathcal{U}$
$\varnothing$		
$\mathcal{U}$		

5. Two sets of elements, each with relations that may be interpreted as $+$ and $\times$, are said to be **isomorphic sets** if there is a one-to-one correspondence of the elements of the sets such that each sum corresponds to the sum of corresponding elements, and each product corresponds to the product of corresponding elements. Show that the set of elements of the binary Boolean algebra considered in Explorations 1 and 2 and the sets of the algebra of sets considered in Explorations 3 and 4 are isomorphic.

Statements may be expressed in many different forms. For example, the words *necessary* and *sufficient* are often used to express conditional statements in alternate forms. Consider the statement

> Working hard is a sufficient condition for passing the course.

Let us use p to mean *work hard* and q to represent *pass the course*. We need to decide whether the given statement means "if p, then q" or "if q, then p." The word sufficient can be interpreted to mean that working hard is adequate or enough, but possibly not always necessary, for passing. That is, there may be other ways to pass the course, but working hard will do it. Thus we interpret the statement to mean

> If you work hard, then you will pass the course.

Then $p \rightarrow q$ may represent each of these statements.

> If p, then q.
>
> p is a sufficient condition for q.

Next consider this statement.

> Working hard is a necessary condition for passing the course.

Here you are told that working hard is necessary or essential in order to pass. That is, regardless of what else you do, you must work hard if you wish to pass. However, there is no assurance that working hard alone will do the trick. It is necessary, but may not be sufficient. (You may also have to get good grades.) Therefore, we interpret the statement to mean

> If you pass the course, then you have worked hard.

Then $q \rightarrow p$ may represent each of these statements.

> If q, then p.
>
> p is a necessary condition for q.

Still another form to consider is the statement "q, only if p." In terms of the example used in this section, we may write this as

> You will pass the course only if you work hard.

Suppose that you will get an A only if you work hard. Does this mean that you will get an A if you work hard? See Exploration 4.

This does *not* say that working hard will insure a passing grade. It does mean that if you have passed, then you have worked hard. That is, "*q*, only if *p*" is equivalent to the statement "if *q*, then *p*." We can also interpret this in another way. This statement "*q*, only if *p*" means "if not *p*, then not *q*." The contrapositive of this last statement, however, is "if *q*, then *p*." In terms of our illustration, this means that if you do not work hard, then you will not pass. Therefore, if you pass, then you have worked hard.

To summarize our discussion, each of the following statements represents a form of the conditional statement $p \rightarrow q$.

If p, then *q*.

q, if *p*.

p is a sufficient condition for *q*.

q is a necessary condition for *p*.

p, only if *q*.

The many distinct ways of expressing a conditional statement illustrate the difficulty of understanding the English language. We shall endeavor to reduce the confusion by expressing conditional statements in the form.

If *p*, then *q*.   Symbolically, $p \rightarrow q$.

Example 1 Write each statement in if-then form.
(**a**) All rainy days are cloudy.
(**b**) $x = 5$, only if $x \neq 0$.
(**c**) Any apple is a piece of fruit.
(**d**) Cats are mammals.

Solution (**a**) If a day is rainy, then it is cloudy.
(**b**) If $x = 5$, then $x \neq 0$.
(**c**) If an object is an apple, then it is a piece of fruit.
(**d**) If an animal is a cat, then it is a mammal.

Example 2 Translate into symbolic form, using *p* and *q* as follows.

p: I shall work hard.

q: I shall get an A.

(**a**) I shall get an A only if I work hard.

(b) Working hard will be a sufficient condition for me to get an A.

(c) If I work hard, then I shall get an A, and if I get an A, then I shall have worked hard.

Solution **(a)** $q \rightarrow p$ **(b)** $p \rightarrow q$ **(c)** $(p \rightarrow q) \wedge (q \rightarrow p)$

$p \rightarrow q$:
p is a sufficient condition for q.

$q \rightarrow p$:
p is a necessary condition for q.

$p \leftrightarrow q$:
p is a necessary and sufficient condition for q.

The statement $(p \rightarrow q) \wedge (q \rightarrow p)$ in Example 2(c) is one form of the **biconditional statement** $p \leftrightarrow q$ (read as "p if and only if q"); the symbol $\leftrightarrow$ is the **biconditional symbol.** Any biconditional statement $p \leftrightarrow q$ is a statement that p is a sufficient condition for q and also p is a necessary condition for q. We may condense this by saying that p is a **necessary and sufficient condition** for q. The biconditional statement may be stated in either of these forms.

> p is a necessary and sufficient condition for q.
>
> p if and only if q; that is, p **iff** q.

Example 3 Complete a truth table for the statement

$$(p \rightarrow q) \wedge (q \rightarrow p)$$

Solution

p	q	$(p \rightarrow q)$	$\wedge$	$(q \rightarrow p)$
T	T	T	T	T
T	F	F	F	T
F	T	T	F	F
F	F	T	T	T

(a) (c) (b)

p	q	$p \leftrightarrow q$
T	T	T
T	F	F
F	T	F
F	F	T

Note that $p \leftrightarrow q$ is true if and only if p and q have the same truth values.

We constructed a truth table for $(p \rightarrow q) \wedge (q \rightarrow p)$ in Example 3. However, we have previously agreed that this conjunction of statements is a form of $p \leftrightarrow q$. This enables us to construct a truth table for $p \leftrightarrow q$. From the truth table, we see that $p \leftrightarrow q$ is true when p and q are both true or both false. Thus each of these biconditional statements is true.

$2 \times 2 = 4$ if and only if $7 - 5 = 2$ (both parts are true)

$2 \times 2 = 5$ if and only if $7 - 5 = 3$ (both parts are false)

Each of the following biconditional statements is false because exactly one part of each statement is false.

$2 \times 2 = 4$ if and only if $7 - 5 = 3$

$2 \times 2 = 5$ if and only if $7 - 5 = 2$

4 SETS AND LOGIC

Example 4 Under what conditions is the following statement true?

I shall get an A if and only if I work hard.

Solution The statement has the form $p \leftrightarrow q$. Such a statement is true only if p and q are both true or p and q are both false. Thus the given statement is true in the following two cases.

You get an A and you worked hard.

You do not get an A and you did not work hard.

EXERCISES *Express each statement in if-then form.*

1. All apples are red.
2. All birds are beautiful.
3. All dogs are good watchdogs.
4. All ovals are round.
5. All squares are polygons.
6. All x's are y's.
7. Any two ball players are competitors.
8. Any large textbook is expensive.
9. Automobiles are expensive.
10. Calculators are useful.
11. You will like this book only if you like mathematics.
12. A necessary condition for liking this book is that you like mathematics.
13. To like this book it is sufficient that you like mathematics.
14. A sufficient condition for liking this book is that you like mathematics.
15. Liking this book is a necessary condition for liking mathematics.
16. A number is a counting number only if its square is a counting number.

Write each statement in symbolic form using p and q as follows.

    p: I feel chilly.

    q: I put on a sweater.

17. If I put on a sweater, then I feel chilly.
18. If I feel chilly, then I put on a sweater.
19. I put on a sweater only if I feel chilly.
20. If I do not feel chilly, then I do not put on a sweater.

21. Feeling chilly is a necessary condition for me to put on a sweater.
22. I put on a sweater if and only if I feel chilly.
23. If I feel chilly, then I do not put on a sweater.
24. For me to put on a sweater it is sufficient that I feel chilly.
25. A necessary and sufficient condition for me to put on a sweater is that I feel chilly.
26. For me to put on a sweater it is necessary that I feel chilly.

Express each statement in if-then form and classify as true or false.

27. $12 - 4 = 7$ if $12 + 4 = 15$.
28. A necessary condition for $2 \times 2 \neq 4$ is $12 - 4 = 8$.
29. For $7 \times 4 = 20$ it is sufficient that $7 + 4 = 11$.
30. $7 \times 5 = 57$ is a sufficient condition for $7 + 5 = 13$.
31. $7 \times 5 = 75$ only if $15 \times 5 \neq 75$.
32. $7 \times 5 = 35$ only if $15 \times 5 \neq 75$.

Under what conditions is the given statement true?

33. I shall be happy if and only if I pass the test.
34. Studying hard is a necessary and sufficient condition for passing the test.
35. A necessary condition for passing the test is that you study hard.
36. A sufficient condition for passing the test is that you study hard.

Explorations

We frequently state things in a variety of ways. In Explorations 1, 2, and 3, restate each sentence in at least four equivalent ways.

1. Learning your part is necessary for you to be in the play.
2. Attending rehearsals is sufficient for you to sing in the chorus.
3. Doing the daily work regularly will enable you to pass the course.

4. Difficulties often arise with the words "only if." For example, the statement "I will get an A, only if I work hard" means "If I get an A, then I have worked hard." However, in everyday language most people tend to interpret the

4 SETS AND LOGIC

first statement (incorrectly) as "If I work hard, then I will get an A." Find several other common examples of such confusions involving "only if."

5. A statement such as "You may leave early if you finish the quiz" logically means "If you finish the quiz, then you may leave early." However, many tend to interpret this permissive use of *if* as meaning *only if*. Thus they would then incorrectly think of the given statement as meaning "If you leave early, then you have finished the quiz."

   An if-then statement that is used to give permission frequently means "don't . . . unless. . . ." With this usage, the statement

   You may leave early if you finish the quiz

   is interpreted as

   Don't leave early unless you have finished the quiz.

   Prepare a set of three "if-then" statements that give permission. Then rewrite each statement in the form "don't . . . unless. . . ."

There exist situations in which statements are neither true nor false and our usual rules of logic cannot be used.

6. Select a plain 3 × 5 card or similar piece of paper. On one side write

   The statement on the other side of this card is true.

   Then on the other side of the card write

   The statement on the other side of this card is false.

   Discuss the possible sets of truth values for the statements on the two sides of the card.

Exploration 7 is a famous paradox. Can you explain it?

7. There is reported to be a town in which the barber is a man who shaves all men who do not shave themselves. Who shaves the barber?

8. Discuss the truth values of the statement

   The sentence you are reading is false.

9. Start a collection of logical paradoxes; that is, sets of statements that do not seem to "make sense" under our usual laws of logic. For example, consider the statements given in Explorations 6, 7, and 8.

10. Read, or reread, Lewis Carroll's *Alice's Adventures in Wonderland* and *Through the Looking Glass* for numerous amusing examples of logical principles and his light touch in handling them.

4-6
Mathematical Proofs

How do you "prove" a statement to a friend? Undoubtedly there are several ways, including these three.

1. In a reference book or from a reliable authority find sufficient support for the statement so that your friend will accept it without further proof.
2. Prove to your friend that the statement is a necessary consequence of some statement that has already been accepted.
3. Prove to your friend that the statement cannot be false.

What does proof mean to you?

In mathematics there are also several ways of "proving" statements. In essence each proof is based upon statements that are accepted as true (assumed). Each *direct proof* consists of a sequence of statements (an argument) such that each statement is either assumed or is a *logical consequence* of the preceding statements, and the statement to be proved is included in the sequence. A statement may also be *proved indirectly* by proving that the statement cannot be false.

Any proof includes, at least informally, some given statements that are assumed to be true statements and one or more statements that are to be proved. The assumed statements are the **"given"** (often called the **premises**) of the proof. The statements that are to be proved are the **conclusions** of the proof. A correct mathematical proof is based on a *valid argument*. Specifically, any argument, mathematical or otherwise, is a **valid argument** if the conjunction of the premises implies the conclusions. In other words, *an argument is valid if, under the assumption of the premises, the conclusions cannot fail to be true*.

One frequently used form of argument is the **law of detachment**, or **modus ponens**.

If a statement of the form "If p, then q" is assumed to be true and p is known to be true, then q must also be true.

Symbolically we write

Given:	p	$\rightarrow$ q	If p, then q; and
Given:	p		p
Conclusion:	q		imply q.

We may also write this argument as

$$[(p \rightarrow q) \wedge p] \rightarrow q$$

p	q	$[(p \rightarrow q)$	$\wedge$	$p]$	$\rightarrow$	q
T	T	T	T	T	T	T
T	F	F	F	T	T	F
F	T	T	F	F	T	T
F	F	T	F	F	T	F

$(a) \qquad (c) \ (b) \ (e) \ (d)$

The truth table shows the validity of the law of detachment. Note that in all possible cases the statement is true, as shown in column (e).

Example 1 Determine whether or not the following argument is valid.

> *Given*: If Mary is a junior, she is taking algebra.
> *Given*: Mary is a junior.
> *Conclusion*: Mary is taking algebra.

Solution Use

> p: Mary is a junior.
> q: Mary is taking algebra.

and think of the argument as

> *Given*: $p \rightarrow q$
> *Given*: p
> *Conclusion*: q

The argument has the form $[(p \rightarrow q) \wedge p] \rightarrow q$ and is therefore valid.

There are other forms of valid arguments. **Argument by contraposition** is valid. (See Exercise 18.) Symbolically, we write this as

Any conditional statement implies its contrapositive, that is, $[p \rightarrow q]$ implies $[(\sim q) \rightarrow (\sim p)]$.

> *Given*: $p \rightarrow q$ If p, then q; and
> *Given*: $\sim q$ not q
> *Conclusion*: $\sim p$ imply not p.

This form of argument may be written as

$$[(p \rightarrow q) \wedge (\sim q)] \rightarrow (\sim p)$$

Example 2 Employ contraposition to give a valid argument using

> p: You work hard.
> q: You pass the course.

If you work hard, then you pass the course. $p \rightarrow q$

You do not pass the course. $\dfrac{\sim q}{\sim p}$

Therefore, you did not work hard.

Not all arguments are valid. Consider the following.

Given: $p \rightarrow q$ If p, then q; and

Given: $\dfrac{q}{p}$ q

Conclusion: imply p.

The argument has the form

$$[(p \rightarrow q) \wedge q] \rightarrow p$$

A truth table will show that a statement of this form is *not* true for all possible cases. Thus the argument is not valid. (See Exercise 19.) As an example of this type of reasoning, consider the statement

Read this misleading advertisement and try to find others.

If you expect to be healthy, eat **KORNIES**.

The advertiser hopes that the consumer will assume, incorrectly, the converse statement

If you eat **KORNIES**, then you expect to be healthy.

The argument is not valid and is called a **fallacy**.

Here is another form of argument that is not valid and is a fallacy. (See Exercise 20.)

Given: $p \rightarrow q$ If p, then q; and

Given: $\dfrac{\sim p}{\sim q}$ not p

Conclusion: imply $\sim q$.

Example 3   Determine whether or not the following argument is valid.

Given: If you worked hard, then you passed the course.

Given: You did not work hard.

Conclusion: You did not pass the course.

Solution   For

p: You worked hard.

q: You passed the course.

the argument is not valid since it has the form

$$[(p \rightarrow q) \wedge (\sim p)] \rightarrow (\sim q)$$

Notice that anyone who uses the argument in Example 3 expects listeners to assume that the inverse $(\sim p) \rightarrow (\sim q)$ of any true statement $p \rightarrow q$ must be true. As in the case of a conditional statement and its converse, we know that a statement does not necessarily imply its inverse. As another example of this type of reasoning, consider the advertisement

> If you brush your teeth with SCRUB, then you will have no cavities.

The advertiser would like you to assume, incorrectly, the inverse statement

> If you do not brush your teeth with SCRUB, then you will have cavities.

Another form of valid reasoning that we shall consider here is of the *chain-reaction* type.

Given:	p	$\rightarrow$	q	If p, then q; and
Given:	q	$\rightarrow$	r	if q then r
Conclusion:	p	$\rightarrow$	r	imply if p then r.

The argument is valid because a truth table will show that the following statement is true in all possible cases. (See Exercise 21.)

$$[(p \rightarrow q) \wedge (q \rightarrow r)] \rightarrow (p \rightarrow r)$$

Here is an example of chain-reaction reasoning.

> If you like this book, then you like mathematics.
> If you like mathematics, then you are intelligent.
> Therefore, if you like this book, then you are intelligent.

The first two statements are premises (assumed to be true) and the third statement is the conclusion. In this and in other cases in which the argument is valid, the conclusion is often called a **valid conclusion** of the premises.

Example 4 Determine whether each of the conclusions is or is not a valid conclusion.

> *Given:* If you study mathematics, then you will be successful.
> *Given:* If you are successful, then you will be rich.
> *Conclusions:*
> (a) If you study mathematics, then you will be rich.
> (b) If you become rich, then you have studied mathematics.
> (c) If you do not become rich, then you have not studied mathematics.

Note that a valid conclusion may be either true or false. Thus in Example 4(c) the conclusion may be false but the argument is valid. *If* you accept the premises, then the conclusion must be accepted as a valid conclusion.

(a) Valid; this argument is of the chain reaction type
$[(p \rightarrow q) \land (q \rightarrow r)] \rightarrow (p \rightarrow r)$.

(b) Not valid; this argument uses the converse for its conclusion.

(c) Valid; this argument uses the contrapositive of part (a).

EXERCISES *In each exercise assume that the premises are true and determine whether or not the argument is valid.*

1. *Given*: If Elliot is a freshman, then Elliot takes mathematics.
   Given: Elliot is a freshman.
   Conclusion: Elliot takes mathematics.

2. *Given*: If you like dogs, then you will live to be 120 years old.
   Given: You like dogs.
   Conclusion: You will live to be 120 years old.

3. If the Braves win the game, then they win the pennant.
   They do not win the pennant.
   Therefore, they did not win the game.

4. If you like mathematics, then you like this book.
   You do not like mathematics.
   Therefore, you do not like this book.

5. If you work hard, then you are a success.
   You are not a success.
   Therefore, you do not work hard.

6. If you are reading this book, then you like mathematics.
   You like mathematics.
   Therefore, you are reading this book.

7. If you are reading this book, then you like mathematics.
   You are not reading this book.
   Therefore, you do not like mathematics.

8. If you work hard, then you will pass the course.
   If you pass the course, then your teacher will praise you.
   Therefore, if you work hard, then your teacher will praise you.

9. If you like this book, then you like mathematics.
   If you like mathematics, then you are intelligent.
   Therefore, if you are intelligent, then you like this book.

10. If you are happy, then you are lucky.
    If you are lucky, then you will be rich.
    Therefore, if you do not become rich, then you are not happy.

In each exercise use all of the given premises and supply a valid conclusion for them.

11. If you drink milk, then you will be healthy.
    You are not healthy.
    Therefore, ...

12. If you eat a lot, then you will gain weight.
    You eat a lot.
    Therefore, ...

13. If you like to fish, then you enjoy swimming.
    If you enjoy swimming, then you are a mathematician.
    Therefore, ...

14. If you do not work hard, then you will not get an A.
    If you do not get an A, then you will have to repeat the course.
    Therefore, ...

15. If you like this book, then you are not lazy.
    If you are not lazy, then you will become a mathematician.
    Therefore, ...

Give a mathematical reason why you would not accept the given conclusion. Then give a nonmathematical reason.

16. *Given:* If you have eyestrain, then you have a headache.
    Given: You have a headache.
    Conclusion: You have eyestrain.

17. *Given:* If you have eyestrain, then you have a headache.
    Given: You do not have eyestrain.
    Conclusion: You do not have a headache.

Complete a truth table for each statement. Use the results to tell whether the given argument is valid or not valid.

18. $[(p \rightarrow q) \wedge (\sim q)] \rightarrow (\sim p)$

19. $[(p \rightarrow q) \wedge q] \rightarrow p$

20. $[(p \rightarrow q) \wedge (\sim p)] \rightarrow (\sim q)$

21. $[(p \rightarrow q) \wedge (q \rightarrow r)] \rightarrow (p \rightarrow r)$

Exercise 22 is a famous nonstandard problem. Can you solve it?

*22. Suppose that you are a prisoner standing alone with the executioner in the execution chamber. You must choose one of two chairs and sit in it. One chair is an electrified chair that kills anyone who sits in it. The other chair is harmless.

   You are allowed to ask the executioner just one question that he or she may answer by "yes" or "no." Furthermore, you know that the executioner either always tells the truth or always lies but you do not know whether he or she tells the truth or lies.

What question could you as the condemned prisoner ask and determine without any doubt which chair is the safe chair?

Suppose that you approach a fork in the road while traveling to the river in a strange country. What single question could you ask to find your way in each of these situations?

*23. You encounter a native who either always tells the truth or always lies but you do not know which.

*24. You encounter two natives, one of whom always tells the truth and one of whom always lies, but you do not know which is which.

Explorations

1. Begin a collection of fallacies in reasoning that you find in newspapers or magazines, or that you hear on radio or television.

2. Examine a newspaper or magazine for if-then statements that are used in advertising. Then analyze the statement to see whether the converse or the inverse of the given statement is intended.

3. Give five valid arguments that could be used with elementary school children.

4. Give five arguments that could be used with elementary school children to illustrate arguments that are not valid.

The following explorations are based upon sets of premises written by Charles Lutwidge Dodgson, who used the name of Lewis Carroll as author of *Alice's Adventures in Wonderland* and *Through the Looking Glass*. Supply a conclusion so that each argument will be valid.

5. Babies are illogical.
   Nobody is despised who can manage a crocodile.
   Illogical persons are despised.
   Therefore, ...

6. No ducks waltz.
   No officers ever decline to waltz.
   All my poultry are ducks.
   Therefore, ...

4 SETS AND LOGIC

7. No terriers wander among the signs of the zodiac.
Nothing that does not wander among the signs of the zodiac is a comet.
Nothing but a terrier has a curly tail.
Therefore, ...

8. Discuss the validity of the arguments used in the following situation.

A man approached the clerk at the checkout counter of a local market and asked the price of a box of blueberries.

Clerk: "65 cents a box."

Customer: "What! They're selling them for 55 cents a box across the street."

Clerk: "Why don't you buy them there?"

Customer: "Because they're all sold out."

Clerk: "Oh! If we were all sold out, our price would be 45 cents a box."

Try to find other descriptions of situations, such as the above, which involve logical concepts.

Chapter 4 Review

Solutions to the following exercises may be found within the text of Chapter 4. Try to complete each exercise without referring to the text.

Section 4-1   Set Notation

1. Give a verbal description for the set $Y = \{1, 3, 5, 7, 9\}$.
2. Show that $\{c, r, a, b\}$ and $\{f, i, s, h\}$ are (a) equivalent sets of letters   (b) not equal sets.
3. Let $\mathcal{U} = \{1, 2, 3, . . . , 10\}$, $A = \{1, 2, 3, 4, 5, 7, 9\}$, and $B = \{2, 4, 6, 8\}$. Find (a) A' (b) B' (c) $A - B$ (d) $B - B$.
4. If $X = \{1, 2, 3, 4, 5\}$ and $Y = \{4, 5, 6, 7\}$, find $X \cap Y$ and $X \cup Y$.

Section 4-2   Venn Diagrams

5. Show by means of a Venn diagram that $(A \cup B)' = A' \cap B'$.
6. Show that $A \cap (B \cup C) = (A \cap B) \cup (A \cap C)$.

7. In a group of 35 students, 15 are studying French, 22 are studying English, 14 are studying Spanish, 11 are studying both French and English, 8 are studying English and Spanish, 5 are studying French and Spanish, and 3 are studying all three subjects. How many are taking only English? How many of these students are not taking any of these subjects?

Section 4-3 Simple and Compound Statements

8. Write in symbolic form: I have not made the last payment on my car or I have not purchased all of the books required for this semester.

Give a truth table for each of the following.

9. $p \wedge q$ 10. $p \vee q$ 11. $p \wedge (\sim q)$

Section 4-4 Conditional Statements

12. Give a truth table for the statement $p \rightarrow q$.
13. Give the truth value of each statement.
    (a) If $5 + 7 = 12$, then $6 + 7 = 13$.
    (b) If $5 \times 7 = 35$, then $6 \times 7 = 36$.
    (c) If $5 + 7 = 35$, then $6 + 7 = 13$.
    (d) If $5 + 7 = 35$, then $6 \times 7 = 36$.
14. Write, in symbolic form, the negation, the converse, the inverse, and the contrapositive for the statement $p \rightarrow q$.
15. Write (a) the negation, and (b) the contrapositive, of the statement: If I work hard, then I shall pass the course.

Section 4-5 Forms of Statements

16. Use p to mean "work hard" and q to represent "pass the course." Then write each of the following in symbolic form.
    (a) Working hard is a sufficient condition for passing the course.
    (b) Working hard is a necessary condition for passing the course.
17. Translate into symbolic form, using p: "I shall work hard" and q: "I shall get an A."
    (a) I shall get an A only if I work hard.
    (b) Working hard will be a sufficient condition for me to get an A.
    (c) If I work hard, then I shall get an A, and if I get an A, then I shall have worked hard.
18. Complete a truth table for the statement $p \leftrightarrow q$.

19. Complete a truth table for this argument:
    $[(p \rightarrow q) \wedge p] \rightarrow q$.

20. Determine whether or not the following argument is valid.
    Given: If you worked hard, then you passed the course.
    Given: You did not work hard.
    Conclusion: You did not pass the course.

21. Determine whether each of the conclusions is or is not a valid conclusion.
    Given: If you study mathematics, then you will be successful.
    Given: If you are successful, then you will be rich.
    Conclusions:
    (a) If you study mathematics, then you will be rich.
    (b) If you become rich, then you have studied mathematics.
    (c) If you do not become rich, then you have not studied mathematics.

Chapter 4 Test

1. Let $\mathcal{U} = \{m, o, r, a, b, l, e\}$ and list A' when A is defined as   (a) $\{m, o, r, e\}$ (b) $\{l, a, b, o, r\}$.

2. List all possible subsets of $\{h, a, t\}$.

Identify as true or as not always true (false).

3. The sets $\{b, a, t\}$ and $\{l, a, b\}$ are   (a) equal sets   (b) equivalent sets.

4. (a) $(\mathcal{U} \cup B) \subseteq \mathcal{U}$ (b) $\mathcal{U} \cap \varnothing \subset \varnothing$

5. (a) $(A \cup B) \subset (B \cup A)$ (b) $(A \cap B) \subset (A \cup B)$

6. (a) $75 \in \{2, 4, 6, \ldots\}$ (b) $75 \in \{1, 3, 5, \ldots\}$

For $\mathcal{U} = \{0, 1, 2, 3, 4, 5, 6, 7, 8, 9\}$, $A = \{2, 4, 6\}$, *and* $B = \{0, 3, 6, 9\}$, *find the following.*

7. $A' \cap B$ 8. $(A \cup B)'$

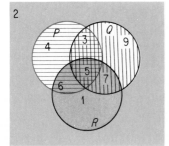

Draw Venn diagrams and shade the indicated regions.

9. $A' \cap B$ 10. $(A \cup B) \cap C$

11. Use the given diagram and find (a) $n(P' \cap Q \cap R)$ (b) $n(Q \cup R)$.

12. Suppose that you operate a newsstand and sell three newspapers: the *Morning Star*, the *Evening Gazette*, and the *Daily Bulletin*. You expect 650 people to buy at least

one paper, 50 to buy all three papers, 130 to buy only the *Star,* 300 to buy the *Bulletin,* 150 to buy the *Bulletin* and the *Gazette,* 175 to buy the *Bulletin* and the *Star,* and 75 to buy only the *Gazette.* How many do you expect to buy both the *Gazette* and the *Star?*

Use

  p: Wendy is happy.

  q: Jan is sad.

Think of "sad" as "not happy" and write each of these statements in symbolic form.

13. (a) Wendy is happy and Jan is sad.
     (b) Wendy is not happy but Jan is happy.
14. (a) Either Wendy or Jan is happy.
     (b) It is not true that Wendy and Jan are both happy.
15. Assume that Wendy and Jan are both happy. Which of the statements in Exercises 13 and 14 are then true?
16. State $p \lor (\sim q)$ in words for
     p: My car is blue.
     q: My car is an American car.

Construct a truth table for the given statement.

17. $\sim[(\sim p) \lor q]$ 18. $(p \land q) \to (\sim p)$
19. $\sim[p \leftrightarrow (\sim q)]$ 20. $(p \land q) \to r$

21. Write the negation, the converse, the inverse, and the contrapositive of the given statement.

     If apples are red, then they are ripe.

22. Give the truth value of each statement.
     (a) If $3 \times 4 = 10$, then $4 \times 3 = 10$.
     (b) If $3 \times 4 = 10$, then $4 \times 3 = 12$.
     (c) If $3 \times 4 = 12$, then $4 \times 3 = 10$.
     (d) If $3 \times 4 = 12$, then $4 \times 3 = 12$.
23. Write each statement in if-then form.
     (a) All horses are quadrupeds.
     (b) Knowing Judy is sufficient reason for liking her.
24. Write each statement in symbolic form. Use

     p: I wear a coat.

     q: Snow is falling.

(a) For me to wear a coat it is necessary that snow be falling.

(b) I wear a coat only if snow is falling.

25. Determine whether each of the conclusions is or is not a valid conclusion.

Given: If you enjoy this test, then you will get an A.

Given: If you get an A, then you will be happy.

Conclusions:

(a) If you enjoy this test, then you will be happy.

(b) If you do not enjoy this test, then you will not be happy.

Sets
of
Numbers

5

5-1
Numbers and Numerals

Symbols for numbers are called **numerals.** For example, we write 1984 using Hindu-Arabic numerals. Four thousand years ago the early Egyptians used groupings of strokes so that they could recognize the number of strokes for each of the numbers 1 through 9.

I,   II,   III,   IIII,   III II,   III III,   IIII III,   IIII IIII,   III III III

1984 can be written using early Egyptian notation in this way:

Probably because we have ten fingers, a new symbol is often introduced for ten. We write this symbol as 10 in our system of notation. The early Egyptians represented 10 by the symbol ∩ and introduced a new symbol for each power of ten.

I	Vertical staff	1
∩	Heel-bone	10
ꝯ	Scroll	100
⚱	Lotus flower	1000
☞	Pointing finger	10,000

The scroll is often called a coil of rope and was sometimes "coiled" counterclockwise instead of clockwise.

The Egyptian system is said to have a **base** of ten because the symbols represent powers of ten. Our system of numeration is called a *decimal system* to emphasize the use of powers of ten for each place value. The Egyptian system has no place value. The absence of a place value means that the position of the symbol does not affect the number represented. For example, in our decimal system of numeration, 23 and 32 represent different numbers. In the Egyptian system ∩∩III and III∩∩ are different representations of the same numeral. Without the concept of place value the early Egyptians needed different symbols for different powers of ten.

Note these comparisons of decimal and early Egyptian number symbols.

3	III
30	∩∩∩
300	ꝯꝯꝯ

25	∩∩IIIII
142	ꝯ∩∩∩∩II
12,321	☞⚱⚱ꝯꝯꝯ∩∩I

Recall that fingers are also called *digits.* Finger reckoning is still used extensively in some parts of the world.

Our **decimal system of numeration** makes use of the ten *decimal digits.*

0 1 2 3 4 5 6 7 8 9

The value of each digit in a numeral depends on the position that the digit occupies. For example, 1984 is read as

one thousand nine hundred eighty-four

$$1984 = (1 \times 1000) + (9 \times 100) + (8 \times 10) + (4 \times 1)$$

The digit 1 represents 1 thousand, the digit 9 represents 9 hundreds, the digit 8 represents 8 tens, and the digit 4 represents 4 ones in the numeral 1984.

It is convenient to use exponents when representing the place values of digits. We define $10^0 = 1$ so that each place value may be written as a power of ten. For example,

$$1000 = 10^3 \qquad 100 = 10^2 \qquad 10 = 10^1 \qquad 1 = 10^0$$

Then we may write 1984 in **expanded notation** as

$$1984 = (1 \times 10^3) + (9 \times 10^2) + (8 \times 10^1) + (4 \times 10^0)$$

Example 1   Use exponents and write 2306 in expanded notation.

Solution

$$2306 = (2 \times 10^3) + (3 \times 10^2) + (0 \times 10^1) + (6 \times 10^0)$$

Example 2   Write in decimal notation.

$$(3 \times 10^5) + (2 \times 10^4) + (7 \times 10^3) + (0 \times 10^2) + (1 \times 10^1) + (3 \times 10^0)$$

Solution   327,013

In contrast to the early Egyptians we have a symbol 0 that can be used to fill a place that is not otherwise occupied. It is this use of zero that enables us to use the same decimal digits for all powers of ten (place values). Our numerals are called **Hindu-Arabic numerals** in recognition of the introduction of 0 by the Hindus and continued development of the numerals by the Arabs.

The symbols used to represent numbers affect the procedures (*algorithms*) used for computations. In the early Egyptian system computations are possible but tedious. For example, we may use these steps to add 27 and 35.

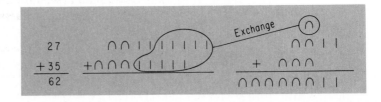

Observe that, in this Egyptian system, an indicated collection of ten ones was replaced by a symbol for ten before the final computation took place. In our decimal system we mentally perform a corresponding exchange of ten ones for a ten when we express $(7 + 5)$ as one ten and two ones. We exchange kinds of units in a similar manner in subtraction.

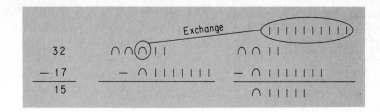

$$
\begin{array}{r}
32 \\
-17 \\
\hline
15
\end{array}
$$

Addition and subtraction may be visualized in terms of re-grouping. Multiplication in terms of repeated additions and division in terms of repeated subtractions are very tedious. The following algorithm for multiplication appeared in one of the first published arithmetic texts in Italy, the *Treviso Arithmetic* (1478). The method was used by early Hindus and Chinese before being widely used by the Arabians, who passed it on to the Europeans during the Middle Ages. We shall refer to the method here as *galley multiplication*, although it was called "Gelosia" multiplication in the original text and is sometimes called "lattice" multiplication. Let us use this method to find the product of the two *factors* 457 and 382.

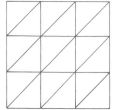

First prepare a "galley" with three rows and three columns, and draw the diagonals, as in the adjacent figure. Our choice for the number of rows and columns is based on the fact that we are to multiply two numbers represented by three-digit numerals.

Place the digits 4, 5, and 7 of one factor in order from left to right at the tops of the columns. Place the digits 3, 8, and 2 of the second factor in order from top to bottom at the right of the rows. Then each product of a digit of 457 and a digit of 382 is called a **partial product** and is placed at the intersection of the column and row of the digits. The diagonal separates the digits of the partial product (ten digit above ones digit). For example, $3 \times 7 = 21$, and this partial product is placed in the upper right-hand corner of the galley; $5 \times 8 = 40$, and this partial product is placed in the center of the galley; $4 \times 2 = 8$, and this partial product is entered as 08 in the lower left-hand corner of the galley. See if you can justify each of the entries in the completed array.

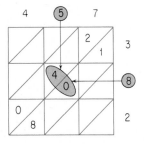

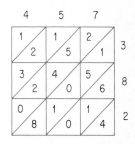

Check the product 457 × 382 using this galley.

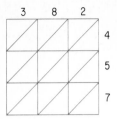

After all partial products have been entered in the galley, we add along diagonals, starting in the lower right-hand corner and carrying to the next diagonal sum where necessary. The next diagram indicates this pattern. The completed problem appears in the figure on the right. We read the final answer, as indicated by the curved arrow in the figure, as 174,574. Note that we read the digits in the opposite order to that in which they were obtained.

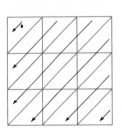

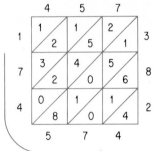

457 × 382 = 174,574

Example 3   Use galley multiplication and multiply 372 by 47.

Solution

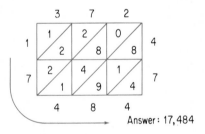

Answer: 17,484

Galley multiplication works because we are really listing all partial products before we add. Compare the following two computations and note that the numerals along the diagonals correspond to those in the columns at the right.

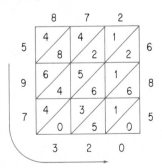

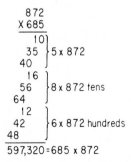

```
    872
  X 685
    10 ⎫
    35 ⎬ 5 x 872
    40 ⎭
    16 ⎫
    56 ⎬ 8 x 872 tens
    64 ⎭
    12 ⎫
    42 ⎬ 6 x 872 hundreds
    48 ⎭
597,320 = 685 x 872
```

5 SETS OF NUMBERS

EXERCISES *Write in early Egyptian notation.*

1. 35
2. 246
3. 3417
4. 60
5. 12,307
6. 21,532

Write in decimal notation.

7. ∩ ∩ | |
8. 𝟡 𝟡 ∩ |
9. 𓎡 𝟡 | |
10. 𓎡 𓎡 𝟡 𝟡 ∩ |
11. 𓎡 𝟡 𝟡 𝟡 ∩ ∩ | | | |
12. 𓎡 𝟡 𝟡 ∩ | |

Use exponents and write in expanded notation.

13. 2504
14. 5210
15. 245,600
16. 301,065
17. 501,200
18. 300,090

Write in decimal notation.

19. $(8 \times 10^3) + (1 \times 10^2) + (6 \times 10^1) + (5 \times 10^0)$
20. $(8 \times 10^6) + (6 \times 10^5) + (4 \times 10^4) + (3 \times 10^3)$
21. (6×10^7) 22. (8×10^9) 23. (7×10^8) 24. (5×10^6)
25. Thirty-seven thousand nineteen.
26. Three million five hundred five.
27. Three hundred thousand twenty-three.
28. Four billion three thousand eleven.
29. One billion three hundred five thousand.
30. Two hundred billion thirty million five.

Read each decimal numeral.

31. 5370
32. 5730
33. 205,030
34. 2,250,300
35. 25,203,500
36. 502,320,035

Multiply, using the galley method.

37. 942
    × 37
38. 586
    × 492
39. 234
    × 762

40. 8764
    × 37
41. 8035
    × 289
42. 8501
    × 3726

Explorations

Roman numerals have been widely used in the past and probably will not be unusual in MCMXCV, that is, 1995. These explorations are intended to help you recognize that it is relatively easy to represent whole numbers as Roman numerals. Some computations with Roman numerals are as easy as with our ordinary numerals, but other computations with Roman numerals are very awkward.

1. Use Roman numerals and list the counting numbers I through XXIX.

2. Use Roman numerals and list by fives the numbers V through L.

3. Describe the use of addition and subtraction in the representation of numbers by Roman numerals, such as XXIX.

4. Roman numerals are often used to state dates of construction of buildings on the cornerstones and in other places as well. Often the pages that precede the introduction to a book are given in Roman numerals. Find as many examples as you can of the use of such numerals.

5. The early Egyptians often used their numerals to form patterns, or pictures. For example, they were able to write 25 in such ways as these.

$$ ||\cap|\cap|| \qquad \cap|||||\cap \qquad |\cap|||\cap| $$

Why can't we do likewise with our numerals? What are the basic advantages of our decimal system of notation that make it desirable to give up such opportunities for "artistic effect"?

6. Refer to a history of mathematics book and determine the origin of the word *algorithm*.

7. Prepare a report on finger reckoning, that is, computations using one's fingers.

Nicaragua honored Napier on one of its stamps and, on the back of the stamp, described his work as follows: "With the invention of logarithms, Napier gave to the world a powerful arithmetic shorthand. It permitted people to do multiplication or division simply by adding or subtracting logarithms of numbers, and it meant that they could rapidly carry out these and more complicated operations for numbers containing many digits. The impact of logarithms on fields such as astronomy and navigation was enormous and comparable to the computer revolution of today."

8. The Scottish mathematician John Napier (1550–1617) made use of galley multiplication as he developed one of the forerunners of modern calculators. His device is referred to as *Napier's rods*, or Napier's bones, named after the material on which he had numerals printed. Find a reference to Napier's rods, make a set on strips of paper or other material, and demonstrate their use.

5-2
Binary Notation

The repesentation of numbers using only two digits, usually 0 and 1, is of special interest because of its applications in modern electronic computers. One of the two digits is represented by the presence of an electric signal; the other digit is represented by the absence of an electric signal. The presence or absence of electric signals may be controlled by the presence or absence of holes in a punched card, the presence or absence of magnetic fields on a magnetic tape, and in other ways. For example, the pictures of the surface of the planet Mars taken by the Mariner spacecraft in 1976 were represented by dots on a coordinate plane. The location, color, and intensity of the dots were transmitted back to Earth in binary notation. Then computers were used to reassemble the pictures so that they could be displayed on television and in other media.

The place value numeration system in which only two digits are used is called **binary notation.** There is some evidence that the basic concepts of binary notation were known to the ancient Chinese about 2000 B.C. However, it is only in relatively recent times that binary notation has been widely applied in card-sorting operations and in computer mathematics. Here are some of the place values in the binary system of numeration.

Here are the first 16 counting numbers written in binary notation.

BASE TEN	BASE TWO
1	1
2	10_2
3	11_2
4	100_2
5	101_2
6	110_2
7	111_2
8	$1\ 000_2$
9	$1\ 001_2$
10	$1\ 010_2$
11	$1\ 011_2$
12	$1\ 100_2$
13	$1\ 101_2$
14	$1\ 110_2$
15	$1\ 111_2$
16	$10\ 000_2$

2^7	2^6	2^5	2^4	2^3	2^2	2^1	2^0
128	64	32	16	8	4	2	1

Binary numerals often involve many digits. We set off these digits in sets of three, as in Example 1, to aid the reading of the numerals.

Example 1 Write $11\ 011\ 101_2$ in base ten notation.

Solution

$$11\ 011\ 101_2 = (1 \times 2^7) + (1 \times 2^6) + (0 \times 2^5) + (1 \times 2^4)$$
$$+ (1 \times 2^3) + (1 \times 2^2) + (0 \times 2^1) + (1 \times 2^0)$$
$$= 128 + 64 + 16 + 8 + 4 + 1 = 221$$

The adjacent tables are for addition and multiplication in binary notation.

+	0	1
0	0	1
1	1	10_2

X	0	1
0	0	0
1	0	1

Example 2   Multiply $101_2 \times 1101_2$.

Solution

$$
\begin{array}{r}
1\ 101_2 \\
\times\ \ \ 101_2 \\
\hline
1\ 101 \\
110\ 10\ \ \ \\
\hline
1\ 000\ 001_2
\end{array}
$$

Check:

$$
\begin{array}{r}
1\ 101_2 = 13 \\
\times\ \ \ 101_2 = 5 \\
\hline
65
\end{array}
$$

$$
1\ 000\ 001_2 = (1 \times 2^6) + (1 \times 2^0)
$$
$$
= 64 + 1 = 65
$$

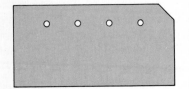

Binary numerals can be shown by means of electric lights, as in Exploration 4. If a light is on, the digit 1 is represented; if the light is off, the digit 0 is represented. Similarly, the binary concept can be used in a card-sorting operation. An exploration of the processes involved in a card-sorting operation provides insight into the processes that occur in a computer. Prepare a set of 16 index cards with four holes punched in each and a corner cut off as in the adjacent figure. At this stage the 16 cards should be exactly alike.

Next represent the numbers 0 through 15 on these cards in binary notation. Cut out the space above each hole to represent 1; leave the hole untouched to represent 0. Several cards are shown in the next figure.

$5 = 0101_2$

$6 = 0110_2$

$10 = 1010_2$

$15 = 1111_2$

After all the cards have been completed in this manner, shuffle them thoroughly and align them, making certain that they remain "face up." (The position of the cut off corner indicates when a card is right side up.) Then, going from right to left, perform the following operation: Stick a pencil or other similar object through the first hole and lift up. Some of the cards will come up, namely, those in which the holes have not been cut through to the edge of the card (those cards representing numbers whose units digit in binary notation is 0).

Place the cards that have been lifted up in front of the other cards and repeat the same operation for the remaining holes in order from right to left. When you have finished, the cards should be in numerical order, 0 through 15.

Note that only four operations are needed to arrange the 16 cards. As the number of cards is doubled, only one additional operation will be needed each time to replace them in order. That is, 32 cards may be placed in numerical order with five of the described card-sorting operations; 64 cards may be arranged with six operations; 128 cards with seven operations;

and so forth. Thus a large number of cards may be arranged in order with a relatively small number of operations. For example, over one billion cards may be placed in numerical order with only 30 sortings.

The seven-digit numeral 1 000 001 is the name for the letter A when that letter is sent over teletype to a computer. The American Standard Code for Information Interchange (ASCII) was adopted in 1967 and is used to convert letters, numerals, and other symbols into binary notation.

Electronic computers are designed to have specified capacities both as to the number of digits that can be represented for a given number and the number of digits that are processed at one time in performing an operation. In the previous example of card sorting, only numbers that required at most four digits in binary notation were represented and the digits were processed one at a time. In "computer language" each binary digit is a *bit* and the *word length* for a particular computer is the number of bits that are processed simultaneously. Some early computers had a word length of three bits and essentially worked in octal notation (see Exercises 39 through 51). As computer technology has improved, the word lengths of computers have increased to 8, 16, 32, and even 64 or more bits.

EXERCISES

Write each number in binary notation.

1.	38	2.	35	3.	29
4.	75	5.	93	6.	129
7.	156	8.	173	9.	200
10.	425	11.	437	12.	511

Change each number to decimal notation.

13.	1110_2	14.	$10\ 100_2$	15.	$11\ 011_2$
16.	$100\ 111_2$	17.	$111\ 011_2$	18.	$101\ 110_2$
19.	$101\ 011_2$	20.	$1\ 001\ 100_2$	21.	$1\ 101\ 010_2$
22.	$11\ 101\ 011_2$	23.	$10\ 111\ 001_2$	24.	$10\ 101\ 010_2$

Perform the indicated operation in binary notation; check in base ten.

25.	1111_2 $+\ 1011_2$	26.	$10\ 001_2$ $+\ 10\ 101_2$	27.	$10\ 011_2$ $+\ 10\ 101_2$
28.	$100\ 101_2$ $+\ 10\ 111_2$	29.	1111_2 $-\ 1011_2$	30.	$11\ 001_2$ $-\ 10\ 110_2$
31.	$100\ 101_2$ $-\ 10\ 111_2$	32.	$100\ 011_2$ $-\ 10\ 101_2$	33.	$1\ 000\ 100_2$ $-\ 111\ 111_2$
34.	1111_2 $\times\ 11_2$	35.	$10\ 111_2$ $\times\ 101_2$	36.	$110\ 110_2$ $\times\ 110_2$

37. In the ASCII code, the numbers for B, D, and G are 66, 68, and 71, respectively. Find the binary representation of (a) B (b) D (c) G.

38. In the ASCII code, letters A, B, C, . . . in alphabetical order are assigned numbers 65, 66, 67, ... What is the word transmitted by the code 1 010 010, 1 010 101, 1 001 110?

39. Write the number 234 in base eight and then in base two notation. Can you discover a relationship between these two notations?

Base eight notation is often called octal notation. *Use the relationship discovered in Exercise 39 and write in octal notation.*

40. $101\ 111\ 001\ 010_2$ 41. $11\ 101\ 011\ 001_2$

42. $1\ 001\ 101\ 110\ 010_2$ 43. $10\ 010\ 101\ 011\ 011_2$

44. $111\ 110\ 101\ 101\ 011\ 001_2$ 45. $1\ 101\ 011\ 001\ 000\ 100\ 110_2$

Use the relationship discovered in Exercise 39 and write in binary notation.

46. 335_8 47. 5023_8 48. 4357_8

49. 4624_8 50. $42\ 345_8$ 51. $36\ 543_8$

Explorations

Many recreational items are based on the binary system of notation. Consider, for example, the boxes shown, within which the numbers 1 to 15 are placed according to the following scheme.

D	C	B	A
8	4	2	1
9	5	3	3
10	6	6	5
11	7	7	7
12	12	10	9
13	13	11	11
14	14	14	13
15	15	15	15

In box A place all numbers that have a 1 in the units place when written in binary notation. In box B place those with a 1 in the second position from the right in binary notation. In C and D are those numbers with a 1 in the third and fourth positions, respectively.

Next ask someone to select a number and tell you in which box or boxes it appears. You then start with zero, add the first number in each of the designated boxes, and state the sum as the selected number. For example, if the number is 11, the designated boxes are A, B, and D. You then find the selected number as $0 + 1 + 2 + 8$.

1. Explain why the method given for finding a number after knowing the boxes in which it appears works as it does.

2. Extend the boxes to include all the numbers through 31.

(A fifth box, *E*, will be necessary.) Then explain for the set of five boxes how to find a number if the boxes in which it appears are known.

	C	B	A
0	0	0	0
1	0	0	1
2	0	1	0
3	0	1	1
4	1	0	0
5	1	0	1
6	1	1	0
7	1	1	1

3. You can use binary notation to identify any one of eight numbers by means of three questions that can be answered by *yes* or *no*. Consider the numbers 0 through 7 written in binary notation. The place values are identified by columns *A*, *B*, and *C*.

   The three questions to be asked are: Does the number have a 1 in position *A*? Does the number have a 1 in position *B*? Does the number have a 1 in position *C*? Suppose the answers are yes, no, yes. Then the number is identified as 101_2; that is, 5. Extend this process to show how you can find any selected number from 0 through 15.

4. Consider five light bulbs in a row on a panel. The bulbs may be labeled so that they can be used in the same manner as the columns in Exploration 3. In the adjacent figure three bulbs are lighted to represent 1110_2, that is, 14.

16 8 4 2 1

   Draw a panel of five bulbs and show how they should be lighted to represent 21. Draw another panel of bulbs on which 57 is represented. How many bulbs would you need in a panel on which the whole numbers 1 through 127 are to be represented?

5-3
Counting Numbers and Their Properties

The famous mathematician Leopold Kronecker (1823–1891) said, "God created the natural numbers; everything else is man's handiwork." He was referring to what we generally call the set of counting numbers.

The counting numbers {1, 2, 3, 4, 5, . . .} are used in at least three different ways. Counting numbers may be used for **identification** such as your social security number, your telephone number, and the number of your driver's license. Counting numbers may be used as **ordinal numbers** to assign an order (first, second, third, . . .) to the elements of a finite set. Counting numbers may be used as **cardinal numbers** to specify the number of elements in a set. For example, if *D* is the set of decimal digits, then $n(D) = 10$, that is, the cardinal number of the set *D* is 10.

Example 1

Tell whether each specified number in the following statement is used for identification, as an ordinal number, or as a cardinal number.

The *second* train through Peoria consisted of *thirty* cars pulled by engine number *534*.

Solution

Second is used as an ordinal number, *thirty* is used as a cardinal number, and *534* is used for identification.

The **equality relation** (is equal to, $=$) has been used to state that two expressions represent the same number. For example, $5 + 7 = 12$. In addition to this *reflexive* property of any number being equal to itself, the equality relation has the two properties illustrated by these examples.

If $5 + 7 = 12$, then $12 = 5 + 7$.

If $5 + 7 = 12$ and $12 = 6 + 6$, then $5 + 7 = 6 + 6$.

We now summarize these properties for any counting numbers a, b, and c.

Reflexive property: $a = a$.
Symmetric property: If $a = b$, then $b = a$.
Transitive property: If $a = b$ and $b = c$, then $a = c$.

The transitive property implies that if two numbers are equal to the same number, then they are also equal to each other. Any relation with these three properties is an **equivalence relation.**

Note, for example, that *having a textbook* is an equivalence relation among students in a class.

Relations are widely used for many types of elements. For example, "is the same age as" is an equivalence relation among people. Consider three students Don, John, and Bill. Note that even though you do not know their ages, you do know that Don is the same age that he is. If Don and John are the same age, then John and Don are the same age. Also, if Don and John are the same age and John and Bill are the same age, then Don and Bill are the same age.

Sums of counting numbers may be introduced using cardinal numbers of sets and unions of sets. Consider, for example, the sum $3 + 4$. Then consider two sets A and B with three and four elements, respectively, and with no elements in common.

Let $A = \{a, b, c\}$; $n(A) = 3$.

Let $B = \{k, l, m, n\}$; $n(B) = 4$.

Then $A \cup B = \{a, b, c, k, l, m, n\}$.

The sum $3 + 4$ is then found to be 7, the number of elements in the set $A \cup B$. In general, if A and B are two sets such that $A \cap B = \varnothing$, then

$$n(A) + n(B) = n(A \cup B)$$

Products of counting numbers may be introduced using Cartesian products of sets. For any two given sets A and B, the

5 SETS OF NUMBERS

set of all possible *ordered pairs* with an element of A as the first element and an element of B as the second element is the **Cartesian product** of A and B, written $A \times B$ and read as "A cross B." For $A = \{p, q, r\}$ and $B = \{1, 2\}$

$$A \times B = \{(p, 1), \quad (p, 2), \quad (q, 1), \quad (q, 2), \quad (r, 1), \quad (r, 2)\}$$

Similarly,

$$B \times A = \{(1, p), \quad (1, q), \quad (1, r), \quad (2, p), \quad (2, q), \quad (2, r)\}$$

The ordered pair $(1, p)$ has first element 1; the ordered pair $(p, 1)$ has first element p. Thus $(p, 1) \neq (1, p)$ and $A \times B \neq B \times A$.

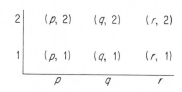

Note that $n(A \times B) = n(B \times A)$ but usually $A \times B \neq B \times A$.

The elements of a Cartesian product $A \times B$ are often represented in an array with the elements of A identifying the columns and the elements of B identifying the rows. In general, if A and B are two sets such that $n(A) = a$ and $n(B) = b$, then $ab = n(A \times B)$.

Example 2   Use a Cartesian product to illustrate $3 \times 2 = 6$.

Solution   Let $P = \{a, b, c\}$ and $Q = \{m, n\}$ where $n(P) = 3$ and $n(Q) = 2$. Then

$$P \times Q = \{(a, m), \quad (b, m), \quad (c, m), \quad (a, n), \quad (b, n), \quad (c, n)\}$$
$$n(P) \times n(Q) = n(P \times Q)$$
$$3 \times 2 = 6$$

Try to illustrate each of these basic properties of our number system with at least two numerical examples.

The properties of the set of counting numbers under addition and multiplication may be summarized for any counting numbers a, b, and c. Each of the operations, addition and multiplication, associates one number with *two* given numbers. For example,

$$4 + 3 = 7 \quad \text{and} \quad 2 \times 3 = 6$$

Since the sum of any two counting numbers is a unique counting number, the set of counting numbers is *closed under addition*.

Closure, $+$:   There is one and only one counting number $a + b$.

Since the product of any two counting numbers is a unique counting number, the set of counting numbers is *closed under multiplication*.

Closure, $\times$:   There is one and only one counting number $a \times b$.

The order of the addends does not affect the sum.

$$2 + 3 = 3 + 2$$

This property is the *commutative property of addition.*

Commutative, $+$: $a + b = b + a$

Similarly, the order of the factors does not affect the product.

$$2 \times 3 = 3 \times 2$$

Commutative, $\times$: $a \times b = b \times a$

There are times when three or more numbers are to be added. We may write, for example, $2 + 3 + 4$ because the same answer is obtained whether we associate the second addend with the first $(2 + 3) + 4$ or associate the second addend with the third $2 + (3 + 4)$.

Associative, $+$: $(a + b) + c = a + (b + c)$

Similarly, whenever we have a product, such as $2 \times 3 \times 4$, of three numbers we may associate the second factor with either the first $(2 \times 3) \times 4$ or the third $2 \times (3 \times 4)$.

Associative, $\times$: $(a \times b) \times c = a \times (b \times c)$

If the counting number 1 is used as a factor, the product is the same as, identical with, the other factor. For example,

$$1 \times 5 = 5 \qquad 7 \times 1 = 7$$

Because of this *multiplication property of one* the counting number 1 is the **multiplicative identity element** and also is called the **identity element for multiplication.**

Identity, $\times$: $a \times 1 = 1 \times a = a$

When a sum is multiplied by another factor we usually find the sum first. For example,

$$5 \times (7 + 11) = 5 \times 18 = 90$$

However, note that if we first distribute the other factor with each of the addends, the same result is obtained.

$$5 \times (7 + 11) = (5 \times 7) + (5 \times 11) = 35 + 55 = 90$$

This property is called the *distributive property.*

$$\boxed{\text{Distributive property:}\quad a \times (b + c) = (a \times b) + (a \times c)}$$

Note that addition is *not* distributive with respect to multiplication since, for example,

$$3 + (5 \times 8) \ne (3 + 5) \times (3 + 8)$$

that is,

$$3 + 40 \ne 8 \times 11$$

Because of the distributive property we may evaluate expressions of the form $a \times (b + c)$ either by adding first and then multiplying or by finding the two products first and then adding. Formally, we call this the **distributive property for multiplication with respect to addition.**

It is the distributive property that allows us, in algebra, to make such statements as

$$2(a + b) = 2a + 2b \qquad 3(x - y) = 3x - 3y$$

Use the distributive property to find the product 7×58.

The distributive property also is often helpful in finding products in arithmetic. For example, the product 8×99 can be found quickly as

$$8 \times 99 = 8 \times (100 - 1) = 800 - 8 = 792$$

Example 3 Apply properties of counting numbers to the left member of the given statement to obtain the right member of the statement. Show each step and name each property. Addition facts and multiplication facts are noted as "closure, +" and "closure, ×" respectively.

(a) $25 \times (11 \times 4) = (25 \times 4) \times 11$
(b) $32 \times (2 + 100) = 3200 + 64$
(c) $(17 + 19) + (3 + 11) = 20 + (19 + 11)$
(d) $(7 \times 5) \times (1 \times 3) = 7 \times (5 \times 3)$

Solution

(a) $25 \times (11 \times 4) = 25 \times (4 \times 11)$ (commutative, ×)
                     $= (25 \times 4) \times 11$ (associative, ×)

(b) $32 \times (2 + 100) = 32 \times (100 + 2)$ (commutative, +)
                     $= (32 \times 100) + (32 \times 2)$ (distributive property)
                     $= 3200 + 64$ (closure, ×)

(c) $(17 + 19) + (3 + 11) = 17 + [19 + (3 + 11)]$ (associative, +)
                          $= 17 + [(19 + 3) + 11]$ (associative, +)
                          $= 17 + [(3 + 19) + 11]$ (commutative, +)
                          $= (17 + 3) + (19 + 11)$ (associative, +)
                          $= 20 + (19 + 11)$ (closure, +)

(d) $(7 \times 5) \times (1 \times 3) = (7 \times 5) \times 3$ (identity, ×)
                         $= 7 \times (5 \times 3)$ (associative, ×)

Often other sequences of steps are also possible in exercises such as those in Example 3. For instance, in part (d) the following steps could be used.

$$(7 \times 5) \times (1 \times 3) = [(7 \times 5) \times 1] \times 3 \quad \text{(associative, } \times)$$
$$= [7 \times (5 \times 1)] \times 3 \quad \text{(associative, } \times)$$
$$= (7 \times 5) \times 3 \quad \text{(identity, } \times)$$
$$= 7 \times (5 \times 3) \quad \text{(associative, } \times)$$

Note that the set of counting numbers does not include an identity element for addition. For example, there does not exist a counting number n such that $n + 5 = 5$, that is, zero is not a counting number. Moreover, the set of counting numbers is *not* closed under subtraction or under division. Note that $2 - 5$ and $12 \div 5$ are meaningless if you are restricted to using counting numbers.

The *order* of two counting numbers is indicated by terms such as *is less than* ($<$) and *is more than* ($>$). For example, $2 < 5$ and $5 > 2$ because there is a counting number 3 such that $2 + 3 = 5$. In general, $a < b$ and $b > a$ if and only if there is a counting number c such that $a + c = b$. The usual **order relations** are

$<$	is less than
$\nless$	is not less than
$\leq$	is less than or equal to
$>$	is greater than
$\ngtr$	is not greater than
$\geq$	is greater than or equal to

This law states that if any two counting numbers a and b are given, then either a is less than b, or a is equal to b, or a is greater than b.

We assume that the **trichotomy law** is a general property of the set of counting numbers, that is, for any counting numbers a and b exactly one of these three relations must hold

$$a < b \qquad a = b \qquad a > b$$

Example 4 Tell whether or not the relation *is greater than* ($>$) is
(**a**) reflexive   (**b**) symmetric   (**c**) transitive.

Solution (**a**) It is not reflexive. For example, the statement $5 > 5$ is not true.
(**b**) It is not symmetric. For example, $5 > 2$, but $2 \ngtr 5$.
(**c**) It is transitive. For example, $8 > 5$, $5 > 2$, and $8 > 2$.

EXERCISES *Tell whether the number specified in the given statement is used for identification, as an ordinal number, or as a cardinal number.*

1. There are *20* volumes in the set of encyclopedia.

2. Mathematics is discussed in the *12th* volume.

3. Dorothy is in the *fourth* row.

5 SETS OF NUMBERS

4. There are *35* students in the class.
5. I am listening to *104* on the FM dial.
6. It takes *9* players to field a baseball team.

Find the cardinal number of each set.

7. $\{p\}$
8. $\{11, 12, \ldots, 18\}$
9. $\{100, 101, \ldots, 110\}$
10. $\varnothing$
11. $\{1, 3, 5, 7, \ldots, 19\}$
12. $\{2, 4, 6, \ldots, 102\}$
13. Find $A \times B$ for $A = \{1, 2\}$ and $B = \{1, 2, 3, 4\}$, thereby showing that $2 \times 4 = 8$.
14. Show that subtraction of counting numbers is not commutative.
15. Show that division of counting numbers is not commutative.
16. Does $8 - (3 - 2) = (8 - 3) - 2$? Is subtraction of counting numbers associative?
17. Does $12 \div (6 \div 2) = (12 \div 6) \div 2$? Is division of counting numbers associative?
18. Show that addition of counting numbers is not distributive with respect to multiplication.

For each arithmetic statement, name the property of counting numbers that is illustrated.

19. $2 + (3 \times 5) = 2 + (5 \times 3)$
20. $2 \times (3 \times 5) = (2 \times 3) \times 5$
21. $4580 = 1 \times 4580$
22. $25 \times (14 + 26) = (25 \times 14) + (25 \times 26)$
23. $17 \times (15 + 21) = 17 \times (21 + 15)$
24. $48 + (19 + 7) = (19 + 7) + 48$

Apply properties of counting numbers to the left member of each equation to obtain the right member. Show each step and name each property.

25. $92 + (50 + 8) = (92 + 8) + 50$
26. $(25 \times 17) \times 4 = 17 \times 100$
27. $37 \times (1 + 100) = 3700 + 37$
28. $(2 + 3) + (8 + 7) = (2 + 8) + (3 + 7)$
29. $(73 + 19) + (7 + 1) = (73 + 7) + (19 + 1)$
30. $(26 \times 1) \times (10 \times 2) = 10 \times (26 \times 2)$

31. In ordinary arithmetic, is addition distributive with respect

to addition? Does $a + (b + c) = (a + b) + (a + c)$ for all possible replacements of a, b, and c?

32. In ordinary arithmetic, is multiplication distributive with respect to multiplication? That is, does $a \times (b \times c) = (a \times b) \times (a \times c)$ for all possible replacements of a, b, and c?

In Exercises 31 and 32 consider the equations with specific replacements for a, b, and c.

Use the distributive property to find the given product by means of a shortcut.

33. 7×79 34. 6×58 35. 8×92 36. 9×63

Tell whether or not each relation is   (a) reflexive   (b) symmetric   (c) transitive   (d) an equivalence relation.

37. For people, *is a daughter of.*
38. For students, *is studying this book in the same class as.*
39. For numbers, *is not equal to,* $\neq$.
40. For numbers, *is greater than or equal to,* $\geq$.
41. For books, *is heavier than.*
42. For meals, *is more expensive than.*

Consider the relations $<$, $=$, $>$ and tell which relation can be used in place of R to make the given statement a true statement. The letters n, p, and q represent counting numbers.

43. (a) 6 R 7   (b) 25 R 19
44. (a) $(10 + 5)$ R $(5 + 10)$ (b) $[5 + (7 \times 6)]$ R $[6 + (7 \times 5)]$
45. (a) p R $(p - 1)$ (b) $(q + n)$ R $(2q + n)$
46. (a) $[p(q + n)]$ R pq (b) $[p(q + n)]$ R $(pq + pn)$

Assume that each letter represents a counting number and classify each statement as true or false. If false, give a specific counterexample to justify your answer.

Recall that a single counterexample is sufficient to show that a statement is not true in general.

47. $n^2 \geq n$
48. $s^2 > s$
49. If $ac < bc$, then $a < b$.
*50. If $s^k = s^n$, then $k = n$.
*51. If $p^k < q^k$, then $p < q$.

Explorations

Arrow diagrams such as the following can be used to show relations.

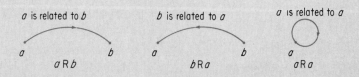

5 SETS OF NUMBERS

If a relation is reflexive, there is a loop at every element of the given set under discussion. If a relation is symmetric, then all arrows that go from an element a to an element b must also have an arrow that goes from b to a. A relation is transitive if whenever there is an arrow from a to b, and one from b to c, then there is also an arrow from a to c.

1. Here is a diagram for the relation *divides* for the set $M = \{2, 3, 6, 12\}$. From the diagram, show that the relation is reflexive and transitive but not symmetric.

Note the arrow from 3 to 6. This arrow indicates that 3 divides 6.

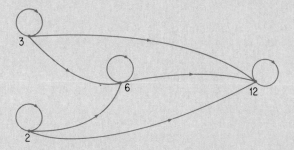

Draw arrow diagrams for each of the following relations on the sets given.

2. *Is greater than* for $A = \{2, 5, 6, 9, 10\}$.
3. *Is greater than or equal to* for $B = \{3, 8, 9, 12\}$.
4. *Is within 3 kilograms of* for the set of weights $C = \{130, 132, 134\}$.

5. Operations may be considered in terms of a **function machine** with three basic parts.

For the rule "add 5" complete the following table.

Input	2	3	7			21	
Output	7			6	11		21

6. In the introduction to this section there appears a quotation from the German mathematician Kronecker. Start a collection of such famous sayings that might be of interest in mathematics classes. For example, Archimedes is supposed to have said, "Give me a place to stand and a lever long enough and I will move the earth."

5-4
The Set of Integers

The counting numbers serve us well and are actually all that we need for many purposes. However, with just this collection of numbers at our disposal, we are unable to say that we have *no* money because 0 is not a counting number.

We need to extend our number system and shall do so by introducing a numeral 0 to represent the number, zero, of elements in the empty set. The number zero is needed to solve sentences such as $5 + n = 5$. If $5 + n = 5$, then $5 + 0 = 5$ and $n = 0$.

When we include 0 with the set of counting numbers, we form a new set W of **whole numbers**.

$$W = \{0, 1, 2, 3, 4, 5, \ldots\}$$

Th only difference between the set of counting numbers and the set of whole numbers is the element zero.

The set of whole numbers is the union of the set of counting numbers and the set containing the single element zero. If the number 0 is used as an addend, the sum is the same as, identical to, the other addend. For example,

$$0 + 5 = 5 \quad \text{and} \quad 7 + 0 = 7$$

Because of this *addition property of zero* the whole number 0 is the **additive identity element** and also is called the **identity element for addition**.

$$\boxed{\text{Identity, } +: \quad a + 0 = 0 + a = a}$$

The basic properties of the set of whole numbers include those of the set of counting numbers and may be listed as follows.

closure, $+$	closure, $\times$
commutative, $+$	commutative, $\times$
associative, $+$	associative, $\times$
identity, $+$	identity, $\times$
distributive property	

Sums of whole numbers may be considered in terms of the addition property of zero and sums of counting numbers. As in the case of counting numbers, subtraction is not always possible with only whole numbers available.

The following relationship between the operations of addition and subtraction enables us to check any addition problem by subtraction and to check any subtraction problem by addition.

$$\boxed{c - b = a \text{ if and only if } a + b = c}$$

194

To check that $90 + 25 = 115$ we may show either that $115 - 25 = 90$ or that $115 - 90 = 25$. To check that $115 - 25 = 90$ we show that $90 + 25 = 115$. Since $n + 0 = n$, the relationship between addition and subtraction also provides the formal justification for the statement $n - 0 = n$.

The product of any whole number n and 0 is 0, for example, $4 \times 0 = 0$. This property is frequently referred to as the *multiplication property of zero*.

$$\text{Zero, } \times: \quad n \times 0 = 0 \times n = 0$$

Multiplication by any whole number different from zero, that is, any counting number, can be done by repeated addition. For example,

$$6 \times 15 = 15 + 15 + 15 + 15 + 15 + 15 = 90$$

Excluding division by zero any division problem can be done by repeated subtraction. For example, $90 \div 15 = 6$ since a remainder of 0 can be obtained by subtracting 15 over and over a total of 6 times.

$$90 - 15 - 15 - 15 - 15 - 15 - 15 = 0$$

If the same process were used for $97 \div 15$, we would subtract 15's until the remainder was less than 15. In this case we would find the quotient to be 6 and the remainder 7 where $0 < 7 < 15$. Then $97 = 6 \times 15 + 7$ and 97 is not divisible by 15, that is, the result of dividing 97 by 15 is not a whole number.

If the subtraction approach to division were tried for division by zero, a computer would work endlessly with no apparent progress. Accordingly, *division by zero is not allowed*. We need to understand that the relationship between the operations of division and multiplication makes it necessary to exclude division by zero.

$$\text{If } b \neq 0, \text{ then } c \div b = a \text{ if and only if } a \times b = c.$$

Multiplication and division are inverse operations.

This discussion shows why division by zero is meaningless, hence not allowed.

To check that $6 \times 15 = 90$ we may show either that $90 \div 15 = 6$ or that $90 \div 6 = 15$. To check that $90 \div 15 = 6$ we show that $6 \times 15 = 90$. If this relationship were considered for $b = 0$ and $c = 7$, then $7 \div 0$ would be the number n such that $0 \times n = 7$. But $0 \times n = 0$ for all whole numbers n. Therefore $7 \div 0$ is meaningless. If any other counting number had been used in place of 7, the result would have been the same, that is, division by zero is not possible. If the relationship between division and multiplication were considered for $b = 0$ and $c = 0$, then $0 \div 0$ would be the number n such that $0 \times n = 0$. Since this equation holds for *all* whole numbers, $0 \div 0$ does not have a *unique* value and is often said to be *indeterminate*.

The whole numbers have many uses. However, with only the set of whole numbers at our disposal we have no numbers to represent such quantities as a deficit of $18 or a temperature of 10 degrees below zero. We need to extend our number system. We start by representing the set of whole numbers on a number line. To obtain a **number line**, draw any line, select any point of that line as the **origin** with *coordinate* 0 (zero), and select any other point of the line as the **unit point** with coordinate 1 (one). Usually the number line is considered in a horizontal position with the unit point on the right of the origin.

The length of the line segment with the origin and the unit point as endpoints is the **unit distance,** or **unit of length,** for marking off a scale on the line. The points representing any given counting numbers may be obtained by marking off successive units to the right of the origin. The numbers are the **coordinates** of the points; the points are the **graphs** of the numbers.

We can now graph different sets of numbers on a number line as in the following examples. We shall use the verb graph to mean draw the graph of.

Example 1 Graph on a number line the set of whole numbers less than 3.

Solution Draw a number line and place a solid dot at each of the points that correspond to 0, 1, and 2.

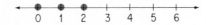

On a number line the point 1 unit to the right of the origin has coordinate 1 and the point 1 unit to the left of the origin is assigned the coordinate -1, *negative* 1, which is often called the *opposite* of 1. The point 2 units to the right of the origin has coordinate 2 and the point 2 units to the left of the origin is assigned the coordinate -2. In general, for any counting number n, the point n units to the right of the origin has coordinate n and the point n units to the left of the origin is assigned the coordinate $-n$, **negative n,** the **opposite of n.**

The graph of the set of integers continues indefinitely both in the positive direction, to the right of the origin, and in the negative direction, to the left of the origin, as shown by the arrows on the number line.

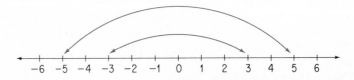

5 SETS OF NUMBERS

We call the set of whole numbers together with the set of the opposites of the whole numbers, the set I of **integers**.

$$I = \{\ldots, -3, -2, -1, 0, 1, 2, 3, \ldots\}$$

The opposite of 0 is defined to be 0, that is, $(-1) \times 0 = 0$.

The set of integers may be considered as the union of three sets of numbers.

If p is a positive integer, then p is a whole number and $p = 0 + p$. If n is a negative integer, then $-n$ is a whole number and $n = 0 - (-n)$.

Positive integers: $\{1, 2, 3, 4, 5, \ldots\}$

Zero: $\{0\}$

Negative integers: $\{\ldots, -5, -4, -3, -2, -1\}$

At times, especially for emphasis, plus signs are used to show positive integers. Thus the set of positive integers may be written in this way.

$$\{+1, +2, +3, +4, +5, \ldots\}$$

We shall consider such expressions as $+4$ and 4 as merely two names for the same number, the number with its graph located at the point four units to the right of the origin on a number line. The positive integers and the negative integers are among the *signed numbers*. Zero is neither a positive integer nor a negative integer and is not a signed number.

Example 2 Graph the set of integers between -3 and 2.

Solution The word *between* indicates that we are not to include the points for -3 and 2. The set to be graphed is $\{-2, -1, 0, 1\}$.

Note that $-b$ may represent either a positive integer or a negative integer. If $b = 5$, then $-b = -5$; if $b = -7$, then $-b = 7$.

The sum of any whole number n and its opposite $-n$ is 0; $n + (-n) = 0$. Consider the equation $n + (-n) = (-n) + n$ and extend the concept of opposites so that each of the numbers n and $-n$ is the opposite of the other. For example, -5 is the opposite of 5 and 7 is the opposite of -7. Then every integer n has an opposite $-n$ such that $n + (-n) = 0$. This property is sometimes called the *addition property of opposites*. Each of the numbers n and $-n$ is the **additive inverse** of the other.

$$\boxed{\text{Inverse, } +: \quad n + (-n) = (-n) + n = 0}$$

The basic properties of the set of integers include those of the set of whole numbers and may be summarized by a list.

closure, +   closure, ×
commutative, +   commutative, ×
associative, +   associative, ×
identity, +   identity, ×
inverse, +

distributive property

A set of elements with the four properties

closure   associative   identity   inverse

The need for additive inverses provided the basis for the extension of the set of whole numbers to the set of integers. The need for multiplicative inverses provides the basis for the extension, in Section 5-6, of the set of integers to the set of rational numbers.

under a specified operation forms a **group** under that operation. A group with the commutative property is a **commutative group.** The set of integers forms a commutative group under addition. Since the set of integers does not, for example, contain an integer n such that $3 \times n = 1$, that is, a multiplicative inverse for 3, the set of integers does not form a group under multiplication. The concept of a group under an operation is widely used in advanced mathematics and its applications.

EXERCISES

For each arithmetic statement, name the property of whole numbers that is illustrated.

1. $5 + 0 = 5$
2. $3 \times 0 = 0$
3. $(17 + 3) \times 0 = 0$
4. $0 \times (28 + 19) = 0$
5. $0 + (28 + 19) = 28 + 19$
6. $2 + (3 \times 5) = 2 + (5 \times 3)$

7. Use repeated addition to find each product.
    (a) 5×20 (b) 6×15 (c) 9×18
8. Use repeated subtraction to find each quotient.
    (a) $40 \div 8$ (b) $120 \div 20$ (c) $105 \div 15$

Graph each set of numbers on a number line.

9.   The set of counting numbers less than 6.
10. The set of whole numbers less than 6.
11. The set of whole numbers between 0 and 5.
12. The set of counting numbers between 1 and 3.
13. The set of counting numbers between 0 and 1.
14. The set of whole numbers greater than or equal to 5 and less than 9.
15. The set of integers between -3 and 5.
16. The set of integers -3 through 5 inclusive.
17. The set of integers that are the opposites of the first six counting numbers.

18. The set of integers that are the opposites of the members of the set $M = \{-4, -3, -2\}$.

Classify each statement as true or false. If the statement is false, give a counterexample to justify your answer.

19. Every counting number is an integer.
20. Every whole number is an integer.
21. Every integer is a whole number.
22. Every integer is either positive or negative.
23. Every integer is the opposite of some integer.
24. The set of integers is the same set as the set of the opposites of the integers.
25. The set of negative integers is the same as the set of the opposites of the whole numbers.
26. The set of integers is closed under multiplication.
27. The set of integers is closed under division.
28. The set of even integers forms a group under addition.
29. The set of odd integers forms a group under addition.
30. The subtraction of integers is commutative.
31. The intersection of the set of positive integers and the set of negative integers is the empty set.
*32. The empty set is the identity element for union of sets.
*33. The universal set is the identity element for intersection of sets.
*34. Complementary sets are inverse elements under union.

Explorations

The concept of infinity as used in mathematics is very difficult for most people to understand. The set of points on a line is infinite. The set of grains of sand on Coney Island is very large but finite. For each of the following finite situations, first make an educated guess. Then find the correct answer.

1. To the nearest day, how long would it take to count to 1 billion at the rate of one number per second?
2. Estimate how many pennies it would take to make a stack 1 inch high. Approximately how high would a stack of 1 million pennies be?
3. One million one-dollar bills are placed end-to-end along the ground. To the nearest 10 miles, how long would this strip of bills be?

4. To the nearest billion, estimate the number of seconds that elapse in a century.
5. To the nearest day how long would it take to spend $1,000,000 at the rate of $1.00 per minute?

One of the largest numbers ever named is a **googol**, which has been defined as 1 followed by 100 zeros:

10 000 000 000 000 000 000 000 000 000 000
000 000 000 000 000 000 000 000 000 000 000
000 000 000 000 000 000 000 000 000 000 000

A googol can be expressed, using exponents, as 10^{100}.

This number is larger than what is considered to be the total number of protons or electrons in the universe!

Even larger than a googol is a **googolplex**, defined as 1 followed by a googol of zeros. One famous mathematician claimed that there would not even be room between the earth and the moon to write all the zeros in a googolplex!

*6. How many zeros are there in a number represented as a googol times a googol?
*7. Express the number obtained in Exploration 6 as a power of ten. Is this number smaller or larger than a googolplex?

The cardinal number of the set W of whole numbers is the same as the cardinal number of the set C of counting numbers since there is a one-to-one correspondence between the elements of the two sets.

$$C = \left\{ 1, \quad 2, \quad 3, \quad 4, \quad \ldots, \quad n, \quad \ldots \right\}$$

$$W = \left\{ 0, \quad 1, \quad 2, \quad 3, \quad \ldots, \quad n-1, \quad \ldots \right\}$$

Show that there is a one-to-one correspondence between the set of counting numbers and the given set.

8. 5, 6, 7, 8, 9, 10, 11, . . .
9. 5, 10, 15, 20, 25, 30, . . .
10. The set 2, 4, 6, 8, 10, . . . of even positive integers.
11. The set 1, 3, 5, 7, 9, . . . of odd positive integers.
12. The set of negative integers.
*13. The set of integers.

The cardinal number of the set of counting numbers is not an integer but is the smallest of the transfinite cardinal numbers, $\aleph_0$, aleph-null. Since the number of whole numbers is one more than the number of counting numbers, the one-to-one correspondence between the set of counting numbers and the set of whole numbers shows that

$$\aleph_0 = \aleph_0 + 1$$

Use a one-to-one correspondence to illustrate each given statement.

14. $\aleph_0 = \aleph_0 + 4$ 15. $\aleph_0 = 5\aleph_0$ 16. $\aleph_0 = \aleph_0 + \aleph_0$

17. Read about the life of Evariste Galois, one of the mathematicians who is given credit for original work on group theory. A fascinating account of his life, and death at age 20, is given in the book by Leopold Infeld, *Whom the Gods Love: The Story of Evariste Galois*, published as a "Classic in Mathematics Education" by the National Council of Teachers of Mathematics.

5-5
Computation with Integers

We assume that the reader has had some past experience performing operations with integers.

Number line representations and the properties of integers provide a basis for addition of integers. Represent the first addend by an arrow that starts at 0, the origin. From the tip of the first arrow, draw a second arrow to show the second addend. Positive numbers are shown by arrows that go to the right, and negative numbers by arrows that go to the left. Finally, the sum is found as the coordinate of the point at the tip of the second arrow. Examples 1 and 2 illustrate this number-line method for addition of integers.

Example 1   Illustrate $(+3) + (-7)$ on a number line.

Solution   Start at 0 and draw an arrow that goes 3 units to the right. Then show an arrow that goes 7 units to the left. The tip of the second arrow is at -4. Thus $(+3) + (-7) = -4$.

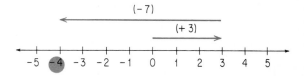

Example 2   Illustrate $(-2) + (-3)$ on a number line.

Solution

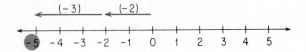

From the figure we see that $(-2) + (-3) = -5$.

Example 3   Find the sum $(-3) + (+3)$.

Solution   Think of a number line. From the origin, move 3 units to the left to represent -3. From this point, move 3 units to the right, back to the origin. Thus, $(-3) + (+3) = 0$. One can also obtain the answer by applying the addition property of opposites, that is, any sum of opposites is 0.

Example 4   Find the sum $(+3) + (-5) + (-2)$.

Solution   Use the associative property and find the sum of two addends at a time. This grouping is essential because of the binary nature of addition; only two numbers can be added at a time.

This answer also can be obtained from the grouping $(+3) + [(-5) + (-2)]$ since $(+3) + (-7) = -4$.

$$(+3) + (-5) + (-2) = [(+3) + (-5)] + (-2)$$
$$= (-2) + (-2)$$
$$= -4$$

The relationship between addition and subtraction holds for integers as well as whole numbers. Thus any subtraction problem may be replaced by an equivalent addition problem. For example,

$$17 - 9 = n \quad \text{if and only if} \quad n + 9 = 17.$$

In general, for any two integers a and b

$$a - b = n \quad \text{if and only if} \quad n + b = a.$$

Example 5   Find the difference $(-2) - (-8)$.

Solution   Let $(-2) - (-8) = n$. Then $n + (-8) = -2$ and we have $(-8) + n = -2$. Thus we need to find a number that must be added to -8 to obtain -2. On a number line, we must move six units to the right to go from -8 to -2. Therefore, $n = 6$, and $(-2) - (-8) = +6$.

The solution of Example 5 illustrates the correspondence of the subtraction problem $(-2) - (-8) = +6$ and the addition problem $(-2) + (+8) = +6$. We use this relationship to define the subtraction of integers a and b.

This is the common rule "to subtract, change the sign and add."

$$\boxed{a - b = a + (-b)}$$

Example 6 Subtract $(+5) - (-3)$.

Solution By the definition of subtraction, $a - b = a + (-b)$. Therefore

The problem in Example 6 can also be written in vertical form.

Subtract: $+5$
            -3

$$(+5) - (-3) = (+5) + [-(-3)]$$
$$= (+5) + (+3) = +8$$

Any sum or product of integers is an integer since the set of integers is closed under addition and also under multiplication. Any difference of integers is an integer since the difference is equivalent to a sum. However, a quotient of integers $a \div b$ is an integer if and only if there is some integer n such that $b \times n = a$. For example, there is no integer n such that $3 \times n = 8$ and the quotient $8 \div 3$ cannot be named by an integer.

The set of integers is not closed under division.

There are a variety of ways in which we can consider the rules for multiplication of integers. Applications of numbers to physical situations can be helpful. Consider, for example, the situation of earning \$5 a day for two days. This can be expressed as the product of two positive integers, $(+2) \times (+5)$. We may think of multiplication by a positive integer as repeated addition.

$$(+2) \times (+5) = (+5) + (+5) = +10$$

Thus the result is a gain of \$10, a positive number. In general, the product of two positive integers is a positive integer.

To evaluate the product of a positive integer and a negative integer, consider the situation of losing \$5 each day for two days. This can be expressed as the product $(+2) \times (-5)$.

$$(+2) \times (-5) = (-5) + (-5) = -10$$

Thus the result is a loss of \$10, a negative number. In general, the product of a positive integer and a negative integer is a negative integer. Then, since multiplication is commutative, the product of a negative integer and a positive integer is also a negative integer. For example,

$$(-5) \times (+2) = (+2) \times (-5) = -10$$

Justifying the product of two negative numbers is somewhat more difficult. Informally, we might say that if you lose $5 a day for two days, then two days *ago* you had $10 more than you have today. Therefore, $(-2) \times (-5) = +10$. However, such arguments are often unconvincing.

The rules for multiplication of integers can be confirmed by mathematical procedures rather than by the somewhat intuitive approach we have just been using.

Let us consider a mathematical justification for the product of two negative integers. We use the convention

$$(-2)(+5) = (-2) \times (+5) \qquad ab = a \times b$$

omitting the multiplication symbol for convenience and begin by exploring a specific example.

$$(-2)[(+5) + (-5)]$$

Because we wish to preserve the distributive property for the set of integers, we agree that there are two ways to obtain an answer to this problem. If we add first, within the brackets, and then multiply, we have

$$(-2)[(+5) + (-5)] = (-2)(0)$$
$$= 0$$

Now we shall use the distributive property: multiply first, and then add. The final result must be the same, namely 0.

$$(-2)[(+5) + (-5)] = [(-2)(+5)] + [(-2)(-5)]$$
$$= (-10) + (?)$$

We have been forced to use a question mark for the product $(-2) \times (-5)$ because this is precisely the product that we are seeking. We do know that the final sum must be 0 if the distributive property is to hold. Furthermore, we know that the sum $(-10) + (+10) = 0$. Therefore, we conclude that $(-2) \times (-5)$ must be equal to $+10$; the product of two negative integers is a positive integer.

We can generalize this approach by considering any positive integers c and b, and the expression $(-c) \times [(b) + (-b)]$. We then evaluate this product in two different ways.

1. $(-c) \times [(b) + (-b)] = (-c) \times 0$
$$= 0$$

2. $(-c) \times [(b) + (-b)] = [(-c) \times (+b)] + [(-c) \times (-b)]$
$$= -cb + (?)$$

Since the final result must be 0, the question mark represents the opposite of $-cb$. Thus $(-c) \times (-b)$ must be the opposite of $-cb$ and $(-c) \times (-b) = cb$; the product of any two negative integers is a positive integer.

5 SETS OF NUMBERS

Example 7 Find each product.
(a) $(+4) \times (+7)$ (b) $(-3) \times (+6)$ (c) $(+5) \times (-8)$
(d) $(-7) \times (-8)$

Solution (a) The product of two positive integers is a positive integer; $(+4) \times (+7) = +28$.
(b) The product of a negative integer and a positive integer is a negative integer; $(-3) \times (+6) = -18$.
(c) The product of a positive integer and a negative integer is a negative integer; $(+5) \times (-8) = -40$.
(d) The product of two negative integers is a positive integer; $(-7) \times (-8) = +56$.

Example 8 Find the product $(-5) \times (-7) \times (-2)$.

Solution Use the associative property and then find the product of two factors at a time.

This answer also can be obtained from the grouping $(-5) \times [(-7) \times (-2)]$ since $(-5) \times (+14) = -70$.

$$(-5) \times (-7) \times (-2) = [(-5) \times (-7)] \times (-2)$$
$$= (+35) \times (-2)$$
$$= -70$$

The relationship between multiplication and division

$$a \div b = c \quad \text{if and only if} \quad a = bc \quad (b \neq 0)$$

provides us with rules for signs of quotients of integers. The quotient of a positive integer and a negative integer is negative. The quotient of two positive integers is positive. The quotient of two negative integers is positive.

Example 9 Find the quotient $(-12) \div (-3)$.

Solution Let $(-12) \div (-3) = n$. Then $(-3) \times n = (-12)$, and $n = +4$. Thus the quotient of these two negative integers is a positive integer.

We are now ready to explore a number of proofs that illustrate the type of reasoning that mathematicians use. First we need several definitions; then we shall consider several proofs and suggest others in the exercises. Some of these proofs will be needed in Section 5-7, where we prove that $\sqrt{2}$ cannot be expressed as a quotient of integers.

An integer is an **even integer** if it is a multiple of 2, that is, if it may be expressed as $2k$, where k stands for an integer. Then the set of even integers is

$$\{\ldots, -6, -4, -2, 0, 2, 4, 6, \ldots\}$$

An integer that is not even is said to be an **odd integer**. Each odd integer may be expressed in the form $2k + 1$, where k stands for an integer. Then the set of odd integers is

$$\{. . . , -7, -5, -3, -1, 1, 3, 5, 7, . . .\}$$

Suppose that we wished to prove that the sum of any two even integers is an even integer. We would first note that any two even integers m and n may be expressed as $2k$ and $2r$ where k and r stand for integers. Then

$$m + n = 2k + 2r = 2(k + r) \qquad \text{(distributive property)}$$

where $k + r$ stands for an integer (closure, $+$). Therefore, $m + n$ may be represented as twice an integer and is an even integer. This completes the proof.

Example 10 Prove that the square of any even integer is an even integer.

Proof Any even integer may be expressed as $2k$, where k stands for an integer. Then the square of the integer may be expressed as $(2k)^2$, where

$$(2k)^2 = (2k)(2k) = 2[k(2k)] \qquad \text{(associative, } \times)$$

and $k(2k)$ stands for an integer (closure, $\times$). Therefore, $(2k)^2$ is an even integer.

EXERCISES *Illustrate each sum on a number line.*

1. $(+5) + (-7)$ 2. $(-3) + (-4)$
3. $(+3) + (+4)$ 4. $(-2) + (-5)$

Find each sum.

5. $(+8) + (-12)$ 6. $(-3) + (-7)$
7. $(-8) + (+12)$ 8. $(-23) + (+23)$
9. $(-15) + (-12)$ 10. $(+13) + (-20)$
11. $(-5) + (-3) + (-7)$ 12. $(+6) + (-8) + (-7)$
13. $(+12) + (-12) + (-9)$ 14. $(-15) + (-7) + (+8)$
15. $(+5) + (-7) + (-6) + (+8)$
16. $(-7) + (-3) + (+12) + (-9)$

Find each difference.

17. $(-12) - (+15)$ 18. $(+13) - (+7)$
19. $(+15) - (+25)$ 20. $(-15) - (-25)$
21. $(+11) - (-5)$ 22. $(+11) - (-20)$

5 SETS OF NUMBERS

Perform the indicated operations.

23. $[(-5) + (-3)] - (+3)$
24. $(-5) + [(-3) - (+3)]$
25. $[(-8) - (-2)] + (-5)$
26. $(-8) - [(-2) + (-5)]$
27. $[(+5) + (-8)] - [(-3) + (-7)]$
28. $[(-7) - (-3)] + [(+5) - (-8)]$

Find each product.

29. $(-5) \times (+9)$
30. $(-5) \times (-9)$
31. $(+5) \times (-9)$
32. $(+5) \times (+9)$
33. $(-8) \times (+12)$
34. $(+12) \times (-12)$
35. $(-25) \times (-25)$
36. $(-15) \times (+15)$
37. $(+10) \times (-17)$
38. $(-11) \times (-11)$
39. $(-3) \times (-7) \times (-5)$
40. $(-3) \times (+7) \times (-7)$

Find each quotient.

41. $(-24) \div (+3)$
42. $(+24) \div (+3)$
43. $(+24) \div (-3)$
44. $(-24) \div (-3)$
45. $(-36) \div (-18)$
46. $(+72) \div (-12)$
47. $(-144) \div (-12)$
48. $(+125) \div (-25)$
49. $(-100) \div 0$
50. $0 \div (-5)$
51. $[(-60) \div (-3)] \div (-4)$
52. $(-100) \div [(-50) \div (+10)]$

Perform the indicated operations.

53. $(-8) + [(-5) + (-7)]$
54. $[(-8) + (-5)] + (-7)$
55. $(-2) \times [4 \times (-5)]$
56. $[(-2) \times 4] \times (-5)$
57. $12 \div [6 \div (-2)]$
58. $(12 \div 6) \div (-2)$

Prove each statement.

59. The sum of any two odd integers is an even integer.
60. The product of any two even integers is an even integer.
61. The square of any odd integer is an odd integer.
*62. If the square of an integer is odd, the integer is odd; if the square of an integer is even, the integer is even.

Explorations

A **nomograph** is a device for performing computations in a simple manner. The following nomograph can be used to find the sums of integers. Just connect the point representing one addend on the *A* scale, with the corresponding point for the other addend on the *B* scale. The point where the line crosses the *S* scale will give the sum. The following figure shows the sum $(+4) + (-6) = -2$.

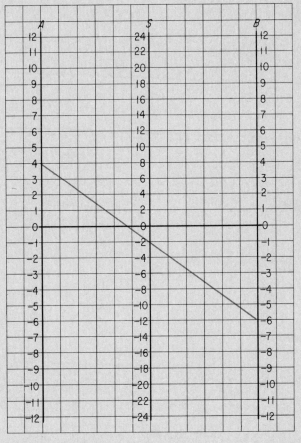

1. Construct your own nomograph on graph paper and use it to find various sums of integers.

2. See if you can use your knowledge of elementary geometry to explain why the nomograph shown works as it does.

3. Try to determine how the nomograph shown can be used for subtraction of integers.

4. Prepare a nomograph that will enable you to find $a + 2b$ directly for any two integers a and b. If possible, prepare a suitable transparency and demonstrate this nomograph on an overhead projector.

5 SETS OF NUMBERS

There are many practical problems that cannot be solved when only the integers are available.

Not one of these problems has a solution if we are restricted to the use of integers only. Think of several examples of your own need for fractions.

1. A cook cannot use a recipe that calls for 3 cups of flour if only one-half the quantity is desired. The problem $3 \div 2$ has no solution in the system of integers.

2. A child cannot find an integer to represent one-half of a candy bar; $1 \div 2$ is not an integer.

3. A driver does not have an integer to represent the average number of miles per gallon of gas for a car that travels 200 miles using 7 gallons; $200 \div 7$ has no solution in the system of integers.

4. When grapefruit are three for a dollar, a different price must be charged for one grapefruit; $100 \div 3$ has no solution in the system of integers.

Furthermore, with only the set of integers at our disposal, we are unable to find replacements for n that make these sentences true.

$$2 \times n = 7 \qquad n + \frac{1}{4} = \frac{7}{8} \qquad 5 \div 3 = n$$

The solution set for each of the preceding sentences is the empty set if only integers may be used as possible replacements for n. To have numbers as solutions of such sentences, as well as to make division always possible (except division by zero), we must extend our set of numbers. In this section we make such an extension and call this new set of numbers the set of rational numbers.

A **rational number** is a number that can be expressed in the form

$$\frac{a}{b}$$

(often written a/b), where a and b are integers and $b \neq 0$.

Each of the following is a rational number.

$$\frac{2}{3} \qquad \frac{1}{2} \qquad \frac{-7}{3} \qquad \frac{0}{1} \qquad \frac{215}{524}$$

We usually refer to such numbers as **fractional numbers**, or **fractions**. Any quotient a/b of integers a and b, $b \neq 0$ is also called a *fraction*. The integer a is the **numerator** and b is the **denominator** of the fraction a/b.

Every whole number n is a rational number $n/1$.

Every integer n is a rational number since $n = n/1$. Any number, such as $2\frac{1}{3}$, that is expressed as the sum of an integer and a fraction is called a **mixed number** and is said to be in **mixed form.** The set of rational numbers includes whole numbers, fractions, mixed numbers, and their opposites.

If we consider fractions as quotients, we obtain

$$1 = \frac{1}{1} = \frac{-1}{-1} = \frac{2}{2} = \frac{-2}{-2} = \frac{3}{3} = \frac{-3}{-3} = \cdots$$

Statements of equality are used when two expressions represent the same number.

> The *product* of two rational numbers is defined as
> $$\frac{a}{b} \times \frac{c}{d} = \frac{ac}{bd}$$

For example,

$$\frac{3}{4} \times \frac{1}{5} = \frac{3 \times 1}{4 \times 5} = \frac{3}{20}$$

Also since any integer a may be expressed as $a/1$,

$$a \times \frac{b}{c} = \frac{a}{1} \times \frac{b}{c} = \frac{a \times b}{1 \times c} = \frac{ab}{c}$$

In particular, for any integer $k \neq 0$

The numerator and the denominator of any fraction a/b may be multiplied by any number $k \neq 0$.

$$\frac{a}{b} = \frac{a}{b} \times 1 = \frac{a}{b} \times \frac{k}{k} = \frac{ak}{bk}$$

This relationship may be used to express any rational number in many ways. For example,

$$\frac{10}{30} = \frac{1 \times 10}{3 \times 10} = \frac{1}{3} = \frac{1 \times 7}{3 \times 7} = \frac{7}{21}$$

Note that since $k = k$ this property of fractions is consistent with the usual "equals multiplied by equals are equal." If $a = b \neq 0$, then

$$1 = \frac{a}{b} = \frac{ak}{bk} \qquad \frac{ak}{bk} = 1 \qquad ak = bk$$

If we think of

$$\frac{ak}{bk} = \frac{a}{b} \quad \text{as} \quad \frac{a\cancel{k}}{b\cancel{k}} = \frac{a}{b}$$

Both sides of an equation may be multiplied, or divided, by any number $k \neq 0$. We call this the *multiplication property of equality.*

we have a very useful basis for *reducing* any fraction by dividing the numerator and the denominator by any number $k \neq 0$. Similarly, for equations

$$\text{if } ak = bk, \qquad \text{then } a = b \qquad (k \neq 0)$$

5 SETS OF NUMBERS

For any integers m and n, $n \neq 0$, the **ratio** of m to n is m/n where

$$m/n = m \div n = \frac{m}{n} = m \times \frac{1}{n}$$

Any equality of ratios is often called a **proportion**. The equality $a/b = c/d$ states a *proportion property* of the integers a, b, c, and d.

The equality of rational numbers represented by fractions is defined as follows.

For any two rational numbers a/b and c/d

$$\frac{a}{b} = \frac{c}{d} \quad \text{if and only if} \quad ad = bc$$

For example,

$$\frac{9}{6} = \frac{15}{10} \quad \text{since} \quad 9 \times 10 = 6 \times 15$$

The equations $ad = bc$ and $9 \times 10 = 6 \times 15$ may be obtained by *cross multiplication* as indicated by these crossed arrows.

$$\frac{a}{b} \diagup\!\!\!\!\!\diagdown \frac{c}{d} \qquad \frac{9}{6} \diagup\!\!\!\!\!\diagdown \frac{15}{10}$$

Since cross multiplication is often attempted by mistake for sums and other pairs of fractions for which it should not be used, many people prefer to use the multiplication property of equality and reduce the fractions.

$$\frac{a}{b} = \frac{c}{d} \qquad \frac{a}{b} \times bd = \frac{c}{d} \times bd$$

$$\frac{a \not{b} d}{\not{b}} = \frac{cb \not{d}}{\not{d}}$$

$$ad = bc$$

For the example $9/6 = 15/10$ we have

$$\frac{9}{\not{6}} \times (\not{6} \times 10) = \frac{15}{\not{10}} \times (6 \times \not{10}) \qquad 9 \times 10 = 15 \times 6$$

Example 1 Find replacements for n to make the given statement true.

(a) $\dfrac{2}{5} = \dfrac{6}{n}$ (b) $\dfrac{2}{n} = \dfrac{6}{9}$ (c) $\dfrac{5}{3} = \dfrac{n}{6}$

Use the definition of the equality of fractions.

(**a**) $2 \times n = 6 \times 5$ (**b**) $2 \times 9 = 6 \times n$ (**c**) $5 \times 6 = 3 \times n$

$$2n = 30 \qquad\qquad 18 = 6n \qquad\qquad 30 = 3n$$
$$2n = 2 \times 15 \qquad 6n = 6 \times 3 \qquad 3n = 3 \times 10$$
$$n = 15 \qquad\qquad n = 3 \qquad\qquad n = 10$$

Recall that the identity element for multiplication is 1 and note that

$$2 \times \frac{1}{2} = 1 \qquad \frac{2}{3} \times \frac{3}{2} = 1 \qquad \left(\frac{-3}{4}\right) \times \left(\frac{4}{-3}\right) = 1$$

Every rational number $a/b \neq 0$ has a *multiplicative inverse* b/a.

Inverse ($\neq 0$), $\times$:   If $\dfrac{a}{b} \neq 0$, then $\dfrac{a}{b} \times \dfrac{b}{a} = 1$.

By extending the set of integers to the set of rational numbers, we have obtained a multiplicative inverse, that is, a **reciprocal,** for each element that is different from zero. The restriction $a/b \neq 0$ is needed because $0 \times n = 0$ for any rational number n and thus zero does not have a reciprocal in the set of rational numbers.

For any integers m and n, $n \neq 0$, the reciprocal of n is $1 \div n$, that is, $1/n$, and

$$m \div n = m \times \frac{1}{n} = \frac{m}{n}$$

The reciprocal of any fraction $(c/d) \neq 0$ is

$$1 \div \frac{c}{d}, \quad \text{that is,} \quad \frac{d}{c} \quad \text{since} \quad \frac{c}{d} \times \frac{d}{c} = \frac{cd}{cd} = 1$$

For any fractions a/b and c/d, $(c/d) \neq 0$,

$$\frac{a}{b} \div \frac{c}{d} = \frac{a}{b} \times \frac{d}{c} = \frac{ad}{bc}$$

This is the common rule "to divide, invert and multiply."

For example,

$$\frac{2}{5} \div \frac{3}{4} = \frac{2}{5} \times \frac{4}{3} = \frac{8}{15} \qquad \frac{4}{3} \div \frac{5}{7} = \frac{4}{3} \times \frac{7}{5} = \frac{28}{15}$$

The equation $(-1) \times n = -n$ holds for any number n. For fractions

There are three signs associated with any fraction. Any two of these signs may be changed without changing the number represented by the fraction.

$$(-1) \times \frac{a}{b} = -\frac{a}{b} = \frac{-a}{b} = \frac{-a}{b} \times \frac{-1}{-1} = \frac{a}{-b}$$

This relationship may be used with the reduction of fractions by dividing numerator and denominator by a number different from zero to express any fraction in **reduced form,** that is, in the form a/b where b is a counting number and 1 is the only counting number that divides both a and b. A fraction in reduced form is often said to be *reduced to lowest terms.*

Example 2   Reduce $\dfrac{6}{-8}$.

Solution   The denominator should be a counting number.

This result is usually written as $-\dfrac{3}{4}$.

$$\frac{6}{-8} = \frac{-6}{8} = \frac{(-3) \times 2}{4 \times 2} = \frac{-3}{4}$$

The word *simplify* is often used here and elsewhere to indicate that all operations are to be performed and the resulting expression reduced to lowest terms.

Example 3   Simplify $\dfrac{2}{3} \times \dfrac{3}{4} \times \left(-\dfrac{1}{8}\right)$.

Solution   There are a number of different ways that this product can be found. The approach shown makes use of the associative property of multiplication, the commutative property, and the reduction of fractions to lowest terms.

Some of the steps shown are often performed mentally and not written. Reasoning such as

$$\frac{\cancel{2} \times \cancel{3} \times (-1)}{\cancel{3} \times \cancel{4} \times 8} = \frac{-1}{16}$$
$$2$$

is useful but note that only factors may be crossed out.

$$\frac{2}{3} \times \frac{3}{4} \times \left(-\frac{1}{8}\right) = \left(\frac{2}{3} \times \frac{3}{4}\right) \times \left(-\frac{1}{8}\right)$$

$$= \frac{2 \times 3}{3 \times 4} \times \left(-\frac{1}{8}\right)$$

$$= \frac{2 \times 3}{2 \times (2 \times 3)} \times \left(-\frac{1}{8}\right)$$

$$= \frac{1}{2} \times \left(-\frac{1}{8}\right)$$

$$= -\frac{1}{16}$$

Addition and subtraction of rational numbers are considered after the introduction of some elementary number theory in Chapter 6. The definitions of *order relations* ($<$, $>$) among rational numbers are similar to the definition of equality of rational numbers.

For any two rational numbers a/b and c/d where b and d are counting numbers

$$\frac{a}{b} < \frac{c}{d} \quad \text{if and only if} \quad ad < bc$$

$$\frac{a}{b} > \frac{c}{d} \quad \text{if and only if} \quad ad > bc$$

In the definition of the order of rational numbers the restriction to counting numbers in the denominators is essential. For example,

$$\frac{2}{-5} < \frac{3}{4} \quad \text{but} \quad 2 \times 4 \not< 3(-5)$$

Then

$$\frac{5}{8} < \frac{2}{3} \quad \text{since} \quad 5 \times 3 < 8 \times 2$$

$$\frac{5}{7} > \frac{2}{3} \quad \text{since} \quad 5 \times 3 > 7 \times 2$$

Draw a number line and locate a point midway between the points with coordinates 0 and 1. The coordinate of the midpoint is called $\frac{1}{2}$. Then locate all points that correspond to "halves," such as $\frac{3}{2} = 1\frac{1}{2}$, $\frac{5}{2} = 2\frac{1}{2}$, $\frac{7}{2} = 3\frac{1}{2}$, and so forth. Also locate points with coordinates $-\frac{1}{2}$, $-1\frac{1}{2}$, $-2\frac{1}{2}$, $-3\frac{1}{2}$, and so forth. The graph of any rational number $n/2$, where n is an integer, may be found in this way.

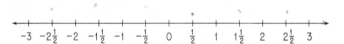

Similarly, the points corresponding to multiples of $\frac{1}{3}$, $\frac{1}{4}$, $\frac{1}{5}$, $\frac{1}{6}$, and so forth, may be located on a number line. Each rational number is the coordinate of some point on the number line. Recall that the set of rational numbers includes the set of integers as a subset.

The set of rational numbers can be classified as being positive (greater than 0), negative (less than 0), or 0. The number 0 is a rational number, but it is neither positive nor negative. As in the case of integers, positive rational numbers are the coordinates of points on the right of the origin and negative rational numbers are coordinates of points on the left of the origin.

Every rational number has a point on the number line as its graph. However, not every point on the number line has a rational *number* as its coordinate.

Our number line has now become dense with points and "resembles" a complete line. The word "resembles" is used because we shall find later that there are still points on the number line that do not have rational numbers as their coordinates. However, the set of rational numbers is said to be **dense,** since between any two elements of the set, there is always another element of the set.

Example 4 Name a rational number between $\dfrac{17}{19}$ and $\dfrac{18}{19}$.

Solution Select a multiple of 19 that is greater than 19, such as 38. Write each given rational number as one with a denominator of 38.

$$\frac{17}{19} \times \frac{2}{2} = \frac{34}{38} \qquad \frac{18}{19} \times \frac{2}{2} = \frac{36}{38}$$

Clearly, 35/38 lies between the two given rational numbers. Note also that

$$\frac{35}{38} = \frac{1}{2}\left(\frac{17}{19} + \frac{18}{19}\right)$$

Example 5 Name a rational number between $\dfrac{17}{19}$ and $\dfrac{35}{38}$.

Solution Select a multiple of 19 and 38 that is greater than 38, such as 76. Write each given rational number as one with a denominator of 76.

$$\frac{17}{19} \times \frac{4}{4} = \frac{68}{76} \qquad \frac{35}{38} \times \frac{2}{2} = \frac{70}{76}$$

Note that 69/76 lies between the two given rational numbers. Also,

$$\frac{69}{76} = \frac{1}{2}\left(\frac{17}{19} + \frac{35}{38}\right)$$

We can summarize the results of the two preceding examples in the following manner.

$$\frac{17}{19} < \frac{35}{38} < \frac{18}{19}$$

$$\frac{17}{19} < \frac{69}{76} < \frac{35}{38}$$

This process can be extended indefinitely. For example, to locate a rational number between 17/19 and 69/76, write each with a denominator of 152. Then find one-half the sum of the two numbers. This existence of a rational number between any two given rational numbers is what is meant by the *density property*.

In addition to the density property the basic properties of the set of rational numbers include those of the set of integers and may be summarized by a list.

closure, +   closure, ×
commutative, +   commutative, ×
associative, +   associative, ×
identity, +   identity, ×
inverse, +   inverse ($\neq$ 0), ×
                distributive property

Mathematicians recognize the presence of these eleven properties by saying that the set of rational numbers forms a **field**.

EXERCISES *Classify the given statement as true or false.*

1. Every rational number is an integer.
2. The reciprocal of every rational number except 0 is a rational number.
3. The number 0 is not a rational number.
4. The multiplicative inverse of a positive rational number is negative.

Each fraction m/n has a numerator m and a denominator n.

5. Identify the numerator of (a) $\dfrac{2}{3}$ (b) $\dfrac{1}{5}$ (c) $\dfrac{2}{7}$

   (d) $\dfrac{0}{7}$.

6. Identify the denominator of each fraction in Exercise 5.

Find a replacement for n to make the given statement true.

7. $\dfrac{5}{6} = \dfrac{n}{30}$ 8. $\dfrac{7}{8} = \dfrac{28}{n}$ 9. $\dfrac{9}{n} = \dfrac{81}{72}$

10. $\dfrac{n}{7} = \dfrac{42}{49}$ 11. $\dfrac{100}{n} = \dfrac{10}{3}$ 12. $\dfrac{144}{12} = \dfrac{n}{1}$

Reduce to lowest terms.

13. $\dfrac{18}{60}$ 14. $\dfrac{45}{70}$ 15. $\dfrac{48}{-64}$ 16. $\dfrac{32}{-36}$

Simplify.

17. $\dfrac{11}{12} \times \dfrac{3}{7}$ 18. $\dfrac{9}{14} \times \dfrac{21}{45}$ 19. $\dfrac{11}{12} \times \dfrac{30}{44}$

20. $\dfrac{8}{9} \div \dfrac{4}{3}$ 21. $\dfrac{-3}{5} \times \dfrac{2}{5}$ 22. $\dfrac{5}{-2} \times \dfrac{-3}{5}$

5 SETS OF NUMBERS

23. $\left(-\dfrac{3}{4}\right) \times \dfrac{2}{5}$ 24. $\left(-\dfrac{7}{8}\right) \times \left(-\dfrac{4}{9}\right)$ 25. $\dfrac{8}{9} \div \dfrac{-2}{3}$

26. $\dfrac{-5}{12} \div \dfrac{10}{-3}$ 27. $\left(-\dfrac{5}{6}\right) \div \dfrac{1}{3}$ 28. $\left(-\dfrac{4}{5}\right) \div \left(-\dfrac{2}{5}\right)$

29. $\dfrac{4}{5} \times \dfrac{3}{4} \times \dfrac{2}{3}$ 30. $\dfrac{7}{8} \times \dfrac{4}{9} \times \left(-\dfrac{3}{2}\right)$ 31. $\dfrac{5}{9} \div \left(\dfrac{2}{3} \div \dfrac{1}{2}\right)$

32. $\left(\dfrac{5}{9} \div \dfrac{2}{3}\right) \div \dfrac{1}{2}$ 33. $\left(\dfrac{7}{9} \times \dfrac{3}{4}\right) \div \dfrac{1}{2}$ 34. $\dfrac{7}{9} \times \left(\dfrac{3}{4} \div \dfrac{1}{2}\right)$

Select a relation R (=, <, >) to make the given statement true.

35. $\dfrac{5}{11} \text{ R } \dfrac{5}{12}$ 36. $\dfrac{1}{8} \text{ R } \dfrac{2}{17}$ 37. $\dfrac{10}{12} \text{ R } \dfrac{25}{30}$

38. $\dfrac{19}{100} \text{ R } \dfrac{18}{100}$ 39. $\dfrac{19}{100} \text{ R } \dfrac{19}{99}$ 40. $\dfrac{19}{100} \text{ R } \dfrac{18}{99}$

41. Name two properties of the set of rational numbers that are not properties of the set of integers.

42. Find three rational numbers between $\dfrac{7}{9}$ and $\dfrac{8}{9}$.

43. Find three rational numbers between 0 and $\dfrac{1}{100}$.

44. Name a rational number between the rational numbers a/b and c/d.

45. Is there a next rational number after 0? Use a property of the set of rational numbers to explain your answer.

46. The results obtained for Exercises 31 and 32 provide a counterexample to show that the set of rational numbers is *not* associative for division. Provide another counterexample of your own to show this fact.

*47. Study the results for Exercises 33 and 34. Then show, in general, that
$$\left(\dfrac{a}{b} \times \dfrac{c}{d}\right) \div \dfrac{e}{f} = \dfrac{a}{b} \times \left(\dfrac{c}{d} \div \dfrac{e}{f}\right)$$

*48. First use your own specific example, and then determine, in general, whether or not the following relationship is true.
$$\dfrac{a}{b} \div \left(\dfrac{c}{d} \times \dfrac{e}{f}\right) = \left(\dfrac{a}{b} \div \dfrac{c}{d}\right) \times \dfrac{e}{f}$$

Copy the following table. Use "√" to show that the set of elements named at the top of a column has the property listed at the side. Use "×" if the set does not have the property.

	Property	Counting numbers	Whole numbers	Integers	Positive rationals	Rational numbers
49.	Closure, +					
50.	Associative, +					
51.	Identity, +					
52.	Inverse, +					
53.	Commutative, +					
54.	Commutative group, +					
55.	Closure, x					
56.	Associative, x					
57.	Identity, x					
58.	Inverse (≠0), x					
59.	Commutative, x					
60.	After excluding 0, commutative group, x					
61.	Distributive					
62.	Field					

Explorations

1. Make a flowchart for reducing fractions. Then use the flowchart to reduce the given fraction.
   (a) $\dfrac{102}{186}$ (b) $\dfrac{84}{120}$ (c) $\dfrac{96}{144}$ (d) $\dfrac{120}{156}$

2. Explain how the density property makes it possible to use rational numbers as measures of line segments.

3. The Greek philosopher Zeno, in the fifth century B.C., proposed several paradoxes that were based on the density property. Prepare a report on these paradoxes using a history of mathematics book as a reference.

Scientific notation is used to represent very large and very small numbers in concise form. To express a number in **scientific notation** we write it as the product of a number between

1 and 10 and some power of 10. These illustrations should clarify this procedure.

$$7,000,000 = 7 \times 10^6$$
$$23,000,000 = 2.3 \times 10^7$$
$$0.000007 = 7 \times 10^{-6}$$
$$0.00000023 = 2.3 \times 10^{-7}$$

This notation is especially helpful in computations with very large or very small numbers. Note, for example, its use in this product.

$$5,000,000 \times 8,000,000 = (5 \times 10^6)(8 \times 10^6)$$
$$= 40 \times 10^{12}$$
$$= 4 \times 10^{13}$$

Unless otherwise instructed, we assume that in ordinary decimal notation zeros that are not otherwise needed are used only to locate the decimal point.

4. Write each number in scientific notation.
   (a) 900,000   (b) 8,000,000,000   (c) 45,000
   (d) 0.00003   (e) 0.00000000005   (f) 0.0000065

5. Use scientific notation to compute each result. Express the answers in scientific notation.
   (a) 300,000 × 30,000   (b) 12,000,000 × 5,000,000
   (c) 45,000,000 ÷ 90,000   (d) 720,000 ÷ 8,000
   (e) 0.000002 × 0.000004   (f) 0.000000045 ÷ 0.0000065

6. Make a collection of very large numbers and very small numbers that are used in science, expressing each in scientific notation.

5-7
The Set of Real Numbers

A line of points with integral coordinates does not fit our concept of a line in the physical universe since there are obvious gaps in a line of integral points. A line of points with rational numbers as coordinates is dense and 3000 years ago appeared to fit the concept of a line in the physical universe, the real world.

There is a legend that the discovery of the need for numbers that are not rational numbers was made by a Pythagorean who was brutally punished for his unorthodox discovery. The Pythagoreans were a secret society of scholars about 2500 years ago. Their concept of a number was as a number of units—a counting number. Numbers such as $\frac{5}{2}$ were considered as five halves; that is, as five units where each unit was $\frac{1}{2}$ instead of 1. Today we still use this concept when we add two fractions with the same number as denominator, that is, *like fractions*. For example,

$$\frac{3}{2} + \frac{5}{2} = 3\left(\frac{1}{2}\right) + 5\left(\frac{1}{2}\right) = (3 + 5)\left(\frac{1}{2}\right) = 8\left(\frac{1}{2}\right) = 4$$

As a mystical sect the Pythagoreans believed that all events and all elements of the universe depended upon numbers (counting numbers). In particular, they believed that the lengths of any two line segments should be expressible as integral multiples of some common unit of length. This belief was shattered when it was discovered that there could not exist a common unit for a diagonal d and a side s of a square.

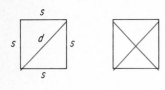

As indicated in the adjacent figures the *square* has four *sides*, each of length s, and four *angles*, each a *right angle*. The two *diagonals* are shown in the second figure. The Pythagoreans discovered that for any given square, there cannot exist a unit line segment such that the lengths of both a diagonal and a side of the square are integral multiples of the length of that unit segment. We consider the case of a unit square, that is, a square with sides 1 unit long.

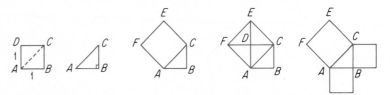

The unit of area is the area of a square on a unit length.

The area of a square on a line segment of length s is s^2, which we read as "s squared."

Let $ABCD$ be a unit square. The area of this unit square is, by definition, 1 square unit. Draw the diagonal AC and consider the **right triangle** (a triangle with a right angle) ABC with sides AB and BC, each 1 unit long. The right angle of the triangle is at B, as indicated by the small square. The side AC is the **hypotenuse** (the side opposite the right angle). Next copy triangle ABC and, as in the figure, draw a square $ACEF$ with side AC. Draw the diagonals AE and CF of this square and label their intersection D, since $ABCD$ in this figure is a copy of the original square $ABCD$. The diagonal AC separates the square $ABCD$ into two *congruent* (same size and shape) right triangles ABC and ADC, each with area $\frac{1}{2}$ square unit. The two diagonals AE and CF separate the square $ACEF$ into four congruent right triangles, each with area $\frac{1}{2}$ square unit. Thus the square $ACEF$

has area 2 square units

has twice the area of the original square $ABCD$

as in the last figure, has area equal to the sum of the areas of the squares on the other two sides of the right triangle ABC

We have demonstrated a special case of the **Pythagorean theorem.**

In modern notation we write
$$a^2 + b^2 = c^2$$

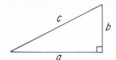

The sum of the areas of the squares on the *legs* (two short sides) of any right triangle is equal to the area of the square on the *hypotenuse*.

Nicaragua honored Pythagoras on one of its stamps and, on the back of the stamp, described the Pythagorean theorem as follows: "The most widely used theorem in geometry is undoubtedly the Pythagorean theorem that refers to the lengths of three sides, *a, b,* and *c,* of a right triangle. It provided for the first time a way of computing lengths by indirect methods, thus permitting people to make surveys and maps. The ancient Greeks used it for measuring the distances between ships at sea, the heights of buildings, and other things. Today scientists and mathematicians constantly use it in developing all kinds of theories."

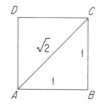

The diagonal AC of the unit square $ABCD$ has length d such that $d^2 = 2$, that is, $d = \sqrt{2}$, the square root of 2, the number whose square is 2. Since $\sqrt{2}$ represents the length of a line segment, $\sqrt{2}$ should be accepted as a number. However, we know that $\sqrt{2}$ is not an integer since

$$1^2 < (\sqrt{2})^2 < 2^2 \quad \text{and} \quad 1 < \sqrt{2} < 2$$

Could $\sqrt{2}$ be a rational number?

The proof that $\sqrt{2}$ is not a rational number depends on these properties of integers (Section 5-5):

An integer n is an even integer if and only if $n = 2k$ where k is an integer.

The square of an even integer is even.

The square of an odd integer is odd.

If the square of an integer is even, then the integer is even.

The proof is an *indirect proof,* that is, we assume that the opposite of what we are trying to prove is true and we show that this assumption leads to an *inconsistency.* (An inconsistency occurs whenever two statements cannot both be true.) Since inconsistencies are not acceptable, the assumption is impossible and its opposite, the statement to be proved, must be true.

PYTHAGORAS

No further extension of the number system is necessary until one wishes to solve such equations as $n^2 + 1 = 0$. Such an extension would require the invention of *imaginary numbers* and the creation of the set of *complex numbers*. However, the use of such numbers is beyond everyday usage and beyond the intended scope of this text. The set of real numbers will suffice for all ordinary consumer needs.

Suppose that $\sqrt{2}$ is a rational number p/q. We may reduce p/q so that $p/q = a/b$, where a and b are not both even integers

$$\sqrt{2} = \frac{a}{b} \qquad \text{Assumed.}$$

$$2 = \frac{a^2}{b^2} \qquad \text{Square both sides.}$$

$$2b^2 = a^2 \qquad \text{Multiply by } b^2.$$

Since $a^2 = 2b^2$ implies that a^2 is even, then a is even and $a = 2k$ for some integer k.

$$2b^2 = (2k)^2 \qquad \text{Substitution.}$$

$$2b^2 = 4k^2 \qquad \text{Multiplication.}$$

$$b^2 = 2k^2 \qquad \text{Divide by 2.}$$

Since $b^2 = 2k^2$ implies that b^2 is even, then b is even. We have shown that if $\sqrt{2} = a/b$, then both a and b are even. However, this is contrary to our assumption that a and b are *not* both even. Thus our assumption must be incorrect. This implies that $\sqrt{2}$ cannot be a rational number.

We assume that any length of a line segment is a **real number**. Then $\sqrt{2}$ is a real number that is not a rational number; in other words, $\sqrt{2}$ is an **irrational number**, a real number that cannot be represented as the quotient a/b of two integers. As a general "rule of thumb" we assume that

If a positive integer n is not the square of a positive integer, then $\sqrt{n}$ is an irrational number.

The assumption that each line segment has a real number as its length is equivalent to an assumption that each point on a number line has a real number as its coordinate. For example, every line segment AB of length d has the same length as the line segment OD, where O is the origin and D is the point with coordinate d on a number line.

There is a one-to-one correspondence between the elements of the set of real numbers and the set of points on the number line. Indeed, this is the distinguishing feature between the set of real numbers and the set of rational numbers. Thus every real number is the coordinate of a point on the number line, and every point on the number line is the graph of a real number. Accordingly, we refer to the number line as the **real number line** and the set of real numbers is said to be **complete**.

The set of real numbers may be obtained either by assuming that all line segments have real numbers as lengths, or by

5 SETS OF NUMBERS

assuming that all decimals represent real numbers. Any integer can be represented by a decimal. Any rational number a/b can be represented by a **terminating decimal** (division is exact at some stage) or by an **infinite repeating decimal.**

$$\frac{1}{4} = 1 \div 4 = 0.25 = 0.25000 \ldots = 0.25\overline{0}$$

$$\frac{3}{8} = 3 \div 8 = 0.375 = 0.375000 \ldots = 0.375\overline{0}$$

$$\frac{1}{3} = 1 \div 3 = 0.333 \ldots = 0.\overline{3}$$

$$\frac{80}{11} = 80 \div 11 = 7.272727 \ldots = 7.\overline{27}$$

In an infinite repeating decimal, we use a bar to indicate the sequence of digits that repeats endlessly. The bar is customarily used over the first repeating digit, or set of digits, on the right of the decimal point. Note that you may, if you wish, think of any terminating decimal as having repeated zeros and thus as a special type of repeating decimal.

Example 1   Write $\dfrac{9}{11}$ as a repeating decimal.

Solution $$\frac{9}{11} = 9 \div 11 = 0.818181 \ldots = 0.\overline{81}$$

Real numbers that are irrational numbers are represented by **nonterminating, nonrepeating decimals,** such as

$$\sqrt{2} = 1.414214 \ldots \qquad \pi = 3.1415926 \ldots$$

In each of these examples the digits do not exhibit a fixed repeating pattern no matter how far they are extended. Some irrational numbers can be represented in decimal notation by a sequence of digits that has a pattern but does not repeat any particular sequence of digits. For example, each of the following decimals names an irrational number.

We consider these decimals rather than numbers such as π, the ratio of the circumference to the diameter of a circle, because advanced mathematical procedures are needed to determine whether π and other such numbers are rational numbers or irrational numbers.

0.20220222022220222220 . . .

0.305300530005300005 . . .

0.404004000400004000004 . . .

Example 2   Write the digit in the sixteenth position to the right of the decimal point in   (a) 0.272272227 . . .   (b) 12.301230012 . . . .

(a) 0.2722722272222722 . . . ; the digit is 2.

(b) 12.3012300123000123 . . . ; the digit is 3.

Example 3 Name two irrational numbers between 0.47 and 0.48.

Solution It is easier to see the solution if the given numbers are written with several decimal places.

$$0.47 = 0.470000\overline{0}$$
$$0.48 = 0.480000\overline{0}$$

We need to write nonterminating, nonrepeating decimals for irrational numbers that are greater than 0.47 and less than 0.48. Here are two of the many possible answers.

$$0.472472247222472222472 . . .$$
$$0.475050050005000050000 . . .$$

The patterns allow us to write as many digits of each decimal as we wish. Such a decimal will never terminate and will never repeat any particular sequence of digits.

We have defined real numbers that are not rational numbers to be irrational numbers. But are irrational numbers useful in any way? Suppose that a and b, where $b \neq 0$, are integers. Then

$$a + b \qquad a - b \qquad a \times b \qquad a \div b$$

are rational numbers. The same is true of rational numbers a and b. Also any number that is obtained as a measure, obtained from a calculator, or obtained from a digital computer, is a rational number. Thus irrational numbers *are not needed* for the four basic operations of arithmetic, for measuring, for use with calculators, or for use with digital computers. Irrational numbers *are needed* as lengths of line segments, as coordinates of points on a number line, as square roots of most positive numbers, and for performing many operations in advanced mathematics.

Operations that are performed on a calculator are based on decimal representations. Any real number that is a rational number may be represented by a terminating decimal or a repeating decimal. Any real number that is not a rational number is an irrational number and may be represented by an infinite nonrepeating decimal.

It is easy to show that every rational number in fractional form can be represented as a repeating (or terminating) deci-

The identification of the set of repeated digits for a rational number can exceed the capacity of a calculator. For example, fractions with denominator 17 have sets of 16 repeated digits.

5 SETS OF NUMBERS

```
   1.714285
7)12.000000
   7
  ⑤0
  4 9
   10
    7
   30
   28
   20
   14
   60
   56
   40
   35
   ⑤
```

mal. Consider, for example, any rational number such as 12/7. When we divide 12 by 7, the possible remainders are 0, 1, 2, 3, 4, 5, 6. If the remainder is 0, the division is exact; if any remainder occurs a second time, the terms after it will repeat also. Since there are only seven possible remainders when you divide by 7, the remainders must repeat or be exact by the seventh decimal place. Consider the determination of the decimal value of 12/7 by long division.

The fact that the remainder 5 occurred again implies that the same steps will be used again in the long division process and the digits 714285 will be repeated over and over; that is, 12/7 = 1.$\overline{714285}$. Similarly, any rational number p/q can be expressed as a terminating or repeating decimal, and at most q decimal places will be needed to identify it.

We can also show that any terminating or repeating decimal can be expressed as a rational number in the form a/b.

Any terminating decimal can be expressed as a fraction, a quotient of integers, with a power of ten as its denominator. For example, if $n = 0.75\overline{0}$, then $100n = 75$, and $n = 75/100$, which reduces to 3/4. Conversely, if a fraction can be expressed as a terminating decimal, its denominator must be a factor of a power of ten.

Any repeating decimal can be expressed as a quotient of integers. For example, if a decimal n repeats one digit, we can find $10n - n$. Suppose that $n = 3.2\overline{4}$; then $10n = 32.\overline{4}$ and we have

$$
\begin{array}{rl}
10n = & 32.4\overline{4} \\
-\quad n = & 3.2\overline{4} \\
\hline
9n = & 29.20
\end{array}
$$

$$n = \frac{29.2}{9} = \frac{29.2}{9} \times \frac{10}{10} = \frac{292}{90} = \frac{146}{45}$$

However, we can avoid the use of decimals in this way.

$$
\begin{array}{rl}
100n = & 234.\overline{4} \\
-\quad 10n = & 32.\overline{4} \\
\hline
90n = & 292.\overline{0}
\end{array}
$$

$$n = \frac{292}{90} = \frac{146}{45}$$

If a decimal n repeats two digits, we find $10^2 n - n$; if it repeats three digits, we find $10^3 n - n$; and so on.

Example 4   Express as a quotient of integers   **(a)** $0.\overline{36}$ **(b)** $0.78\overline{346}$.

Solution   **(a)** Let $n = 0.\overline{36}$; then $100n = 36.\overline{36}$.

$$\begin{array}{r} 100n = 36.\overline{36} \\ -\quad n = \ \ 0.\overline{36} \\ \hline 99n = 36.\overline{0} \end{array}$$

$$n = \frac{36}{99} = \frac{4}{11}$$

(b) Let $n = 0.78\overline{346}$ and $100n = 78.\overline{346}$ so that only the repeating digits occur on the right of the decimal point. Then multiply by 1000.

$$\begin{array}{r} 100{,}000n = 78{,}346.\overline{346} \\ -\quad 100n = \ \ \ \ \ 78.\overline{346} \\ \hline 99{,}900n = 78{,}268.\overline{000} \end{array}$$

$$n = \frac{78{,}268}{99{,}900} = \frac{19{,}567}{24{,}975}$$

The representation of rational numbers as decimals may be summarized as follows.

Every rational number can be named by a terminating or a repeating decimal.

Every terminating or repeating decimal is the name of a rational number.

The real numbers may be classified in several different ways. For example, any real number is

positive, zero, or negative;

a rational number or an irrational number;

expressible as a terminating, a repeating, or a nonterminating, nonrepeating decimal.

At each step of the development of the set of real numbers we may think of the original set of numbers as a proper subset of the new set.

| Counting numbers | ⊂ | Whole numbers | ⊂ | Integers | ⊂ | Rational numbers | ⊂ | Real numbers |

The numbers of each set are *ordered* in the same manner as their graphs on a number line, that is, for the number line in its usual position, $a < b$ and $b > a$ if and only if the point with

5 SETS OF NUMBERS

coordinate a is on the left of the point with coordinate b. The set of real numbers with the operations $+$ and $\times$ forms a **complete linearly ordered field**.

Example 5 Identify each number as a rational number or an irrational number.

(a) $\sqrt{19}$ (b) $2 + \sqrt{7}$ (c) $3 + \sqrt{9}$

(d) $5\sqrt{11}$ (e) $\sqrt{13} - 2$ (f) $5 - \sqrt{15}$

Solution (a) Irrational number, since 19 is not the square of an integer.
(b) Irrational number, since 7 is not the square of an integer.
(c) Rational number, since 9 is the square of an integer.
(d), (e), (f) Irrational numbers, since 11, 13, and 15 are not squares of integers.

EXERCISES *Classify each statement as true or false. Give a counterexample for each false statement.*

1. Every rational number is a real number.

2. Every real number is a rational number.

3. Every real number is either a rational number or an irrational number.

4. Every point on a number line has a real number as its coordinate.

5. Every real number has a point of the number line as its graph.

6. For a given unit of length every positive real number may be represented by the length of a line segment.

7. Every real number is either positive or negative.

8. Every irrational number is either positive or negative.

9. The additive inverse of an irrational number is an irrational number.

10. The multiplicative inverse of an irrational number is an irrational number.

Identify the given number as a rational number or an irrational number.

11. (a) $\sqrt{11}$ (b) $5\sqrt{11}$

12. (a) $-4 + \sqrt{16}$ (b) $11 - \sqrt{63}$

13. (a) $7\sqrt{36}$ (b) $5\sqrt{37}$

14. (a) $\sqrt{20}$ (b) $\sqrt{63} + 1$

15. (a) $\sqrt{2}/\sqrt{98}$ (b) $-2\sqrt{25}$

16. (a) $\sqrt{25} - \sqrt{9}$ (b) $\sqrt{20} + \sqrt{45}$

17. (a) $2\sqrt{2} + (5 - \sqrt{8})$ (b) $2\sqrt{5} + (3 - \sqrt{20})$

18. (a) $(\sqrt{12} + \sqrt{27})/\sqrt{3}$ (b) $\sqrt{7}(\sqrt{28} + \sqrt{63})$

Classify the given decimal as the name of a rational number or an irrational number.

19. 0.745   20. $0.7\overline{45}$

21. 0.745455455545 . . .   22. $0.\overline{745}$

23. 0.745455455545   24. 0.745445444544 . . .

Tell whether the given number can be represented by a terminating decimal, a nonterminating, repeating decimal, or a nonterminating, nonrepeating decimal. Give the decimal representation of each number that can be expressed as a terminating or repeating decimal.

25. $\dfrac{3}{8}$ 26. $\dfrac{5}{12}$ 27. $\sqrt{8}$

28. $\sqrt{3}$ 29. $\sqrt{100}$ 30. $\dfrac{13}{16}$

List the numbers of each set in order from smallest to largest.

31. 0.45,   0.454554555 . . . ,   0.45455, $0.\overline{45}$, $0.4\overline{5}$

32. 2.525,   2.5252, $2.5\overline{2}$,   2.525225222 . . . ,   2.5

33. 0.067, $0.06\overline{7}$,   0.067677677767 . . . ,   0.06, $0.0\overline{6}$

34. State whether or not the indicated number is a rational number between 0.37 and 0.38.

    (a) 0.375   (b) $0.3\overline{7}$ (c) 0.373773777 . . .   (d) $0.37\overline{8}$

35. State whether or not the indicated number is an irrational number between 0.234 and 0.235.

    (a) 0.2345   (b) $0.2\overline{34}$

    (c) 0.234040040004 . . .   (d) 0.235545545554 . . .

36. Name two rational numbers between 0.524244244424 . . . and 0.52525525525 . . . .

37. Name two irrational numbers between 0.48 and 0.49.

38. Name two irrational numbers between $0.\overline{78}$ and $0.\overline{79}$.

39. Which of the sets of numbers that we have studied in this chapter are (a) dense (b) complete?

Give at least three examples that satisfy the given condition.

40. A sum of irrational numbers that is 0.

41. A sum of irrational numbers that is 2.

42. A difference of irrational numbers that is 1.

43. A product of irrational numbers that is a rational number.

*44. A product of irrational numbers that is an irrational number.

*45. A quotient of irrational numbers that is a rational number.

*46. A quotient of irrational numbers that is an irrational number.

Explorations

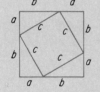

1. The construction for $\sqrt{2}$ as the length of a diagonal of a unit square can be extended to find segments whose measures are $\sqrt{3}$, $\sqrt{4}$, $\sqrt{5}$, and so on. Merely continue to construct right triangles, using the hypotenuse of the preceding triangle as one leg and a segment of one unit as the other leg. The adjacent figure shows the construction for $\sqrt{3}$. Continue this method of construction to find three more line segments with lengths that are irrational numbers.

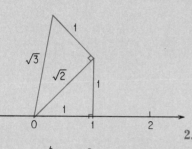

2. Construct any right triangle and designate the lengths of its sides as a, b, and c, where c is the hypotenuse. Make three copies of this triangle to obtain four congruent right triangles. Cut out the four triangles. Arrange the four triangles as in the adjacent figure to represent

$$(4 \text{ triangles}) + c^2 = (a + b)^2$$

Arrange the four triangles as in the next adjacent figure to represent

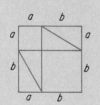

$$(4 \text{ triangles}) + a^2 + b^2 = (a + b)^2$$

Explain the sense in which the Pythagorean theorem has been demonstrated.

3. If the denominator of a fraction can be expressed as the product of powers of 2 and 5 only, then the fraction can be written as a terminating decimal.
   (a) Show that this is so for several examples, such as $\dfrac{3}{50}$ and $\dfrac{33}{60}$.

   (b) Explain why $\dfrac{n}{2^q 5^p}$ for any whole numbers n, p, and q can be represented by a terminating decimal.

   (c) Explain why a fraction that cannot be reduced to the form given in part (b) also cannot be represented by a terminating decimal.

Chapter 5 Review *Solutions to the following exercises may be found within the text of Chapter 5. Try to complete each exercise without referring to the text.*

Section 5-1   Numbers and Numerals

1. Write 142 in early Egyptian notation.

2. Write $\nearrow\int\limits^{\!\!\!\curvearrowright}\,\int\limits^{\!\!\!\curvearrowright}\, 999 \cap\cap |$ in decimal notation.

3. Use exponents and write 2306 in expanded notation.

4. Use galley multiplication and multiply 372 by 47.

Section 5-2   Binary Notation

5. Write 11 011 101$_2$ in base ten notation.

6. Multiply   101$_2$ × 1101$_2$.

Section 5-3   Counting Numbers and Their Properties

7. Tell whether each specified number in the following statement is used for identification, as an ordinal number, or as a cardinal number.

   The *second* train through Peoria consisted of *thirty* cars pulled by engine number *534*.

8. Use a Cartesian product to illustrate 3 × 2 = 6.

9. Apply properties of counting numbers to the left member of the equation 25 × (11 × 4) = (25 × 4) × 11 to obtain the right member of the equation. Show each step and name each property.

10. Tell whether or not the relation *is greater than* (>) is
    (**a**) reflexive   (**b**) symmetric   (**c**) transitive.

Section 5-4   The Set of Integers

11. For each arithmetic statement name the property of whole numbers that is illustrated.
    (**a**) 7 + 0 = 7   (**b**) 4 × 0 = 0

12. Graph the set of integers between −3 and 2.

13. State five properties of the set of integers under addition.

Section 5-5   Computation with Integers

14. Find the sum   (+3) + (−5) + (−2).

15. Subtract   (+5) − (−3).

5 SETS OF NUMBERS

16. Find each product.
(**a**) $(+4) \times (+7)$ (**b**) $(-3) \times (+6)$ (**c**) $(+5) \times (-8)$
(**d**) $(-7) \times (-8)$

17. Show that the set of integers is not closed with respect to the operation of division.

18. Prove that the sum of any two even integers is an even integer.

Section 5-6 *The Set of Rational Numbers*

19. Reduce $\frac{6}{-8}$.

20. Simplify $\frac{2}{3} \times \frac{3}{4} \times \left(-\frac{1}{8}\right)$.

21. Name a rational number between $\frac{17}{19}$ and $\frac{18}{19}$.

Section 5-7 *The Set of Real Numbers*

22. Write $\frac{9}{11}$ as a repeating decimal.

23. Name two irrational numbers between 0.47 and 0.48.

24. Express as a quotient of integers (**a**) $0.\overline{36}$ (**b**) $0.78\overline{346}$.

25. Identify as a rational number or an irrational number
(**a**) $2 + \sqrt{7}$ (**b**) $3 + \sqrt{9}$.

26. State the relationship between rational numbers and decimals.

Chapter 5 Test

1. Write
(**a**) 235 in early Egyptian notation.
(**b**)  in decimal notation.

Compute in binary notation.

2. $\begin{array}{r} 11\ 011_2 \\ +\ \underline{10\ 111_2} \end{array}$

3. $\begin{array}{r} 11\ 011_2 \\ \times\ \underline{101_2} \end{array}$

Classify each statement as true or false.

4. (**a**) The opposite of each whole number is a negative integer.
(**b**) The opposite of each integer is an integer.

5. (**a**) Between any two rational numbers there is always another rational number.

(b) There is a one-to-one correspondence between the set of rational numbers and the points of the number line.

6. Name the property of whole numbers that is illustrated by each statement.
   (a) $7 \times (5 \times 8) = (7 \times 5) \times 8$
   (b) $9 \times (7 + 6) = 9 \times (6 + 7)$

7. Graph the set of integers between -4 and 2 on a number line.

Perform the indicated operations.

8. $(-20) + (+3)$ 9. $(+7) + (-18)$

10. $(-5) \times (+8)$ 11. $(-9) \times (-8)$

12. $(+36) \div (-4)$ 13. $(-42) \div (-7)$

14. $(-3) + [(+2) + (-7)]$ 15. $[(-9) - (-2)] - (+3)$

16. $(-5) \times [(-2) \times (-1)]$ 17. $[(-40) \div (+5)] \div (-2)$

18. $\left(-\dfrac{3}{4}\right) \times \left(-\dfrac{8}{9}\right)$ 19. $\left(\dfrac{-5}{12}\right) \div \left(\dfrac{3}{-4}\right)$

20. Reduce $\dfrac{30}{72}$ to lowest terms.

21. Express each decimal as a fraction.
    (a) $0.\overline{63}$ (b) $0.6\overline{12}$

22. Express each fraction as a repeating decimal.
    (a) $\dfrac{7}{12}$ (b) $\dfrac{7}{13}$

23. Classify each statement as true or false. Give a counter-example for each false statement.

    (a) The sum of any two irrational numbers is an irrational number.
    (b) The product of any two irrational numbers is an irrational number.

24. List the given numbers in order from smallest to largest.
    $2.\overline{56}$ $2.5\overline{6}$ 2.56 2.566 2.565

25. Name two irrational numbers between $0.\overline{56}$ and $0.5\overline{6}$.

5 SETS OF NUMBERS

Elements
of
Number
Theory

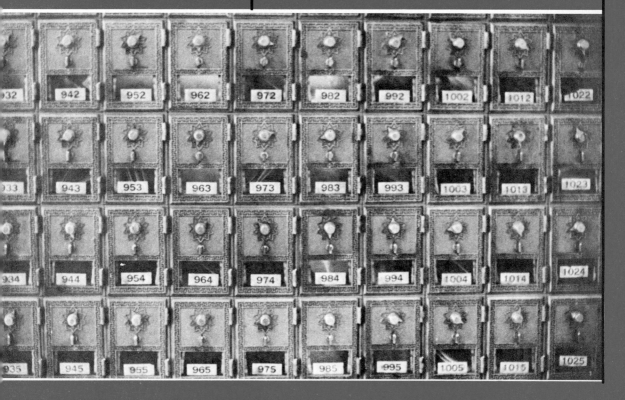

6-1

Factors, Multiples, and Divisibility Rules

Elementary number theory is concerned with relationships among counting numbers. One of the basic relationships is divisibility. The counting number 24 is divisible by the counting number 3 since there is another counting number, 8, such that $24 = 3 \times 8$. Similarly, 24 is divisible by each member of the set of counting numbers {1, 2, 3, 4, 6, 8, 12, 24}.

> A counting number n is **divisible** by a counting number s if and only if there is a counting number k such that $n = s \times k$.

Because 24 is divisible by each of the numbers in the set {1, 2, 3, 4, 6, 8, 12, 24}, each of these numbers is a *factor* of 24 and 24 is a *multiple* of each of them. On the other hand, 24 is not divisible by 5 since there is no counting number b such that $24 = 5 \times b$. Therefore 5 is not a factor of 24 and 24 is not a multiple of 5.

> If n is divisible by s, then s is a **factor** of n and n is a **multiple** of s.

Only counting numbers are considered as factors of other counting numbers. The set of counting numbers that are factors of a given counting number may be found by listing pairs of numbers whose product is the given number. This is illustrated for 12 and 18.

	12		18
	1×12		1×18
	2×6		2×9
	3×4		3×6
Factors	12: {1, 2, 3, 4, 6, 12}	18:	{1, 2, 3, 6, 9, 18}

The counting numbers 1, 2, 3, . . . are tried, in numerical order, as possible factors. This process is continued until the two factors are alike or, as in 3×6 for 18, the second factor is the next larger counting number that is a factor of the given number.

Example 1 Find the set of counting numbers that are factors of 16.

Solution   {1, 2, 4, 8, 16}

Each of these counting numbers is a factor of 16, and 16 is divisible by each of these numbers.

234

6 ELEMENTS OF NUMBER THEORY

Example 2   Find the factors of 19.

Solution   There is one and only one pair of counting numbers that have 19 as their product. These numbers are 1 and 19. Thus the set of factors of 19 is {1, 19}.

Any counting number has been defined to be a multiple of, and divisible by each of its factors. For example, 16 is a multiple of 8 but 19 is not a multiple of 8 since 8 is not a factor of 19.

Example 3   Show that   **(a)** 245 is a multiple of 5   **(b)** 257 is not a multiple of 3.

Solution   **(a)** Since $245 = 5 \times 49$, the counting number 245 is divisible by 5 and is a multiple of 5.
   (b) There is no counting number k such that $257 = 3 \times k$. By division, $257 \div 3 = 85\frac{2}{3}$, which is not a counting number.

Multiples of a number can be obtained by multiplying the given number by members of the set of counting numbers. The first ten multiples of 9 are listed in this display.

Note some of the properties of the digits of these multiples of 9. In each case the sum of the digits is 9. As the multiples increase the units digits decrease and the tens digits increase.

$$
\begin{array}{ll}
1 \times 9 = 9 & 6 \times 9 = 54 \\
2 \times 9 = 18 & 7 \times 9 = 63 \\
3 \times 9 = 27 & 8 \times 9 = 72 \\
4 \times 9 = 36 & 9 \times 9 = 81 \\
5 \times 9 = 45 & 10 \times 9 = 90
\end{array}
$$

The multiples of 9 can be written in the form $9n$ where n is a counting number.

The set of multiples of 9 is an infinite set and can be listed in set notation.

{9, 18, 27, 36, 45, 54, 63, 72, 81, 90, . . .}

Each member of this set is divisible by 9, and 9 is a factor of each member.

Example 4   List the first ten multiples of 5.

Solution   5, 10, 15, 20, 25, 30, 35, 40, 45, 50

Because we are restricting our discussion in this section to the set of counting numbers, no mention has been made of the number zero. It should be noted that in some texts zero is considered to be a multiple of every number. For example, zero is a multiple of 9 since $0 \times 9 = 0$ and $0 \div 9 = 0$. However, because of problems with zero that would be encountered in later sections, we shall continue to use only counting numbers as we consider multiples and factors of numbers.

Any counting number *is divisible by* another counting number if the division produces a zero remainder. Divisibility rules are obtained from patterns that arise when the numbers are expressed in decimal notation. Any counting number N less than 10,000 can be expressed in the form

$$N = 1000T + 100h + 10t + u$$

Note that T represents the number of thousands while t represents the number of tens. Usually $T \neq t$.

where T, h, t, and u are elements, not necessarily distinct, of the set $\{0, 1, 2, 3, 4, 5, 6, 7, 8, 9\}$. The letter u is used for the ones (units) digit to avoid confusion of o and 0. The expanded notation for N can be used to establish rules for the divisibility of N.

$$\frac{N}{2} = \frac{1000T}{2} + \frac{100h}{2} + \frac{10t}{2} + \frac{u}{2} = 500T + 50h + 5t + \frac{u}{2}$$

Thus $N/2$ is a whole number if and only if $u/2$ is a whole number; that is, N *is divisible by 2 if and only if u is divisible by 2.*

$$\frac{N}{3} = \frac{(999 + 1)T}{3} + \frac{(99 + 1)h}{3} + \frac{(9 + 1)t}{3} + \frac{u}{3}$$
$$= 333T + 33h + 3t + \frac{T + h + t + u}{3}$$

Thus $N/3$ is a whole number if and only if $(T + h + t + u)/3$ is a whole number; that is, N *is divisible by 3 if and only if the sum of its decimal digits is divisible by 3.*

$$\frac{N}{4} = \frac{1000T}{4} + \frac{100h}{4} + \frac{10t}{4} + \frac{u}{4} = 250T + 25h + \frac{10t + u}{4}$$

Thus $N/4$ is a whole number if and only if $(10t + u)/4$ is a whole number; that is, N *is divisible by 4 if and only if the number represented by its tens and ones digits is divisible by 4.*

According to these rules:

1984 is divisible by 2 since its ones digit, 4, is divisible by 2.

1984 is not divisible by 3 since the sum of its digits is 22 and 22 is not divisible by 3. (Note that $2 + 2 = 4$ and 4 is not divisible by 3.)

Other rules for divisibility are considered in the exercises.

1984 is divisible by 4 since 84 is divisible by 4.

6 ELEMENTS OF NUMBER THEORY

List the first five multiples of the given number.

1. 2	2. 3	3. 4	4. 6
5. 7	6. 8	7. 10	8. 11
9. 15	10. 20	11. 25	12. 32

Find the factors of the given number.

13. 30	14. 32	15. 49	16. 36
17. 20	18. 48	19. 64	20. 80
21. 92	22. 72	23. 17	24. 54
25. 40	26. 51	27. 37	28. 99
29. 84	30. 60	31. 95	32. 111
33. 125	34. 150	35. 225	36. 450

For decimal notation, find a rule for divisibility by the given whole number.

37. 5	38. 6	39. 8	40. 9

Test each number for divisibility by **(a)** 2 **(b)** 3 **(c)** 4 **(d)** 5 **(e)** 6 **(f)** 8 **(g)** 9.

41. 5280	42. 225	43. 1728	44. 16,275
45. 17,540	46. 19,678	47. 36,000	48. 27,600
49. 45,460	50. 80,124	51. 100,200	52. 100,100

Explorations

Consider as many numerical examples as necessary and explain why each statement must be true for all integers.

1. The sum of any two consecutive integers, in general n and $n + 1$, is an odd number.

2. The product of any two consecutive integers is an even number.

3. The product of any three consecutive integers is divisible by 3 and by 6.

4. The product of any four consecutive integers is divisible by 24.

5. A positive integer N has an odd number of positive integers as factors if and only if N is the square of an integer.

6. For any positive integer k the sum of the first k positive odd integers is k^2.

7. Any sum of even numbers is an even number.

8. The sum of any even number of odd numbers is an even number.

9. The sum of any odd number of odd numbers is an odd number.

10. This problem appeared on a recent college aptitude test. Try it.

One student protested that there were two possible answers. Show that such is the case. The cube of an integer s is s^3, that is, $s \times s \times s$.

Row A:	7	2	5	4	6
Row B:	3	8	6	9	7
Row C:	5	4	3	8	2
Row D:	9	5	7	3	6
Row E:	5	6	3	7	4

Which row in the list above contains both the square of an integer and the cube of a different integer?

6-2 Prime Numbers

Study the set of factors for each of the following numbers.

2: {1, 2}   3: {1, 3}   5: {1, 5}   7: {1, 7}

In each case the number shown has exactly two distinct counting numbers as factors, the number itself and 1, and is said to be a *prime number*.

> A **prime number** is a counting number that has exactly two distinct factors.

As shown, 2, 3, 5, and 7 are examples of prime numbers. On the other hand, 6 is not a prime number because it can be factored as 1×6 and 2×3. That is, 6 has more than two distinct factors. The set of factors of 6 is {1, 2, 3, 6}.

The counting numbers that are greater than 1 and are not prime are **composite numbers**. Note that every counting number greater than 1 is either prime or composite. The number 1 is neither prime nor composite because it does not have exactly two different factors; the only way to factor 1 is as 1×1. Thus we may classify any counting number as belonging to one of the following sets.

The set whose only element is the number 1.
The set of prime numbers.
The set of composite numbers.

Example 1   Classify as prime or composite   (a) 24   (b) 31.

6 ELEMENTS OF NUMBER THEORY

(a) Composite. The number 24 can be factored as 1×24, 2×12, 3×8, or 4×6 and has more than two factors.

(b) Prime. The number 31 can be factored only as 1×31 and has exactly two distinct factors.

Eratosthenes (ca. 276–194 B.C.) was a gifted mathematician, astronomer, geographer, historian, philosopher, poet, and athlete. Ptolemy III invited him to Alexandria, Egypt to tutor his son and take charge of the library of the University.

The following method for identifying prime numbers was discovered over two thousand years ago by a Greek mathematician named Eratosthenes. This method, known as the **Sieve of Eratosthenes**, is illustrated for the set consisting of the counting numbers through 100.

First prepare a table as shown, and cross out 1 since 1 is not classified as a prime number. Draw a circle around 2, the smallest prime number. Then cross out every multiple of 2 that follows, since each one is divisible by 2 and thus is not prime. That is, cross out the numbers in the set $\{4, 6, 8, \ldots, 100\}$.

1̶	②	3	4̶	5	6̶	7	8̶	9	1̶0̶
11	1̶2̶	13	1̶4̶	15	1̶6̶	17	1̶8̶	19	2̶0̶
21	2̶2̶	23	2̶4̶	25	2̶6̶	27	2̶8̶	29	3̶0̶
31	3̶2̶	33	3̶4̶	35	3̶6̶	37	3̶8̶	39	4̶0̶
41	4̶2̶	43	4̶4̶	45	4̶6̶	47	4̶8̶	49	5̶0̶
51	5̶2̶	53	5̶4̶	55	5̶6̶	57	5̶8̶	59	6̶0̶
61	6̶2̶	63	6̶4̶	65	6̶6̶	67	6̶8̶	69	7̶0̶
71	7̶2̶	73	7̶4̶	75	7̶6̶	77	7̶8̶	79	8̶0̶
81	8̶2̶	83	8̶4̶	85	8̶6̶	87	8̶8̶	89	9̶0̶
91	9̶2̶	93	9̶4̶	95	9̶6̶	97	9̶8̶	99	1̶0̶0̶

Draw a circle around 3, the next prime number in the list. Then cross out each succeeding multiple of 3. Some of these numbers, such as 6 and 12, will already have been crossed out because they are also multiples of 2.

Draw a circle around 5, the next prime number. Then exclude each fifth number after 5. The next prime number is 7; exclude each seventh number after 7. The next prime number is 11. Since all multiples of 11 in the table have already been excluded, the remaining numbers that have not been excluded are prime numbers and may be circled, as on page 240.

Notice that 49 is the first number that is divisible by 7 and is not also divisible by a prime number less than 7. In other words, each composite number less than 7^2 has at least one of its factors less than 7. Similarly, each composite number less

To determine whether a counting number N is a prime number try the prime numbers p such that $p^2 \leq N$ as factors of N. For example, to check 217 try 2, 3, 5, 7, 11, and 13.

~~1~~	**2**	**3**	~~4~~	**5**	~~6~~	**7**	~~8~~	~~9~~	~~10~~
11	~~12~~	**13**	~~14~~	~~15~~	~~16~~	**17**	~~18~~	**19**	~~20~~
~~21~~	~~22~~	**23**	~~24~~	~~25~~	~~26~~	~~27~~	~~28~~	**29**	~~30~~
31	~~32~~	~~33~~	~~34~~	~~35~~	~~36~~	**37**	~~38~~	~~39~~	~~40~~
41	~~42~~	**43**	~~44~~	~~45~~	~~46~~	**47**	~~48~~	~~49~~	~~50~~
~~51~~	~~52~~	**53**	~~54~~	~~55~~	~~56~~	~~57~~	~~58~~	**59**	~~60~~
61	~~62~~	~~63~~	~~64~~	~~65~~	~~66~~	**67**	~~68~~	~~69~~	~~70~~
71	~~72~~	**73**	~~74~~	~~75~~	~~76~~	~~77~~	~~78~~	**79**	~~80~~
~~81~~	~~82~~	**83**	~~84~~	~~85~~	~~86~~	~~87~~	~~88~~	**89**	~~90~~
~~91~~	~~92~~	~~93~~	~~94~~	~~95~~	~~96~~	**97**	~~98~~	~~99~~	~~100~~

than 5^2 has at least one factor less than 5. In general, *for any prime number p each composite number less than p^2 has a prime number less than p as a factor.* A number N is a prime number if and only if N is not divisible by any prime number p where p^2 is less than or equal to N.

The set of prime numbers is an infinite set, and to this date the search continues for larger and larger prime numbers. The advent of the computer has made such searches possible, and the discovery of a new prime number invariably finds its way as a news article into the daily papers.

We use this property to tell us when we have excluded all composite numbers from a set. In the set {1, 2, . . . , 100} we have considered the prime numbers 2, 3, 5, and 7. The next prime number is 11 and 11^2 is greater than 100. Thus we have already excluded all composite numbers and identified all prime numbers that are less than or equal to 100.

Example 2 List the set of prime numbers less than 70.

Solution From the table, this set is

{2, 3, 5, 7, 11, 13, 17, 19, 23, 29, 31, 37, 41, 43, 47, 53, 59, 61, 67}

EXERCISES *Classify the given number as prime or composite.*

1. 42	2. 51	3. 71	4. 92
5. 89	6. 101	7. 147	8. 203
9. 257	10. 707	11. 727	12. 729

Express the given number as the product of as many different pairs of counting numbers as possible.

13. 16	14. 17	15. 21	16. 24
17. 31	18. 51	19. 54	20. 60
21. 100	22. 120	23. 125	24. 150

List the prime numbers between the given numbers.

25. 1 and 20	**26.** 30 and 40	**27.** 40 and 50
28. 60 and 70	**29.** 80 and 100	**30.** 100 and 120

List the composite numbers between the given numbers.

31. 1 and 10	**32.** 20 and 30	**33.** 30 and 50
34. 50 and 70	**35.** 80 and 90	**36.** 110 and 120

37. Is every odd number a prime number? Is every prime number an odd number?

38. Exhibit a pair of prime numbers that differ by 1 and show that there is only one such pair possible.

Christian Goldbach (1690–1764) was a German mathematician who left his mathematical work for a career in the Russian civil service. His conjecture has been viewed as a wild guess but no one has been able to prove that it is wrong.

39. Here is a famous theorem that has not yet been proved: Every even number greater than 2 is expressible as the sum of two prime numbers, not necessarily distinct. (This theorem is often called **Goldbach's conjecture.**) Express each of the even numbers 4 through 40 as a sum of two prime numbers.

40. Here is another famous theorem. Two prime numbers such as 17 and 19 that differ by 2 are called **twin primes.** It is believed but has not yet been proved that there are infinitely many twin primes. Find a pair of twin primes that are between **(a)** 35 and 45 **(b)** 65 and 75 **(c)** 95 and 105.

41. A set of three prime numbers that differ by 2 is called a **prime triplet.** Exhibit a prime triplet and explain why it is the only possible triplet of prime numbers.

42. It has been conjectured but not proved that every odd number greater than 5 is expressible as the sum of three prime numbers, not necessarily distinct. Verify this for the numbers 7, 9, 11, 13, and 15.

43. What is the largest prime number that you need to consider to be sure that you have excluded all composite numbers less than or equal to **(a)** 200 **(b)** 500 **(c)** 1000?

*44. Extend the Sieve of Eratosthenes to find the prime numbers less than or equal to 200.

Explorations

1. No one has ever been able to find a formula that will produce only prime numbers. At one time it was thought that the expression $n^2 - n + 41$ would give only prime numbers for the set of counting numbers as replacements for

Pierre de Fermat (1601–1665) was a lawyer who spent his spare time studying mathematics. He corresponded with many of the leading mathematicians of his time, helped spread the discoveries of others, and made major contributions of his own. He was the founder of modern number theory.

For $n = 4$ the Fermat number 65,537 is prime. For $n = 5$ the number $4{,}294{,}967{,}297 = 641 \times 6{,}700{,}417$. No one has found a prime Fermat number for n greater than 4 and it is not known whether or not any exist.

n. Show that prime numbers are obtained when n is replaced by 1, 2, 3, 4, and 5. Show that the formula fails for $n = 41$.

2. A theorem about prime numbers first stated in 1640 by Pierre de Fermat (1601–1665) states that if p is a prime number, then for every integer a, $a^p - a$ is divisible by p. For example, if $p = 2$, then $a^2 - a$ is divisible by 2 for all integers a. Test that this is so for at least five different values of a. Then let $p = 3$ and test again. Then let $p = 4$, not a prime, and show by a single counterexample that the theorem does not hold for composite numbers.

3. For $n = 0, 1, 2, 3, \ldots$, numbers of the form

$$2^{(2^n)} + 1$$

are known as **Fermat numbers.** Fermat conjectured, incorrectly, that all such numbers would be prime. For $n = 0$, we have

$$2^{(2^0)} + 1 = 2^1 + 1 = 3$$

a prime number. Show that prime numbers are obtained for $n = 1$, $n = 2$, and $n = 3$.

4. It can be proved that every counting number greater than 11 is the sum of two composite numbers. Show, by example, that this is true for the counting numbers 12 through 25.

5. Euclid proved that the set of prime numbers is infinite. Refer to a textbook or a book on the history of mathematics and study this simple, yet elegant, proof.

6. On October 30, 1978, two 18-year-old students discovered the largest prime number known at that time. On February 9, 1979, one of these students found a larger prime number. Try to find out what is the largest prime number known today and by whom it was discovered.

7. Assume that you have a list P of the prime numbers that are less than 100. Describe in words, by a flowchart, or by a computer program a procedure for determining whether or not any given counting number N, where $N \le 10{,}000$, is a prime number.

6-3
Prime Factorization

We have seen that every counting number greater than 1 is either a prime number or a composite number. Now we shall find that, except for the order of the factors, every counting

6 ELEMENTS OF NUMBER THEORY

number greater than 1 can be expressed as the product of one and only one set of powers of prime numbers.

In this sense the prime numbers are the "building blocks" of the counting numbers.

Consider these factorizations of 24.

$$24 = 1 \times 24$$
$$24 = 2 \times 12$$
$$24 = 3 \times 8$$
$$24 = 4 \times 6$$
$$24 = 2 \times 2 \times 6$$
$$24 = 2 \times 3 \times 4$$
$$24 = 2 \times 2 \times 2 \times 3 = 2^3 \times 3$$

In the last factorization the set $\{2^3, 3\}$ of powers of primes is unique. The actual factorization in terms of prime numbers can be written in other ways such as $2 \times 3 \times 2^2$. However, these ways are equivalent, since the order of the factors does not affect the product. Thus 24 can be expressed as the product of a unique set of powers of its prime factors.

One of the easiest ways to find the prime factors of a number is to consider the prime numbers

$$2, \quad 3, \quad 5, \quad 7, \quad 11, \quad 13, \quad 17, \quad 19, \quad 23, \quad 29, \quad 31, \quad \ldots$$

in order and use each one as a factor as many times as possible. Then for 24 we would have

$$24 = 2 \times 12$$
$$= 2 \times 2 \times 6$$
$$= 2 \times 2 \times 2 \times 3$$

These steps may be performed by successive division.

```
2)24
2)12
 2)6
   3
```

Since 3 is a prime number, no further steps are needed and $24 = 2^3 \times 3$.

Example 1   Express 3850 in terms of its prime factors.

Solution

```
2)3850
5)1925
 5)385      3850 = 2 × 5² × 7 × 11
  7)77
    11
```

$$3850 = 2 \times 5^2 \times 7 \times 11$$

In general, if a counting number n is greater than 1, then n has a prime number p_1 as a factor. Suppose that

$$n = p_1 n_1$$

Then if n is a prime number, $n = p_1$ and $n_1 = 1$. If n is not a prime number, then n_1 is a counting number greater than 1. In this case n_1 is either a prime number or a composite number. Suppose that

$$n_1 = p_2 n_2 \quad \text{and thus} \quad n = p_1 p_2 n_2$$

where p_2 is a prime number. As before, if $n_2 \neq 1$, then

$$n_2 = p_3 n_3 \quad \text{and thus} \quad n = p_1 p_2 p_3 n_3$$

where p_3 is a prime number, and so forth. We may continue this process until some $n_k = 1$, since there are only a finite number of counting numbers less than n and

$$n > n_1 > n_2 > n_3 > \cdots > n_k = 1$$

Then we have an expression for n as a product of prime numbers,

$$n = p_1 p_2 p_3 \cdots p_k$$

It is customary to write the prime factors in increasing order, as in Examples 1 and 2.

We call this the **prime factorization** of n, that is, the factorization of n into its prime factors. Except for the order of the factors, the prime factorization of any counting number greater than 1 is unique; that is, *any counting number greater than 1 may be expressed as a product of a unique set of powers of its prime factors.*

Example 2   Find the prime factorization of 5280.

Solution

$$
\begin{array}{r}
2)\overline{5280} \\
2)\overline{2640} \\
2)\overline{1320} \\
2)\overline{660} \\
2)\overline{330} \\
3)\overline{165} \\
5)\overline{55} \\
11
\end{array}
\qquad 5280 = 2^5 \times 3 \times 5 \times 11
$$

In the sections that follow, we shall find that the prime factorization of a number can be used to develop techniques for simplifying fractions, as well as for computing with fractions.

6 ELEMENTS OF NUMBER THEORY

The prime factorization of a number also may be used to determine the set of all factors of that number. Consider, for example, the prime factorization of 30.

$$30 = 2 \times 3 \times 5$$

Select the prime factors of 30 one at a time, two at a time, and three at a time as in the array.

ONE AT A TIME	TWO AT A TIME	THREE AT A TIME
2	2×3	$2 \times 3 \times 5$
3	2×5	
5	3×5	

The number 1, which is a factor of every number, and the numbers shown in the array form the set {1, 2, 3, 5, 6, 10, 15, 30} of factors of 30.

EXERCISES

Find the number represented by each prime factorization.

1. $3 \times 5^2 \times 7$
2. $2^3 \times 3^2$
3. $2 \times 3 \times 5^2$
4. $2 \times 7 \times 11^2$
5. $2^5 \times 3^3$
6. $2 \times 5^2 \times 7 \times 13$
7. $3^3 \times 5^3$
8. 11^2
9. $11 \times 13 \times 17^2$
10. $7^3 \times 13 \times 17$
11. $2^4 \times 11^3$
12. $3^2 \times 13 \times 17^2 \times 19$

Find the prime factorization of each number.

13. 96
14. 64
15. 415
16. 213
17. 938
18. 2425
19. 257
20. 618
21. 3000
22. 4800
23. 4895
24. 5780

Use the prime factorization shown to write the set of factors of each number.

25. $2 \times 5 \times 7$
26. $3 \times 7 \times 11$
27. $2^2 \times 5$
28. $3^2 \times 7$
29. $2 \times 3^2 \times 5$
30. $2^2 \times 3 \times 5$

Find the given number N, where each of the other letters represents a prime number.

*31. $N = 2 \times a^2 \times 5 = b \times 3^2 \times c$
*32. $N = d^3 \times 3 \times e = 2^3 \times f \times 5$
*33. $N = g^2 \times h \times 7 = 2^2 \times 5 \times j$
*34. $N = k^2 \times 5^2 \times m \times 13 = 2^2 \times p^2 \times 11 \times q$

Explorations

Some textbooks use the idea of a branching tree to help students think about factors. If the branches terminate with prime numbers, then we have a **prime-factor tree.** Two such trees are shown in the following figure.

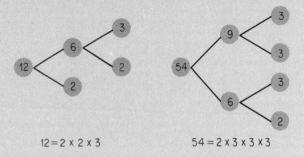

$$12 = 2 \times 2 \times 3 \qquad\qquad 54 = 2 \times 3 \times 3 \times 3$$

Draw a prime-factor tree for the given number.

1. 30   2. 60   3. 96

Here is a trick that depends upon the prime factorization of a number for its explanation. Begin with a box of colored objects such as beads, blocks, or marbles. Assign numerical values to each color. For example, suppose that a box contains red beads, green beads, and blue beads and suppose that the beads are assigned these values:

Red—2   Green—3   Blue—5

Next ask someone to select any number of beads from the box and to find the product of their values. Suppose the person selects three red, two green, and four blue beads. Then the product of their values is

$$\underbrace{2 \times 2 \times 2}_{\text{red}} \times \underbrace{3 \times 3}_{\text{green}} \times \underbrace{5 \times 5 \times 5 \times 5}_{\text{blue}} = 45{,}000$$

At this point you ask for the product of the values of the selected beads. From this product, you are able to determine the number of beads of each color that have been chosen. The trick is to write the prime factorization of the product. Since

$$45{,}000 = 2^3 \times 3^2 \times 5^4$$

the beads chosen are three red (2^3), two green (3^2), and four blue (5^4). Note that the exponent of the associated prime number value indicates the number of beads that have been chosen of each color.

Consider the given number as the product of the values of

the beads in a selection and find the number of beads that have been chosen of each color.

4. 2400 5. 10,000 6. 6750

7. Assume that you have a list P of the prime numbers less than 100. Describe in words, by a flowchart, or by a computer program a procedure for finding the prime factorization of any given counting number N where $N \leq 10,000$.

6-4
Greatest Common Factor

We can make effective use of prime factorizations in many arithmetic situations.

Let us consider the various factors of counting numbers. Recall that a counting number s is said to be a factor of another counting number n if and only if there is a counting number k such that $n = s \times k$. For example, the number 3 is a factor of 18 because there exists a number, namely 6, such that $18 = 3 \times 6$.

Consider the sets of factors of 12 and 18 as shown.

12: {①, ②, ③, 4, ⑥, 12}

18: {①, ②, ③, ⑥, 9, 18}

The circled numbers are those that are factors of both 12 and 18 and are called the **common factors** of the two numbers.

Common factors of 12 and 18: {1, 2, 3, 6}

The largest member of this set, 6, is called the *greatest common factor* (GCF) of the two numbers. In general, the **greatest common factor** of two or more counting numbers is the largest counting number that is a factor of each of the given numbers. The GCF is sometimes called the *greatest common divisor* (GCD).

We may use the prime factorization of two counting numbers to find their greatest common factor. First represent each number by its prime factorization. Then consider the prime numbers that are factors of both of the given numbers, and take the product of those prime numbers with each raised to the highest power that is a factor of *both* of the given numbers. For $12 = 2^2 \times 3$ and $18 = 2 \times 3^2$, we have GCF $= 2 \times 3$, that is, 6.

Example 1 Find the greatest common factor of 60 and 5280.

Solution

$$
\begin{array}{r}
2\overline{)60} \\
2\overline{)30} \\
3\overline{)15} \\
5
\end{array}
\qquad 60 = 2^2 \times 3 \times 5
$$

We have $5280 = 2^5 \times 3 \times 5 \times 11$ as in **Example 2** of Section 6-3. Then the highest power of 2 that is a common factor of 60 and 5280 is 2^2; 3 is a common factor; 5 is a common factor; 11 is not a common factor. The greatest common factor of 60 and 5280 is $2^2 \times 3 \times 5$, that is, 60.

Example 2   Find the GCF of 3850 and 5280.

Solution
$$3850 = 2 \times 5^2 \times 7 \times 11$$
$$5280 = 2^5 \times 3 \times 5 \times 11$$

The greatest common factor of 3850 and 5280 is $2 \times 5 \times 11$, that is, 110.

Example 3   Find the GCF of 12, 36, and 60.

Solution   First write the prime factorization of each number.
$$12 = 2^2 \times 3$$
$$36 = 2^2 \times 3^2$$
$$60 = 2^2 \times 3 \times 5$$

The GCF is $2^2 \times 3$, that is, 12.

The greatest common factor can be used to reduce (simplify) a fraction.

For any counting numbers a, b, and c

$$\frac{ac}{bc} = \frac{a}{b} \times \frac{c}{c} = \frac{a}{b} \times 1 = \frac{a}{b}$$

If c is the greatest common factor of ac and bc, then the fraction a/b has been reduced to *lowest terms*. For example,

$$\frac{60}{4880} = \frac{2^2 \times 3 \times 5}{2^4 \times 5 \times 61}$$

The greatest common factor of 60 and 4880 is $2^2 \times 5$. Thus

$$\frac{60}{4880} = \frac{(2^2 \times 5) \times 3}{(2^2 \times 5) \times (2^2 \times 61)} = \frac{3}{2^2 \times 61} = \frac{3}{244}$$

Since 3 is the only prime factor of the numerator and 3 is not a factor of the denominator, the fraction 3/244 is in lowest terms. That is, the numerator and denominator do not have any common prime factors and are said to be **relatively prime**.

Example 4   Reduce the fraction 60/168 to lowest terms.

Solution
$$60 = 2^2 \times 3 \times 5$$
$$168 = 2^3 \times 3 \times 7$$

6 ELEMENTS OF NUMBER THEORY

The greatest common factor is $2^2 \times 3$.

$$\frac{60}{168} = \frac{(2^2 \times 3) \times 5}{(2^2 \times 3) \times (2 \times 7)} = \frac{5}{14}$$

EXERCISES *Find the set of common factors for the given pair of numbers.*

1. 8 and 10
2. 12 and 30
3. 16 and 40
4. 9 and 30
5. 5 and 7
6. 17 and 24
7. 10 and 30
8. 12 and 60
9. 25 and 75
10. 93 and 155
11. 202 and 606
12. 103 and 600

Write the prime factorizations and find the GCF of the given numbers.

13. 42 and 60
14. 68 and 96
15. 123 and 287
16. 96 and 1425
17. 123 and 615
18. 285 and 1425
19. 68 and 112
20. 112 and 480
21. 600 and 800
22. 1850 and 7400
23. 2450 and 3500
24. 2025 and 5400
25. 12, 18, and 21
26. 18, 24, and 45
27. 15, 25, and 40
28. 12, 30, and 48
29. 10, 20, and 35
30. 24, 40, and 64

Use the GCF to reduce the given fraction.

31. $\dfrac{8}{20}$
32. $\dfrac{12}{18}$
33. $\dfrac{18}{45}$

34. $\dfrac{30}{45}$
35. $\dfrac{60}{72}$
36. $\dfrac{54}{90}$

37. $\dfrac{68}{112}$
38. $\dfrac{112}{480}$
39. $\dfrac{128}{320}$

40. $\dfrac{378}{405}$
41. $\dfrac{2450}{3500}$
42. $\dfrac{2025}{5400}$

Classify each statement as true or false. If false, give a specific counterexample to justify your answer.

43. If two numbers are relatively prime, then each of the numbers must be prime.
44. If two numbers are relatively prime, they must both be odd numbers.
45. If two numbers are relatively prime, then one number must be even and the other number must be odd.
46. Every two counting numbers have at least one common factor.
47. The greatest common factor of any two prime numbers is 1.

In 1866 Nicolo Paganini, a 16-year-old Italian boy, discovered that 1184 and 1210 are amicable numbers. More than 1000 pairs of amicable numbers have been identified.

48. The factors of a given number N that are less than N are often called **proper divisors** of the given number. Two numbers are called **amicable**, or *friendly*, **numbers** if each is the sum of the proper divisors of the other. Show that 220 and 284 are amicable numbers.

Explorations

Extend the first three columns of this table for counting numbers 1 through 17.

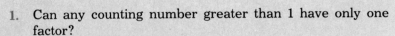

Counting number	Factors	Number of Factors	Sum of Factors
1	1	1	1
2	1, 2	2	3
.	.	.	.
.	.	.	.
.	.	.	.

1. Can any counting number greater than 1 have only one factor?

2. Can the number of its factors be used to identify any given number as a prime number?

3. Can the number of its factors be used to identify any given number as a composite number?

4. Give a rule for determining whether or not a counting number is a prime number if the number of its factors is known.

5. Find a way to distinguish in terms of the number of factors the counting numbers that are squares of counting numbers.

Extend the fourth column of the previous table as needed for Explorations 6 through 8.

6. A number is a **deficient number** if the sum of its factors is less than twice the number. List the first five deficient numbers.

7. A number is a **perfect number** if the sum of its factors is equal to twice the number. Find at least one perfect number.

In the sixth century B.C. the Pythagoreans believed that the counting numbers provided the basis for all qualities of man and matter. Their motto was "all is number." Beliefs in the creation of the world in 6 days and the circling of Earth by the moon in 28 days led to special recognition of these numbers. Much later Alcuin (735–804) observed that the second creation of the human race from the eight people on Noah's ark was imperfect since the number 8 is deficient.

6 ELEMENTS OF NUMBER THEORY

8. A number is an **abundant number** if the sum of its factors is more than twice the number. Find at least one abundant number.

9. Assume that you have a list P of the prime numbers that are less than 100. Describe in words, by a flowchart, or by a computer program a procedure for finding the greatest common factor of any two given counting numbers N and D where $N \leq 10{,}000$ and $D \leq 10{,}000$.

10. As in Exploration 9 describe a procedure for reducing any fraction N/D to lowest terms where N and D are counting numbers such that $N \leq 10{,}000$ and $D \leq 10{,}000$.

6-5 Least Common Multiple

Let us now turn our attention to the concept of a multiple of a number. Consider, for example, the multiples of 5 and 6 as shown in this diagram.

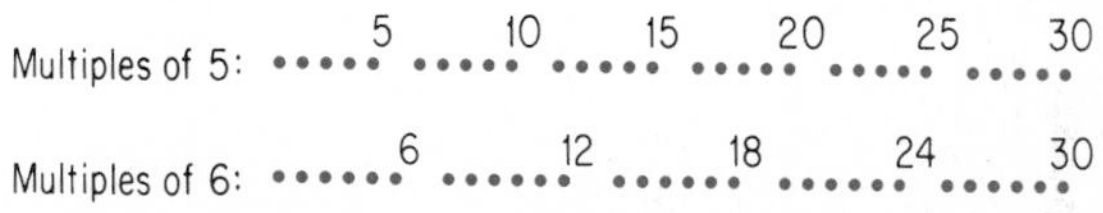

The first place where the multiples of 5 and the multiples of 6 coincide is 30, the *least common multiple* of the two numbers.

As another example, consider the set of multiples of 12 and the set of multiples of 18.

12: {12, 24, ㊱, 48, 60, (72), 84, 96, (108), 120, . . .}
18: {18, ㊱, 54, (72), 90, (108), 126, . . .}

The numbers that are circled are those that are multiples of both 12 and 18 and are called the *common multiples* of the two numbers.

Common multiples of 12 and 18: {36, 72, 108, . . .}

As indicated, this is an infinite set; there is no greatest common multiple. The smallest member of this set, 36, is the least common multiple of the two numbers. In general, the **least common multiple** (LCM) of two or more counting numbers is the smallest number that is a multiple of each of the given numbers.

Example 1 List the first three common multiples for 6 and 9.

Solution First list the set of multiples for each number.

6: {6, 12, ⑱, 24, 30, ㊱, 42, 48, (54), 60, . . .}
9: {9, ⑱, 27, ㊱, 45, (54), 63, 72, . . .}

Now select the multiples that appear in both sets. The first three common multiples are 18, 36, 54 as shown. The LCM of 6 and 9 is 18.

We may use the prime factorization of two numbers to find their least common multiple. First represent each number by its prime factorization. Then consider the prime factors that are factors of either of the given numbers, and take the product of these prime numbers with each raised to the highest power that occurs in *either* of the prime factorizations. In the case of $12 = 2^2 \times 3$ and $18 = 2 \times 3^2$, we have the LCM $= 2^2 \times 3^2$, that is, 36.

Example 2   Find the LCM of 3850 and 5280.

Solution   Write the prime factorization of each number.

$$3850 = 2 \times 5^2 \times 7 \times 11$$
$$5280 = 2^5 \times 3 \times 5 \times 11$$

The least common multiple of 3850 and 5280 is

$$2^5 \times 3 \times 5^2 \times 7 \times 11, \quad \text{that is} \quad 184{,}800.$$

Example 3   Find the LCM of 12, 18, and 20.

Solution   Write the prime factorization of each number.

$$12 = 2^2 \times 3$$
$$18 = 2 \times 3^2$$
$$20 = 2^2 \times 5$$

The LCM is $2^2 \times 3^2 \times 5$, that is, 180.

We use the least common multiple of the denominators of any two fractions first to express the fractions as *like fractions* and then to add or subtract the fractions. Consider, for instance, 7/12 and 5/18. The least common multiple of 12 and 18 is 36.

$$\frac{7}{12} = \frac{7}{12} \times \frac{3}{3} = \frac{21}{36}$$

$$\frac{5}{18} = \frac{5}{18} \times \frac{2}{2} = \frac{10}{36}$$

$$\frac{7}{12} + \frac{5}{18} = \frac{21}{36} + \frac{10}{36} = \frac{31}{36}$$

$$\frac{7}{12} - \frac{5}{18} = \frac{21}{36} - \frac{10}{36} = \frac{11}{36}$$

The LCD is the LCM of the denominators of two or more fractions.

In general, the least common multiple of the denominators of two or more fractions is the **least common denominator** (**LCD**) of the fractions. For any two like fractions a/c and b/c the least common denominator is c.

$$\frac{a}{c} + \frac{b}{c} = \frac{a + b}{c} \qquad \frac{a}{c} - \frac{b}{c} = \frac{a - b}{c}$$

These definitions may be used for any two rational numbers since any rational number may be expressed as a fraction. In adding or subtracting fractions it is customary to use the least common denominator. However, any common multiple of the denominators may be used.

For any two fractions a/b and c/d

$$\frac{a}{b} = \frac{a}{b} \times \frac{d}{d} = \frac{ad}{bd} \qquad \frac{c}{d} = \frac{b}{b} \times \frac{c}{d} = \frac{bc}{bd}$$

$$\frac{a}{b} + \frac{c}{d} = \frac{ad}{bd} + \frac{bc}{bd} = \frac{ad + bc}{bd}$$

$$\frac{a}{b} - \frac{c}{d} = \frac{ad}{bd} - \frac{bc}{bd} = \frac{ad - bc}{bd}$$

Example 4   Subtract $\dfrac{3}{4} - \dfrac{2}{5}$.

Solution   The least common multiple of 4 and 5 is 20 which is the least common denominator of the given fractions. To subtract these fractions we first rewrite them using this common denominator.

$$\frac{3}{4} - \frac{2}{5} = \frac{3}{4} \times \frac{5}{5} - \frac{2}{5} \times \frac{4}{4}$$

$$= \frac{3 \times 5}{4 \times 5} - \frac{2 \times 4}{5 \times 4}$$

$$= \frac{15}{20} - \frac{8}{20} = \frac{7}{20}$$

Example 5   Simplify $\dfrac{37}{5280} - \dfrac{19}{3850}$.

Solution   As in Example 2,

Recall that the instruction "simplify" is used to mean "perform the indicated operations and express the answer in simplest form." In the case of fractions, "express in simplest form" means "reduce to lowest terms."

$$5280 = 2^5 \times 3 \times 5 \times 11$$

$$3850 = 2 \times 5^2 \times 7 \times 11$$

The least common multiple, $2^5 \times 3 \times 5^2 \times 7 \times 11$, of 5280 and 3850 may be thought of as $5280 \times (5 \times 7)$ or as $3850 \times (2^4 \times 3)$.

$$\frac{37}{5280} - \frac{19}{3850} = \frac{37 \times (5 \times 7)}{(2^5 \times 3 \times 5 \times 11) \times (5 \times 7)} - \frac{19 \times (2^4 \times 3)}{(2 \times 5^2 \times 7 \times 11) \times (2^4 \times 3)}$$

$$= \frac{37 \times 5 \times 7}{2^5 \times 3 \times 5^2 \times 7 \times 11} - \frac{19 \times 2^4 \times 3}{2^5 \times 3 \times 5^2 \times 7 \times 11}$$

$$= \frac{1295 - 912}{2^5 \times 3 \times 5^2 \times 7 \times 11} = \frac{383}{184{,}800}$$

Example 6 Simplify $\left(\dfrac{2}{3} - \dfrac{1}{2}\right) \div \left(\dfrac{1}{4} + \dfrac{1}{6}\right).$

Solution The least common multiple of 3 and 2 is 6; the least common multiple of 4 and 6 is 12.

Expressions involving fractions are simplified using the same sequence of operations as for counting numbers. The first step is to simplify expressions in parentheses.

$$\left(\frac{2}{3} - \frac{1}{2}\right) \div \left(\frac{1}{4} + \frac{1}{6}\right) = \left(\frac{2 \times 2}{3 \times 2} - \frac{1 \times 3}{2 \times 3}\right) \div \left(\frac{1 \times 3}{4 \times 3} + \frac{1 \times 2}{6 \times 2}\right)$$

$$= \left(\frac{4}{6} - \frac{3}{6}\right) \div \left(\frac{3}{12} + \frac{2}{12}\right) = \frac{1}{6} \div \frac{5}{12}$$

$$= \frac{1}{6} \times \frac{12}{5} = \frac{1 \times 12}{6 \times 5} = \frac{6 \times 2}{6 \times 5} = \frac{2}{5}$$

Example 7 Express as a single quotient $\dfrac{a}{b} \div \left(\dfrac{c}{d} + e\right).$

Solution $\dfrac{a}{b} \div \left(\dfrac{c}{d} + e\right) = \dfrac{a}{b} \div \left(\dfrac{c}{d} + \dfrac{e}{1}\right) = \dfrac{a}{b} \div \dfrac{c + ed}{d}$

Expressions involving letters are simplified using the same procedures as for numbers.

$$= \frac{a}{b} \times \frac{d}{c + ed} = \frac{ad}{b(c + de)}$$

EXERCISES *List the first three common multiples of each pair of numbers.*

1. 3 and 4	2. 4 and 5	3. 3 and 5
4. 5 and 7	5. 3 and 9	6. 4 and 9
7. 10 and 30	8. 12 and 20	9. 14 and 20
10. 2 and 17	11. 32 and 48	12. 48 and 60

Write the prime factorization and find the LCM of the given numbers.

13. 14 and 40	14. 68 and 96	15. 123 and 287
16. 96 and 1425	17. 123 and 615	18. 285 and 1425
19. 68 and 112	20. 112 and 480	21. 600 and 800

6 ELEMENTS OF NUMBER THEORY

22. 1850 and 7400	23. 2450 and 3500	24. 2025 and 5400
25. 12, 18, and 21	26. 18, 24, and 45	27. 15, 25, and 40
28. 12, 30, and 48	29. 10, 20, and 35	30. 24, 40, and 64

Classify each statement as true or false. If false, give a specific counterexample to justify your answer.

31. The least common multiple of two prime numbers is their product.

32. If a number m is a multiple of another number n, then their least common multiple is m.

33. There is no greatest common multiple for two counting numbers.

34. The least common multiple of a prime number and a composite number is their product.

Simplify.

35. $\dfrac{5}{8} + \dfrac{7}{12}$ 36. $\dfrac{5}{12} + \dfrac{9}{16}$ 37. $\dfrac{9}{10} + \dfrac{1}{15}$

38. $\dfrac{11}{12} + \dfrac{7}{18}$ 39. $\dfrac{11}{12} - \dfrac{7}{15}$ 40. $\dfrac{7}{10} - \dfrac{5}{18}$

41. $\dfrac{11}{14} - \dfrac{3}{40}$ 42. $\dfrac{7}{68} + \dfrac{13}{96}$ 43. $\dfrac{9}{123} - \dfrac{7}{615}$

44. $\dfrac{5}{68} + \dfrac{7}{112}$ 45. $\dfrac{17}{800} - \dfrac{13}{600}$ 46. $\dfrac{7}{285} + \dfrac{2}{1425}$

47. $\left(-\dfrac{2}{3}\right) + \left(-\dfrac{1}{5}\right)$ 48. $\left(-\dfrac{1}{4}\right) + \left(+\dfrac{1}{3}\right)$

49. $\dfrac{2}{3} + \left(\dfrac{3}{4} + \dfrac{7}{8}\right)$ 50. $\left(\dfrac{2}{3} + \dfrac{3}{4}\right) + \dfrac{7}{8}$

51. $\dfrac{7}{8} - \left(\dfrac{3}{4} - \dfrac{1}{2}\right)$ 52. $\left(\dfrac{7}{8} - \dfrac{3}{4}\right) - \dfrac{1}{2}$

53. $\left(\dfrac{3}{4} + \dfrac{1}{2}\right) \div \left(\dfrac{7}{8} - \dfrac{1}{4}\right)$ 54. $\left(\dfrac{11}{12} - \dfrac{2}{3}\right) \div \left(\dfrac{5}{6} + \dfrac{1}{2}\right)$

Simplify by multiplying numerator and denominator by the least common denominator of each fraction involved.

55. $\dfrac{\frac{2}{3}}{\frac{5}{6}}$ 56. $\dfrac{\frac{1}{2}}{\frac{3}{8}}$ 57. $\dfrac{\frac{4}{5}}{\frac{2}{3}}$ 58. $\dfrac{\frac{7}{8}}{\frac{7}{8}}$ 59. $\dfrac{\frac{4}{7}}{\frac{5}{6}}$ 60. $\dfrac{\left(-\frac{3}{5}\right)}{\left(-\frac{7}{10}\right)}$

Express as a single quotient.

*61. $\dfrac{a}{b} + \left(\dfrac{c}{d} + \dfrac{e}{f}\right)$

*62. $\left(\dfrac{a}{b} + \dfrac{c}{d}\right) + \dfrac{e}{f}$

*63. $\dfrac{a}{b}\left(\dfrac{c}{d} + \dfrac{e}{f}\right)$

*64. $\left(\dfrac{a}{b} \times \dfrac{c}{d}\right) + \left(\dfrac{a}{b} \times \dfrac{e}{f}\right)$

*65. $\dfrac{\dfrac{a}{b} + \dfrac{c}{d}}{\dfrac{e}{f}}$

*66. $\dfrac{\dfrac{a}{b}}{\dfrac{c}{d} + \dfrac{e}{f}}$

Explorations

Solve each problem. Identify a possible use of common factors, greatest common factors, common multiples, or least common multiples in solving problems of each type.

1. An interior decorator would like a wallpaper pattern that would fit exactly in a room with walls 8 feet high and also fit exactly under window sills that are 30 inches from the floor. What is the height of the largest pattern that can be considered?

2. Describe the sizes of square tiles that could be used for the floor of a room 10 feet by 15 feet without cutting any tiles.

3. Describe some of the shapes of rooms whose floors can be tiled completely using tiles 9 inches square without cutting any tiles.

4. A manufacturer ships one product in packages 4 by 6 by 6 inches and another in packages 2 by 8 by 9 inches. Cartons are to be made that can be used without extra space for one dozen of either product. Find the dimensions of at least three shapes of cartons that are at most 2 feet on their longest side.

5. Venn diagrams with the prime factors, including repeated factors, as elements may be used to represent the procedures for finding the greatest common factor and the least common multiple of any two or more given counting numbers. Consider for example

$12 = 2 \times 2 \times 3$ $18 = 2 \times 3 \times 3$

GCF = 2 × 3
LCM = 2 × 2 × 3 × 3

6 ELEMENTS OF NUMBER THEORY

Explore this use of Venn diagrams for finding the greatest common factor and the least common multiple of the given numbers.

(a) 8 and 18   (b) 27 and 36   (c) 12 and 35

(d) 9, 12, and 15   (e) 18, 24, and 60   (f) 15, 24, and 50

6. Assume that you have a list P of the prime numbers less than 100. Describe in words, by a flowchart, or by a computer program a procedure for finding the least common multiple of any two given counting numbers N and D where $N \le 10,000$ and $D \le 10,000$.

7. As in Exploration 6 describe a procedure for adding any two given fractions A/B and C/D where $B \le 10,000$ and $D \le 10,000$.

6-6 Modular Arithmetic

Integers are often classified into subsets according to their remainders when divided by a specified number. For example, consider numbers of hours after noon today. The hands on a 12-hour clock will be in the same positions 9, 21, 33, 45, . . . hours after noon and 3, 15, 27, . . . hours before noon. Each of the numbers

$$. . . , \quad -27, \quad -15, \quad -3, \quad 9, \quad 21, \quad 33, \quad 45, \quad . . .$$

can be expressed in the form $9 + 12k$ for some integer k, that is, each has remainder 9 when divided by 12. Note that any two of these numbers differ by a multiple of 12.

The arithmetic on a 12-hour clock is a *modular arithmetic* with *modulus* 12. The hands on a 12-hour clock are in the same position at two different numbers of hours after noon if and only if the numbers differ by a multiple of 12, that is, the numbers are *congruent modulo 12*. We may write for example

$$45 \equiv 9 \ (\text{mod } 12) \qquad \text{read "45 is congruent to 9 modulo 12."}$$

We may use the arithmetic on a 12-hour clock as a model for arithmetic modulo 12. Suppose that Jack works an 8-hour shift starting at 11 P.M. We know that his shift is over at 7 A.M., but note the arithmetic.

$$11 + 8 = 19 \equiv 7 \ (\text{mod } 12)$$

$$11 + 8 = 19 = 12 + 7 = 7 \qquad \text{(on a 12-hour clock)}$$

Any problem in arithmetic modulo 12 may be completed as a problem on a 12-hour clock with 0 used for 12.

Also in arithmetic modulo 12

$$7 \times 9 = 63 \equiv 3 \ (\text{mod } 12)$$

$$7 \times 9 = (12 \times 5) + 3 = 3 \qquad \text{(on a 12-hour clock)}$$

Example 1 Solve in arithmetic modulo 12.
(a) $8 + 7$ (b) $3 - 7$ (c) 3×10

Solution

In arithmetic modulo 12 every integer is congruent to one of the numbers 0, 1, 2, 3, 4, 5, 6, 7, 8, 9, 10, 11. To solve a statement in arithmetic modulo 12 is to find the number(s) from this set that satisfy the given statement.

(a) $8 + 7 = 15 \equiv 3 \pmod{12}$
   Alternatively, $8 + 7 = 15 = 12 + 3 = 3$ (on a 12-hour clock).
(b) $3 - 7 = -4 \equiv 8 \pmod{12}$
   Alternatively, $3 - 7 = -4 = -4 + 12 = 8$ (on a 12-hour clock).
(c) $3 \times 10 = 30 \equiv 6 \pmod{12}$
   Alternatively, $3 \times 10 = 30 = (2 \times 12) + 6 = 6$ (on a 12-hour clock).

On a 12-hour clock division problems such as $10 \div 2$ are solved as in ordinary arithmetic; $10 \div 2 = 5$ on a 12-hour clock. Division problems such as $2 \div 10$ are considered in terms of the related multiplication problem

$$2 \div 10 = n \quad \text{if and only if} \quad 2 = 10 \times n \text{ on a 12-hour clock}$$

To solve such problems it is helpful to consult a multiplication table modulo 12 (see Exercise 43). In such a table the multiples of 10 would include 10×5 as 2 since

$$10 \times 5 = 50 = (4 \times 12) + 2 \equiv 2 \pmod{12}$$

Therefore, $2 \div 10 = 5$ on a 12-hour clock as well as modulo 12. We use arithmetic modulo 5 to illustrate this procedure and as another example of modular arithmetic.

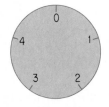

Any problem in arithmetic modulo 5 may be completed as a problem on a 5-hour clock.

There are five elements 0, 1, 2, 3, and 4 in arithmetic modulo 5. Verify each of the entries in these addition and multiplication tables modulo 5.

+	0	1	2	3	4
0	0	1	2	3	4
1	1	2	3	4	0
2	2	3	4	0	1
3	3	4	0	1	2
4	4	0	1	2	3

×	0	1	2	3	4
0	0	0	0	0	0
1	0	1	2	3	4
2	0	2	4	1	3
3	0	3	1	4	2
4	0	4	3	2	1

Example 2 Solve in arithmetic modulo 5. (a) $2 - 3$ (b) $2 \div 4$

Solution   **(a)** $2 - 3 \equiv x$ (mod 5) where $2 \equiv 3 + x$. We use the addition table and in the row for 3 find 2 in the column headed by 4. Therefore we have $3 + 4 \equiv 2$ (mod 5) and $2 - 3 \equiv 4$ (mod 5).

(b) $2 \div 4 \equiv x$ (mod 5) where $2 \equiv 4 \times x$ (mod 5). We use the multiplication table and in the row for 4 find 2 in the column headed by 3. Therefore we have $4 \times 3 \equiv 2$ (mod 5) and $2 \div 4 \equiv 3$ (mod 5).

Example 3   Solve in arithmetic modulo 5. **(a)** $x + 2 \equiv 1$ (mod 5)
(b) $2 - 3 \equiv x$ (mod 5)   **(c)** $4 \div 3 \equiv x$ (mod 5)

Solution   **(a)** In the addition table in the row for 2 we find 1 in the column headed by 4. Therefore, $2 + 4 \equiv 1$ (mod 5), $4 + 2 \equiv 1$ (mod 5), and $x \equiv 4$ (mod 5).

(b) $2 - 3 \equiv x$ (mod 5) if and only if $2 \equiv 3 + x$ (mod 5). In the addition table in the row for 3 we find 2 in the column headed by 4. Therefore $3 + 4 \equiv 2$ (mod 5) and $x \equiv 4$ (mod 5).

(c) $4 \div 3 \equiv x$ (mod 5) if and only if $4 \equiv 3 \times x$ (mod 5). In the multiplication table in the row for 3 we find 4 in the column headed by 3. Therefore, $3 \times 3 \equiv 4$ (mod 5) and $x \equiv 3$ (mod 5).

Example 4   Suppose that you leave at 9 A.M. on a business flight around the world that takes 74 hours. What time will it be when you return?

Solution   Since time is measured in 24-hour days, we think in terms of arithmetic modulo 24.

$$74 = 24 \times 3 + 2 \equiv 2 \text{ (mod 24)}$$

The time of your return will be two hours later than the time of your departure, that is, 11 A.M.

If New Year's Day is on Sunday this year, what day of the week is July 4, the 185th day of the year? The days of the week may be considered in arithmetic modulo 7.

$$185 = 7 \times 26 + 3 \qquad 185 \equiv 3 \text{ (mod 7)}$$

If the first day is Sunday, then the third day (and the 185th day) is Tuesday. The fact that Tuesday, July 4, is 26 weeks later is not significant. Only the day of the week was requested; only arithmetic modulo 7 is needed.

In general, two integers are **congruent modulo** m if the integers differ by a multiple of m. The *arithmetic of congruences* is very similar to our usual arithmetic.

The theory of modular congruences was developed by Carl Friedrich Gauss. This theory was included in *Disquisitiones Arithmeticae,* which Gauss completed in 1801 at the age of 24.

If $a \equiv b$ (mod m) and $c \equiv d$ (mod m), then for any integer k

$$a + c \equiv b + d \text{ (mod } m)$$
$$ak \equiv bk \text{ (mod } m)$$
$$ac \equiv bd \text{ (mod } m)$$

We assume these properties and leave their proofs for more advanced courses.

EXERCISES

Solve in arithmetic modulo 12.

1. $9 + 8$	2. $7 + 11$	3. $6 + 10$
4. 4×9	5. 3×8	6. 9×9

Solve in arithmetic modulo 5.

7. $3 + 4$	8. $2 + 3$	9. $4 + 4$	10. 2×3
11. 2×4	12. 4×3	13. 3×3	14. 4×4
15. $2 - 4$	16. $1 - 2$	17. $1 - 4$	18. $1 - 3$
19. $3 - 4$	20. $2 \div 3$	21. $3 \div 2$	22. $3 \div 4$
23. $4 \div 3$	24. $1 \div 3$		

Solve in the indicated modular arithmetic.

25. $3 + x \equiv 1$ (mod 5)	26. $x + 4 \equiv 1$ (mod 5)
27. $4 \times x \equiv 2$ (mod 5)	28. $3 \times x \equiv 2$ (mod 5)
29. $1 \div 3 \equiv x$ (mod 5)	30. $2 \div 3 \equiv x$ (mod 5)
31. $x + 5 \equiv 0$ (mod 7)	32. $x - 3 \equiv 2$ (mod 4)
33. $3x \equiv 1$ (mod 7)	34. $x \times x \equiv 1$ (mod 8)
35. $x \div 4 \equiv 3$ (mod 9)	36. $2 \div x \equiv 3$ (mod 7)
37. $1 - x \equiv 4$ (mod 6)	38. $4 + x \equiv 1$ (mod 7)
39. $x + 5 \equiv 1$ (mod 8)	40. $2 - x \equiv 3$ (mod 6)
41. $2x \equiv 3$ (mod 6)	42. $3 \div x \equiv 3$ (mod 9)

43. Make a multiplication table for arithmetic modulo 12.

44. Use the multiplication table from Exercise 43 and solve in arithmetic modulo 12.
    (a) $11 \div 5$ (b) $1 \div 7$ (c) $4 \div 5$
    (d) $7 \div 11$ (e) $6 \div 7$ (f) $6 \div 11$

45. We describe the property illustrated by the equation $3 \times 4 \equiv 0$ in arithmetic modulo 12, where the product is zero but neither number is zero, by saying that 3 and 4 are **zero divisors**. Are there other zero divisors in arithmetic modulo 12? If so, list them.

46. Consider a system modulo 7 where the days of the week correspond to numbers as follows: Monday—0; Tuesday—1; Wednesday—2; Thursday—3; Friday—4; Saturday—5; Sunday—6. Memorial Day, May 30, is the 150th day of a certain year and falls on a Thursday. In that same year, on what day of the week does July 4, the 185th day of the year, fall? On what day does Christmas, the 359th day of the year, fall?

Explorations

1. Note $10 \equiv 1$ (mod 9) and $10^n \equiv 1$ (mod 9). Show that any integer N expressed in decimal notation is congruent modulo 9 to the sum of its digits. Restate this fact as a rule for divisibility by 9.

2. Use $10^n \equiv 1$ (mod 9) as in Exploration 1 and develop a check modulo 9 for addition, subtraction, multiplication, and division of integers. Look up the method of **casting out nines** and compare your procedure with this method. Note that since, for example, $1984 \equiv 1894$ (mod 9), a check modulo 9 is not a "complete" check.

3. Note that $10 \equiv -1$ (mod 11) and $10^n \equiv (-1)^n$ (mod 11). As in Exploration 1 develop a check modulo 11 for addition, subtraction, and multiplication of integers. Note that since, for example, $1984 \equiv 1489$ (mod 9) and $1984 \equiv 1489$ (mod 11), answers that check both modulo 9 and modulo 11 may still not be correct.

4. One of these equations has the empty set as its solution set on a four-hour clock.

   $t = 3 \qquad 2t = 3 \qquad 3t = 3$

   Make a multiplication table for arithmetic modulo 4, identify the given impossible equation, and give another impossible equation in the arithmetic on a 4-hour clock.

5. Give at least four impossible equations in the arithmetic on a 12-hour clock.

6. Explore the possible types of impossible equations in the arithmetic on a 12-hour-clock.

7. Use congruences as in Exploration 1 and develop rules for the divisibility by 2, 3, 4, 5, and 8 of any counting number N in terms of its decimal digits.

Chapter 6 Review   *Solutions to the following exercises may be found within the text of Chapter 6. Try to complete each exercise without referring to the text.*

Section 6-1 Factors, Multiples, and Divisibility Rules

1. Find the set of counting numbers that are factors of 16.
2. Show that **(a)** 245 is a multiple of 5 **(b)** 257 is not a multiple of 3.
3. List the first ten multiples of 5.
4. Explain why 1984 is divisible by 2 and by 4 but is not divisible by 3.

Section 6-2 Prime Numbers

5. Define a prime number.
6. Classify as prime or composite **(a)** 24 **(b)** 31.
7. List the set of prime numbers less than 70.

Section 6-3 Prime Factorization

8. Express 3850 in terms of its prime factors.
9. Find the prime factorization of 5280.

Section 6-4 Greatest Common Factor

10. List the common factors of 12 and 18.
11. Find the greatest common factor of 60 and 5280.
12. Find the greatest common factor of 12, 36, and 60.
13. Use the concept of a greatest common factor to reduce 60/168 to the lowest terms.

Section 6-5 Least Common Multiple

14. List the first three common multiples for 6 and 9.
15. Find the least common multiple of 3850 and 5280.
16. Find the LCM of 12, 18 and 20.
17. Use the concept of a least common multiple to simplify $\dfrac{37}{5280} - \dfrac{19}{3850}$.

Section 6-6 Modular Arithmetic

18. Solve in arithmetic modulo 12.
    (a) $8 + 7$ **(b)** $3 - 7$ **(c)** 3×10
19. Solve in arithmetic modulo 5.
    (a) $2 - 3$ **(b)** $2 \div 4$ **(c)** $4 \div 3 \equiv x \pmod{5}$
20. If New Year's Day is on Sunday this year, what day of the week is July 4, the 185th day of the year?

6 ELEMENTS OF NUMBER THEORY

Chapter 6 Test

Classify each statement as true or false. If false, give a specific counterexample to justify your answer.

1. Every prime number is odd.
2. The product of any two prime numbers is a composite number.
3. The smallest prime number is 1.
4. The *least* common factor of two counting numbers is 1.
5. The least common multiple of two prime numbers is their product.
6. If $N \equiv 0 \pmod 7$, then N is divisible by 7.

7. Find the factors of   (a) 42   (b) 71.
8. Find the first five multiples of   (a) 4   (b) 11.
9. List the set of prime numbers that are less than 19.
10. List the set of prime numbers between 20 and 30.
11. List the set of composite numbers between 10 and 20.

Find the number represented by the given prime factorization.

12. $2^3 \times 3^2 \times 5^3$

13. $2 \times 5^2 \times 7 \times 11$

Find the prime factorization of the given number.

14. 60

15. 1500

Find the set of common factors for the given pair of numbers.

16. 9 and 15

17. 24 and 40

Find the greatest common factor of the given numbers.

18. 120 and 140

19. 24, 30, and 42

20. List the first three common multiples for 5 and 9.

Find the least common multiple of the given numbers.

21. 90 and 1500

22. 12, 20, and 24

23. Solve either as on a 12-hour clock or in arithmetic modulo 12.
    (a) $5 - 7$ (b) 5×7

Solve in the indicated modular arithmetic.

24. $5 + x \equiv 3 \pmod 7$

25. $4 \times x \equiv 3 \pmod 5$

An Introduction to Algebra

7

7-1
Sentences and Statements

In arithmetic specific numbers are represented by such numerals as

$$2 \quad 11 \quad 3/4 \quad 0 \quad 1.4 \quad -17 \quad \sqrt{2}$$

In algebra an arbitrary element of a specified set of numbers is represented by a **variable,** usually a letter from our alphabet; for example,

Algebra is an extension of the arithmetic of integers, rational numbers, and real numbers that we considered in Chapter 5. In this chapter we also consider graphs of sets of numbers, and thus explore the interrelationship of algebra and geometry.

any counting number n
any prime numer p
any real number x

The set of possible replacements for (*values of*) a variable is its **replacement set.** If no replacement set is specified for a variable, then the replacement set is assumed to be the set of real numbers.

Statements are sentences that can be identified as true or identified as false. A statement of equality, whether true or false, is an **equation.** A statement using, as in Section 5-3, one of the relations $\neq$, $<$, $\nless$, $\leq$, $>$, $\ngtr$, or $\geq$ is a **statement of inequality.**

In arithmetic the truth values of statements can be identified.

$7 + 3 = 3 + 7$ is a true statement of equality
$7 - 3 = 3 - 7$ is a false statement of equality
$7 + 2 > 7 - 2$ is a true statement of inequality
$5 + 3 \neq 3 + 5$ is a false statement of inequality
$8 - 2 < 2 - 8$ is a false statement of inequality

In algebra a sentence in one variable may be true for all, none, or some but not all of the possible replacements of the variable. A sentence such as

$$x + 2 = 2 + x \quad \text{or} \quad x - 2 < x$$

that is true for all possible values of the variable is a *true statement* (**identity**). A sentence such as

$$x - 2 = x \quad \text{or} \quad x + 2 \neq 2 + x$$

that is false for all possible values of the variable is a *false statement.* A sentence such as

$$x + 5 = 3 \quad \text{or} \quad x^2 = 9$$

that is true for at least one value of the variable and false for at least one value of the variable is an **open sentence.**

For integers x the sentence $x + 5 = 3$ is an open sentence since it is true for $x = -2$ and false for $x \neq -2$. For whole numbers x the sentence $x + 5 = 3$ is a false statement of equality. That is, there is no whole number x for which $x + 5 = 3$. To *solve* an open sentence in one variable is to identify the set of possible replacements for which the sentence is a true statement, that is, to find the *solution set* of the given sentence. The solution set may depend on the replacement set of the variable. In practice the selection of a replacement set is usually determined by the conditions of the problem being solved.

SENTENCE	REPLACEMENT SET	SOLUTION SET
$x - 1 < 3$	Counting numbers	$\{1, 2, 3\}$
$x - 1 < 3$	Whole numbers	$\{0, 1, 2, 3\}$
$x - 1 < 3$	Integers	$\{. . . , -2, -1, 0, 1, 2, 3\}$
$x - 1 < 3$	Real numbers	All real numbers less than 4

Example 1   Solve.
(a) $x + 1 > 4$ for whole numbers x.
(b) $n + 2 = 2 + n$ for real numbers n.
(c) $x + 2 < x$ for integers x.
(d) $x^2 = 2$ for rational numbers x.

Solution   (a) For $x = 3$ we have $3 + 1 = 4$. The solution set S consists of all whole numbers greater than 3.
$$S = \{4, 5, 6, . . .\}$$

(b) The given sentence is a true statement of equality for all possible replacements of the variable, since the statement is an application of the commutative property of addition. The solution set is the set of all real numbers.

(c) Regardless of the integer selected as the replacement for x, the given sentence is *always* false. There are no solutions and the solution set is the empty set.

In Example 1(d) note that if $x^2 = 2$, then $x = \sqrt{2}$ or $x = -\sqrt{2}$. However, $\sqrt{2}$ and $-\sqrt{2}$ are *not* rational numbers. (See Section 5-7.)

(d) There is no rational number x for which the given statement is a true statement. The solution set is the empty set.

An equation such as $x + 3 = 5$ can be thought of as a **set-selector**; it selects from the replacement set just those numbers that make the sentence true when used as replacements for x. The selected set is the solution set of the equation. The solution

Set-builder notation is often convenient to use. However, we will usually omit it and simply write a sentence such as $x + 3 = 5$.

set $\{2\}$ is "the set of all x such that $x + 3 = 5$." We may designate this solution set in *set-builder notation* as
$$\{x \mid x + 3 = 5\}$$

7 AN INTRODUCTION TO ALGEBRA

The replacement set of the variable may be indicated in the set-builder notation as

$$\{x \mid x + 3 = 5, x \text{ an integer}\}$$

We may also write

$$\{x \mid x + 3 = 5\} = \{2\}$$

The solution set of any sentence in one variable may be graphed on a real number line. We draw a graph to represent the set of points that correspond to the solution set of a sentence. We often refer to this graph simply as the *graph of the equation* or inequality.

Example 2 Graph the given sentence.
(a) $x + 3 = 5$ for integers x.
(b) $x + 3 \leq 5$ for whole numbers x.
(c) $x + 2 = 2 + x$ for real numbers x.

Solution (a) The solution set is $\{2\}$. We graph this solution set on a number line by drawing a solid dot at 2. The graph consists of this single point.

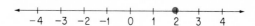

(b) The given sentence is read "$x + 3$ is less than or equal to 5." For whole numbers x, the solution set is $\{0, 1, 2\}$ and is graphed as a set of three points.

(c) The given sentence is true for all replacements of x; it is an identity. The solution set is the set of all real numbers. Thus the graph is the entire number line.

On a real number line the graphs of inequalities are common geometric figures. Indeed, the inequalities may be used to define the geometric figures.

Example 3 Graph the given sentence for real numbers x.
(a) $x + 3 > 5$ (b) $x + 3 \geq 5$ (c) $x + 3 \nless 5$

Solution (a) The solution set consists of all real numbers greater than 2. The graph of the solution set is drawn by first placing a hollow dot at 2 on the number line to indicate that this point is not a member of the solution set. Then a heavily shaded arrow is drawn to show that all real numbers greater than 2 satisfy the given inequality.

(b) The given sentence is read "$x + 3$ is greater than or equal to 5." Therefore the given sentence is true if $x > 2$ and also if $x = 2$, that is, if x is greater than or equal to 2. In symbols the solution may be written as $x \geq 2$. The graph of the solution set is indicated by a solid dot at 2, to show that this point is a member of the solution set, and a shaded arrow as in the figure.

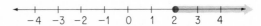

(c) The given sentence is read "$x + 3$ is not less than 5." This is equivalent to saying that $x + 3$ is greater than or equal to 5; $x + 3 \geq 5$. Thus the solution set and its graph are the same as for part (b) of this example.

For any real number b the graph of an inequality of the form $x > b$ (see Example 3(a)) is a **half-line**. The graph of an inequality of the form $x < b$ is also a half-line, the opposite half-line of the graph of $x > b$. Similarly, the graph of an inequality of the form $x \geq b$ (see Example 3 (b)) is a **ray**; also the graph of $x \leq b$ is a ray, the opposite ray of the graph of $x \geq b$. The point with coordinate b is the **endpoint** of each of these rays.

Example 4 Graph the given sentence for real numbers x.
(a) $-1 \leq x \leq 3$ (b) $-1 < x < 3$

Solution (a) The given sentence is read "-1 is less than or equal to x, which is less than or equal to 3." In other words, "-1 is less than or equal to x and x is less than or equal to 3." That is, "x is greater than or equal to -1 and x is less than or equal to 3." Thus the solution set is the set of real numbers -1 *through* 3.

(b) The solution set consists of all the real numbers *between* -1 and 3.

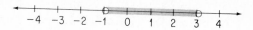

For any real numbers a and b where $a < b$, the graph of the sentence $a \le x \le b$ is a **line segment.** In Example 4(a), the points with coordinates -1 and 3 are **endpoints** of the line segment. The points of a line segment that are not endpoints are **interior points** of the line segment. The graph in Example 4(b) can be obtained from the graph in Example 4(a) by removing the endpoints of the line segment. When we wish to name each of these graphs, we call a line segment with its endpoints a **closed line segment** and a line segment without either endpoint, an **open line segment;** a line segment with one endpoint but not both is neither closed nor open. For example, the following graph of $-1 \le x < 3$ is neither closed nor open; it is sometimes called a **half-open line segment.**

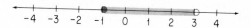

EXERCISES *For integers x identify the given sentence as a true statement, a false statement, or an open sentence.*

1. $x + 3 = 5$
2. $x + 5 = 3$
3. $x + 3 \ne 5$
4. $x + 5 \ne 3$
5. $x + 3 > x$
6. $x - 3 < x$
7. $x - 3 > x$
8. $x + 3 < x$
9. $x < x + 1$
10. $x + 2 < x + 4$
11. $x + 2 < x + 1$
12. $x > x - 1$

Solve the given sentence for whole numbers x.

13. $x + 3 = 8$
14. $x - 3 = 2$
15. $x + 2 < 6$
16. $x - 2 < 6$
17. $x - 2 < 5$
18. $x - 2 > 5$

Solve the given sentence for integers x.

19. $x + 1 < 5$
20. $x - 1 < 5$
21. $x - 2 < 1$
22. $x + 2 < 1$
23. $x + 5 > 2$
24. $x - 2 > 3$

Describe in words the solution set of the given sentence for real numbers x.

25. $x + 1 > 3$
26. $x + 1 < 3$
27. $x - 2 < 4$
28. $x - 2 > 4$
29. $x + 2 \ne 7$
30. $x + 3 \ne 3$

Identify the graph of the solution set of the given sentence as a point, a half-line, a ray, a line segment, or a line.

31.	$x + 3 = 7$	32.	$x - 2 > 7$	33.	$x + 2 \geq 5$
34.	$x < x + 3$	35.	$-2 \leq x \leq 5$	36.	$x + 3 \leq 5$
37.	$x + 1 > x$	38.	$x - 2 = 7$	39.	$x - 1 < 5$
40.	$-3 \leq x \leq 0$	41.	$x + 2 \not> 5$	42.	$x + 2 \not\leq x + 1$

Graph the given sentence for real numbers x.

43.	$x + 1 > 3$	44.	$x - 2 \leq 5$	45.	$-3 \leq x \leq 4$
46.	$-2 < x < 2$	47.	$x + 3 \geq 5$	48.	$x - 2 < 4$
49.	$x + 2 > 4$	50.	$x + 2 \not< 4$	51.	$x + 3 > x$
52.	$x + 3 = x$	53.	$x + 2 \not> 5$	54.	$x + 3 \neq 5$
*55.	$3x - 2 \geq -8$	*56.	$2x + 5 \leq 1$	*57.	$x^2 = 36$
*58.	$x^2 + 3 = 28$	*59.	$x^2 \leq 9$	*60.	$x^2 > 16$

Explorations

The use of two variables in a mathematical expression can be introduced on an intuitive basis through the use of flowcharts. Consider, for example, these flowcharts.

About 1650 B.C., an Egyptian scribe named Ahmes copied an earlier manuscript which he described as "the entrance into knowledge of all existing things and all obscure secrets." The copy by Ahmes is often called the Rhind Mathematical Papyrus. It contains 85 problems. Here is one of those problems.

A number and its one-fourth added together become 15. What is the number?

Variables were not known 3000 years ago. Algebra did not exist. Problems were solved by arithmetic with a procedure that we now call the *method of false position*. This method can be used for some problems but is not useful for many others. For the preceding problem, we note that we need to take one-fourth of the number, so we try 4.

$$4 + \tfrac{1}{4}(4) = 4 + 1 = 5$$

If we try 4, we get 5. But we need 15, that is, 3×5. Therefore, the answer is 3×4, that is, 12.

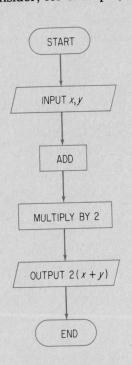

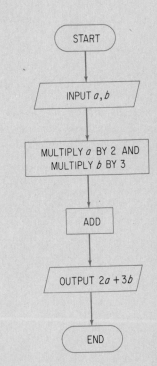

7 AN INTRODUCTION TO ALGEBRA

Copy each table of values and complete using the flowcharts.

1.

x	3	5	-2	3	-1
y	4	1	7	-2	-4
$2(x+y)$					

2.

a	2	3	-5	-9	-7
b	3	2	4	-1	-5
$2a+3b$					

Draw a flowchart for each of these expressions.

3. $3x + 5y$

4. $3(x + y)$

5. $\frac{1}{2}x + 3y$

6. $3(2x + 3y)$

7-2 Compound Sentences

Many compound sentences occur in algebra. For example, consider the following compound sentence.

$$x + 1 > 2 \quad \text{and} \quad x - 2 < 1$$

Since no replacement set for x is specified, we assume that the replacement set is the set of real numbers. The sentence $x + 1 > 2$ is true for all x greater than 1; the sentence $x - 2 < 1$ is true for all x less than 3. Recall that a **compound sentence** of the form $p \wedge q$ (p and q) is true only when both parts are true. Thus the given compound sentence is true for the set of elements in the *intersection* of the two sets. Graphically, we can show this as follows.

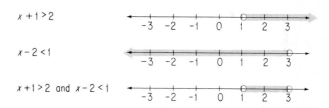

The graph of the compound sentence consists of an *open interval* (open line segment) and can be described as

$$1 < x < 3$$

The solution set of this compound sentence can be written in set-builder notation as $\{x \mid 1 < x < 3\}$ or as

$$\{x \mid x > 1\} \cap \{x \mid x < 3\}$$

This is read as "the intersection of the set of all x such that x is greater than 1 and the set of all x such that x is less than 3."

Example 1   Solve for integers x.

$$x \geq -2 \quad \text{and} \quad x + 1 \leq 4$$

Solution   Here we want the set of integers that are greater than or equal to -2 but also are less than or equal to 3. (If $x + 1 \leq 4$, then $x \leq 3$.) The solution set is $\{-2, -1, 0, 1, 2, 3\}$.

Example 2   Solve for real numbers x.

$$x + 3 < 5 \quad \text{and} \quad 3 < 1$$

Solution   Note that the second part of this sentence $(3 < 1)$ is false. If part of a sentence of the form $p \wedge q$ is false, then the entire sentence is false. Thus the solution set is the empty set.

Next we consider a compound sentence involving the connective *or*.

$$x + 1 < 2 \quad \text{or} \quad x - 2 > 1$$

Note the distinction between the connectives *and* and *or* in a compound sentence. A sentence of the form *p and q* implies the *intersection* of two sets; a sentence of the form *p or q* implies the *union* of two sets. (See Section 4-3).

The sentence $x + 1 < 2$ is true for all x less than 1; the sentence $x - 2 > 1$ is true for all x greater than 3. Recall that a sentence of the form $p \vee q$ *(p or q)* is true unless both parts are false. Thus the given compound sentence is true for the set of elements in the *union* of the two sets. Graphically, we have the following.

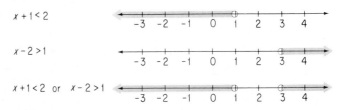

The graph of the compound sentence consists of the union of two half-lines. The solution set can be written in set-builder notation as

$$\{x \mid x < 1\} \cup \{x \mid x > 3\}$$

This is read "the union of the set of all x such that x is less than 1 with the set of all x such that x is greater than 3."

Example 3   Solve for real numbers x.
(a) $5 > 1 \quad$ or $\quad x + 2 < 5$
(b) $x + 2 \neq 2 + x \quad$ or $\quad x + 2 < x$

7 AN INTRODUCTION TO ALGEBRA

Solution   (a) The compound sentence $p \lor q$ is true if at least one of the parts is true. Since the first part of the given sentence is *always* true, the whole sentence is always true. That is, the compound sentence is true for *all* real numbers x; the solution set is the entire set of real numbers.

(b) Both parts of the sentence are always false; the solution set is the empty set.

Example 4   Graph   (a) $x \leq -1$ or $x \geq 2$ (b) $x = -2$ or $x \geq 1$.

Solution   (a) The graph is the union of two rays.

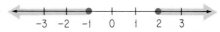

(b) The graph is the union of a point and a ray.

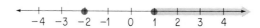

EXERCISES   *Solve for integers x.*

1. $x \geq 1$ and $x + 1 \leq 6$ 2. $x \leq -2$ and $x > -5$
3. $x > 1$ and $x < 5$ 4. $x \geq 2$ or $x \leq 1$
5. $x < 0$ or $x > 0$ 6. $x < 1$ or $x > -1$

Graph for real numbers x.

7. $x \geq 0$ or $x < 0$ 8. $x < 5$ or $x + 1 > 5$
9. $x = -1$ or $x \geq 0$ 10. $x = 0$ or $x \geq 1$
11. $x + 1 < 5$ and $x > 5$ 12. $x > 2$ and $x + 2 < 0$
13. $x + 2 = 5$ and $x < 3$ 14. $x > 3$ and $x + 2 = 5$
15. $x \geq 2$ and $x \leq 5$ 16. $x \leq 0$ or $x \geq 3$
17. $x + 2 < 5$ and $x \geq 0$
18. $x - 2 > 5$ or $x < 0$
19. $x + 3 \geq 5$ or $x - 1 < 0$
20. $x + 1 \geq 3$ or $x = -2$
21. $x + 3 \geq 5$ and $x - 1 < 0$
22. $x + 2 \geq 2$ and $x - 1 < 3$
23. $x - 1 > x$ and $x + 2 = 7$
24. $x < x + 1$ and $x - 3 > 5$
25. $x + 2 \neq x$ or $x + 2 \leq 5$
26. $x + 1 = x$ or $x^2 < 0$
27. $x^2 < 0$ or $x^2 + 1 > 0$ 28. $x^2 \geq 4$
*29. $1 \leq x^2 \leq 9$ *30. $4 \leq x^2 \leq 25$

Explorations

The absolute value of a number x (written as $|x|$) may be defined as the distance from the origin of the point with coordinate x on the number line. Thus $|3| = 3$ and $|-3| = 3$ because the point with coordinate 3 and the point with coordinate -3 are both at a distance of 3 units from 0 on the number line. In general, for any real number x

$$|x| = x, \qquad \text{if } x \text{ is positive}$$
$$|x| = 0, \qquad \text{if } x = 0$$
$$|x| = -x, \qquad \text{if } x \text{ is negative}$$

Note that if x is negative, then $-x$ is the opposite of x and therefore is a positive number. Thus if $x = -3$, then $|-3| = -(-3) = 3$. Accordingly, the absolute value of any real number different from zero is a positive number; $|0| = 0$.

Here is a flowchart for finding the absolute value of a number x.

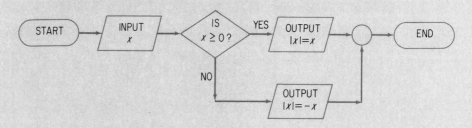

Solve for integers x.

1. $|x| = 2$
2. $|x| \leq 3$
3. $|x| < 5$
4. $|x| = -3$
5. $|x - 1| = 3$
6. $|x + 1| = 5$
7. $|x + 2| = 1$
8. $|x - 2| = 3$

Evaluate.

9. $|-7|$
10. $|-2| + |-3|$
11. $|(-2) + (-3)|$
12. $(|-5|)^2$
13. $|(-5)^2|$
14. $|-5| \times |-3|$

The following sentences may be graphed on a real number line. Identify each graph as two points, a line segment, or the union of two rays.

15. $|x| \leq 7$
16. $|x - 2| = 5$
17. $|x| \geq 1$
18. $|x + 3| \leq 1$

Graph on a real number line.

19. $|x| \leq 2$ 20. $|x| \geq 1$
21. $|x| > 3$ 22. $|x| < 5$
23. $|x - 1| \geq 2$ 24. $|x + 1| \leq 3$
25. $|x| = -2$ 26. $|x| \geq 0$
27. $|x + 2| \geq 3$ 28. $|x - 2| \leq 1$
29. $|x - 3| < 2$ 30. $|x + 1| > 2$
*31. $2 \leq |x| \leq 3$ *32. $3 \leq |x - 1| \leq 5$
*33. $|x| + 2 = |x + 2|$ *34. $|x| = -x$
*35. $|x| + |x - 3| = 3$ *36. $|x^2 - 10| \leq 6$

7-3

Equations and Inequalities of the First Degree

A sentence that involves the variable but does not involve products or quotients of the variable is a **sentence of the first degree**. Properties of numbers, equalities, and inequalities are used in the solution of such sentences. For example, consider the equation

$$x - 3 = 7$$

Note that

$$(x - 3) + 3 = x + (-3 + 3) = x + 0 = x$$

Therefore we may solve the given equation as follows.

$$x - 3 = 7$$
$$(x - 3) + 3 = 7 + 3$$
$$x = 10$$

We have added 3 to both members (sides) of the equation to obtain an **equivalent sentence**, that is, a sentence with the same solution set as the given sentence. In fact, we have used a property of equality. Here are some of the common **properties of statements of equality** involving real numbers a, b, and c.

REFLEXIVE, $=$: $a = a$.

ADDITION, $=$: If $a = b$, then $a + c = b + c$.

MULTIPLICATION, $=$: If $a = b$, then $ac = bc$.

Note that subtraction is defined in terms of addition; division is defined in terms of multiplication.

Subtraction may be performed by adding the opposite of a number, and division may be performed by multiplying by the reciprocal.

$$a - b = a + (-b) \qquad a \div b = a \times \frac{1}{b}$$

Example 1   Solve $x + 3 = 7$ and explain each step.

Solution

STATEMENTS	REASONS
$x + 3 = 7$	Given.
$(x + 3) + (-3) = 7 + (-3)$	Addition, $=$.
$x + [3 + (-3)] = 7 + (-3)$	Associative, $+$.
$x + 0 = 4$	Inverse, $+$; addition.
$x = 4$	Zero, $+$. (Addition property of zero.)

Alternate Solution   Frequently we omit some of the steps and think:

$x + 3 = 7$	Given.
$x + 3 - 3 = 7 - 3$	Subtract 3 from both sides.
$x = 4$	Subtraction.

Example 2   Solve $2x - 3 = 7$ and explain each step.

Solution

$2x - 3 = 7$	Given.
$2x = 10$	Add 3 to both sides.
$x = 5$	Divide both sides by 2.

These methods may be used to solve any equation of the **first degree in one variable** x, that is, any sentence that can be expressed in the form $ax + b = 0$, $a \neq 0$. The expression $ax + b$ is called an **expression of the first degree in x.**

$ax + b = 0$	Given.
$ax = -b$	Subtract b from both sides.
$x = -\dfrac{b}{a}$	Divide both sides by a.

Inequalities are closely related to equalities. For all real numbers a and b

$a < b$ and $b > a$ if and only if $a + c = b$ for some positive number c.

This interrelation between inequalities and equalities may be taken as a definition of the inequalities. Then we may establish order relations by considering specific cases as follows. In each case verify that the first inequality listed is true. Then note that the accompanying inequality is also true.

$$2 < 5 \qquad 2 + 3 < 5 + 3$$
$$8 > 3 \qquad 8 + 5 > 3 + 5$$
$$-3 < 5 \qquad -3 + 1 < 5 + 1$$

These properties indicate that if the same number is added to both members of a statement of inequality, the "sense" of the inequality is preserved.

In general, we may list these **addition properties of order** for all real numbers a, b, and c:

ADDITION, $<$: If $a < b$, then $a + c < b + c$.

ADDITION, $>$: If $a > b$, then $a + c > b + c$.

If both members of a true statement of inequality are multiplied by the same positive number, the resulting sentence is also true. If both members of an inequality are multiplied by a negative number, it is necessary to reverse the sense of the inequality to obtain an equivalent sentence. Thus in the following examples each member of the first inequality is multiplied by the same positive number, and the order of the inequality is maintained.

$$3 < 7 \qquad 2 \times 3 < 2 \times 7$$
$$5 > -1 \qquad 3 \times 5 > 3 \times (-1)$$

In the next two examples each member of the first inequality is multiplied by the same negative number, and the order of the inequality is reversed.

$$2 < 8 \qquad -3 \times 2 > -3 \times 8$$
$$3 > -1 \qquad -2 \times 3 < -2 \times (-1)$$

We summarize our discussion by listing the **multiplication properties of order** for real numbers:

MULTIPLICATION, $<$: If $a < b$ and $c > 0$, then $ac < bc$.
If $a < b$ and $c < 0$, then $ac > bc$.

MULTIPLICATION, $>$: If $a > b$ and $c > 0$, then $ac > bc$.
If $a > b$ and $c < 0$, then $ac < bc$.

Several uses of properties of order relations are illustrated in the examples that follow.

Example 3 Solve $\frac{1}{2}x - 2 > 5$ and explain each step.

Solution
$$\frac{1}{2}x - 2 > 5 \qquad \text{Given.}$$

$$\frac{1}{2}x > 7 \qquad \text{Add 2 to both sides.}$$

$$x > 14 \qquad \text{Multiply both sides by 2.}$$

Example 4   Solve $-2x + 3 < 7$ and explain each step.

Solution

$$-2x + 3 < 7 \qquad \text{Given.}$$

$$-2x < 4 \qquad \text{Subtract 3 from both sides.}$$

Note that dividing by -2 is the same as multiplying by $-1/2$.

$$x > -2 \qquad \text{Divide both sides by } -2.$$

EXERCISES   *Explain each step in the following solutions.*

1. (a) $2x + 3 = 7$
   (b) $2x = 4$
   (c) $x = 2$

2. (a) $3x - 4 = 11$
   (b) $3x = 15$
   (c) $x = 5$

3. (a) $2x - 4 < 10$
   (b) $2x < 14$
   (c) $x < 7$

4. (a) $5x + 2 > 22$
   (b) $5x > 20$
   (c) $x < 4$

5. (a) $-3x + 5 > 17$
   (b) $-3x > 12$
   (c) $x < -4$

6. (a) $17 - 2x < 11$
   (b) $-2x < -6$
   (c) $x > 3$

Solve for x.

7. $3x + 2 = 14$
8. $2x - 5 = 9$
9. $4x - 3 = 17$

10. $5x + 7 = 17$
11. $-3x + 1 = 10$
12. $-2x + 3 = 7$

13. $5 - 2x = 13$
14. $1 - 3x = 16$
15. $7 - 3x = 1$

16. $9 - 4x = 5$
17. $3x - 2 < 10$
18. $3x - 1 > 8$

19. $2x + 1 > 7$
20. $5x + 3 < 18$
21. $-2x + 1 < 9$

22. $-3x + 2 < 8$
23. $6 - 2x > 0$
24. $20 - 3x > 8$

25. $\frac{1}{2}x - 3 = 8$
26. $\frac{2}{3}x - 7 = 5$
27. $\frac{3}{4}x + 1 = 10$

28. $\frac{5}{6}x + 18 = 3$
29. $-\frac{1}{2}x + 3 < 7$
30. $-\frac{2}{3}x - 4 < 8$

Determine whether each statement is always true for integers a, b, c, and d, where $a < b$ and $c < d$. If the statement is not always true, give an example for which it is false.

*31. $a + c < a + d$
*32. $a + c < b + d$

*33. $a - c < b - d$
*34. $a + d > b + c$

*35. $ac < bc$
*36. $ad < bc$

Explorations

Note the equations described by these flowcharts.

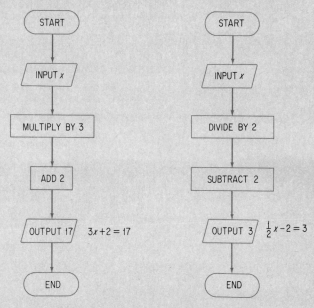

Draw a flowchart to describe each equation.

1. $2x + 3 = 7$

2. $3x - 5 = 10$

3. $\frac{1}{2}x - 1 = 7$

4. $\frac{1}{4}x + 2 = 6$

Reverse flowcharts may be used in the solutions of equations. Start at the end and work backward, using opposite (inverse) operations, as in these examples.

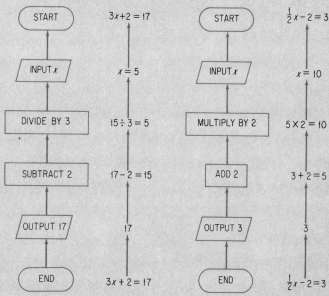

5. Draw reverse flowcharts to solve each of the equations in Explorations 1 through 4.

7-4
Problem Solving

Many problems can be solved using either arithmetic or algebra. Often sentences of equality provide the bases for the solutions. When necessary, the algebraic approach can be used to explain the method used in arithmetic. A systematic approach is helpful in the solution of any problem.

Example 1 The sum of two numbers is 20; their difference is 6. Find the numbers.

Solution Let the smaller number be n. Then $n + 6$ represents the larger number so that their difference is 6. The statement "the sum of the numbers is 20" can be expressed as an equation.

$$n + (n + 6) = 20$$
$$n + n = 14 \qquad \text{Subtract 6 from both sides.}$$
$$2n = 14 \qquad \text{Add; } n + n = 2n.$$
$$n = 7 \qquad \text{Divide both sides by 2.}$$

If the smaller number is n, then the larger number is $n + 6$. If $n = 7$, then $n + 6 = 13$. The numbers are 7 and 13.

Check: $7 + 13 = 20$
$13 - 7 = 6$

Example 2 Find each number using algebra and then using arithmetic.
(a) If a number is doubled and then increased by 5, the result is 17.
(b) If a number is increased by 5 and that sum is doubled, the result is 18.

Solution (a) If the number is x, then

$2x$ is the double of (twice) the number

$2x + 5$ is the double increased by 5

$$2x + 5 = 17$$
$$2x = 12$$
$$x = 6 \qquad \text{The number is 6.}$$

If only arithmetic is used, we have:

Twice the number is increased by 5 and the sum is 17.
Twice the number is 12.
The number is 6.

7 AN INTRODUCTION TO ALGEBRA

(b) If the number is x, then

If only arithmetic is used, we have:

Twice the sum of the number and 5 is 18.
The sum of the number and 5 is 9.
The number is 4.

$x + 5$ is the number increased by 5
$2(x + 5)$ is twice that sum
$2(x + 5) = 18$
$x + 5 = 9$
$x = 4$ The number is 4.

Note that representations for numbers have been used in each of the previous examples.

n	$n + 6$	$2n + 6$
x	$2x$	$2x + 5$
x	$x + 5$	$2(x + 5)$

Such translations from verbal statements to expressions in the language of algebra are an important part of problem solving.

Each of the preceding problems also involved an equation in which two expressions (names) for the same number are equal to each other.

$$n + (n + 6) = 20$$
$$2x + 5 = 17$$
$$2(x + 5) = 18$$

Frequently such equalities are translations of verbal statements in which *is* means *is equal to*.

Problems arise in many different forms. The following four **steps for problem solving** are helpful in solving almost all problems. We illustrate the steps for this problem involving a rectangle.

The perimeter of a rectangle is 32 centimeters. The length of the rectangle is 1 centimeter greater than twice the width. Find the length and the width of the rectangle.

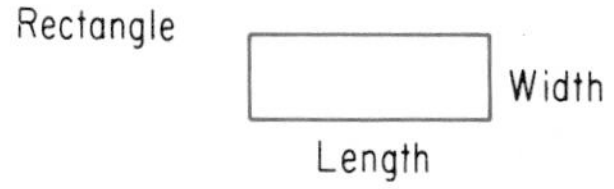

1. *Read the problem several times.* Be sure that you understand the terms that are used. Try to visualize what is given and what must be found. Where appropriate, draw a sketch as in the adjacent figure.

Rectangle

Width

Length

2. *Plan the solution.* Use a variable to represent one of the unknown quantities—usually one of the quantities that you are trying to find. Represent other unknown quantities in terms of the variable.

Let x represent the width.
Then $2x + 1$ represents the length.

3. *State and solve an equation or inequality.* This usually requires that a relation be found among the known and unknown quantities in the problem.

The perimeter is the distance around the rectangle:

$$\text{(width)} + \text{(length)} + \text{(width)} + \text{(length)} = \text{perimeter}$$

$$x + (2x + 1) + x + (2x + 1) = 32$$

$$6x + 2 = 32$$

$$6x = 30$$

$$x = 5 \qquad \text{The width is 5 cm.}$$

$$2x + 1 = 11 \qquad \text{The length is 11 cm.}$$

Steps in problem solving:

READ
PLAN
SOLVE
CHECK

Note that it is often necessary to find the value of more than one unknown quantity.

4. *Return to the original problem to check your answers.* Answers that have been correctly obtained may not be meaningful in the original problem.

The length: $11 = 2(5) + 1$

The perimeter: $5 + 11 + 5 + 11 = 32$

Example 3 Doris' class is renting a bus for a short field trip. If the members of the class pay $1 each, there is a surplus of $2. If the members pay 90¢ each, there is a shortage of $1. How many members are in the class and what is the charge for renting the bus?

Solution Let x be the number of members in the class. Then the charge for renting the bus is $(1)x - 2$ and also $(0.90)x + 1.00$.

$$x - 2 = 0.9x + 1$$

$$0.1x = 3$$

$$x = 30 \qquad \text{There are 30 members in the class.}$$

$$x - 2 = 28 \qquad \text{The charge for renting the bus is \$28.}$$

Check: $30(1) = 28 + 2 \qquad 30(0.90) + 1 = 27 + 1 = 28$

Example 4 Ginny and Bill each drive small cars. Ginny averages 40 miles per gallon with her car; Bill averages 30 miles per gallon with his. They attend the same college and live the same distance from campus. If Bill needs 1 gallon more fuel than Ginny for five round trips to campus, how far does each live from campus?

Solution Let d be the distance in miles that each lives from campus.

$2d$ is the length of one round trip
$10d$ is the length of five round trips

7 AN INTRODUCTION TO ALGEBRA

At 40 miles per gallon, Ginny's car requires 10d/40 gallons of fuel to travel 10d miles; at 30 miles per gallon, Bill's car requires 10d/30 gallons of fuel.

$$\frac{10d}{40} + 1 = \frac{10d}{30}$$

$$\frac{d}{4} + 1 = \frac{d}{3}$$

$3d + 12 = 4d$ Multiply both sides by 12.

$12 = d$ Each lives 12 miles from campus.

Check: 12 miles one way means 120 miles for five round trips. Ginny requires 3 gallons of fuel and Bill requires 4 gallons for 120 miles; $3 + 1 = 4$.

EXERCISES   *For a given number n, represent the given information as an algebraic expression.*

1. Three more than twice the given number.
2. Twice the sum of the given number and three.
3. The given number decreased by two and that difference multiplied by five.
4. The given number multiplied by five and that product decreased by two.

Find each number using algebra and then using only arithmetic.

5. Three times a number is 99.
6. A number increased by 5 is 25.
7. Three times the sum of a number and 5 is 21.
8. Four times the difference of a given number decreased by 2 is 12.
9. Four times the difference of 7 decreased by a given number is 12.
10. Five times the difference of 11 decreased by a given number is 35.
11. If the product of a number and 5 is increased by 2, the sum is 22.
12. If the quotient of a number divided by 7 is increased by 3, the sum is 24.
13. If the quotient of a number divided by 9 is decreased by 1, the difference is 17.
14. If the product of a number and 9 is decreased by 1, the difference is 17.

Solve each problem.

15. The perimeter of a rectangle is 60 centimeters. The length is twice the width. Find the length and the width.

16. The perimeter of a rectangle is 10 meters. The width is 1 meter less than the length. Find the length and the width.

17. The perimeter of a rectangle is three times the length. The width is 5 centimeters. Find the length and the perimeter.

18. The perimeter of a rectangle is five times the width. The length is 6 meters. Find the width and the perimeter.

19. Find a number such that the sum of twice the number and 5 is the same as the sum of the number and 17.

20. Find a number such that 50 more than three times the number is the same as the sum of the number and 80.

21. John's team needed cash for a project. At $5 each there was a surplus of $1; at $4.75 each there was a shortage of $1. How many people were on the team and how much money was needed?

22. The children of José and Dolores shared equally in the cost of a present for their mother. If each gave $2, there was a surplus of $1. If each gave $1.50, there was a shortage of $1. How many children were there and what was the cost of the present?

23. Joe and Ruth drive the same distance each week. Joe averages 25 miles per gallon and needs 6 gallons of fuel more than Ruth, who averages 35 miles per gallon. How far does Ruth drive each week?

24. Susan drives twice as far as Tom each week and uses 5 gallons of fuel more than Tom. If Susan averages 30 miles per gallon and Tom averages 35 miles per gallon, how far does Susan drive each week?

Explorations

Compose a trick of your own similar to the one shown in Exploration 1.

1. Many mathematical "tricks" can be explained and developed through the use of algebraic techniques. Consider, for example, this set of directions.

Think of a number.
Multiply that number by 2.
Add 9.
Subtract 3.
Divide by 2.
Subtract the original number.
The final answer will always be 3.

7 AN INTRODUCTION TO ALGEBRA

Use n to represent the original number, and write a mathematical phrase to represent each step in the set of directions just given. For example, the first three steps are n, $2n$, and $2n + 9$. From this, show why the result is always 3, regardless of the number originally selected.

2. As in Exploration 1 write a set of directions for obtaining in order the numbers

$$n \qquad 5n \qquad 5n + 7 \qquad 5n + 5 \qquad n + 1 \qquad 1$$

7-5
Equations in One Variable

The Celsius thermometer is named after the Swedish astronomer and scientist, Anders Celsius (1701–1744). He was the first to consider the separation of the distance between the freezing point and the boiling point of water into 100 equal parts.

Any letter may be used as the variable when writing an equation. Often the letter is selected as the first letter of the word representing the quantity for which the variable is used. For example, in the metric system temperatures are measured on a scale of degrees Celsius where water freezes at 0 degrees and boils at 100 degrees. As of January 1, 1973, the American Society for Testing and Materials adopted the notation °C for degrees **Celsius** and °F for degrees **Fahrenheit**. A normal body temperature of 98.6 °F is about 37 °C.

It is often helpful to consider degrees Celsius in terms of commonly recognized temperatures, as in the adjacent figure. Corresponding temperatures in degrees Fahrenheit are also shown in the figure.

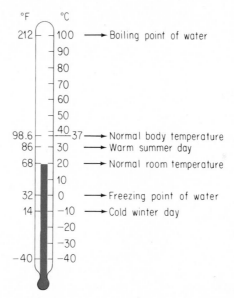

There is a formula that relates temperatures in degrees Fahrenheit *(F)* to those in degrees Celsius *(C)*. For temperatures above freezing, the number of degrees Fahrenheit above freezing is $F - 32$; the number of degrees Celsius is C. Furthermore, the interval from freezing to boiling is 180 °F but only 100 °C. Accordingly, a degree Fahrenheit is a smaller unit than a degree Celsius and the ratio of the number of degrees Fahrenheit above freezing to the number of degrees Celsius above freezing is given by this proportion:

$$\frac{F - 32}{C} = \frac{180}{100}$$

$$F - 32 = \frac{9}{5}C$$

$$F = \frac{9}{5}C + 32$$

This formula holds for all temperatures and involves the two variables F and C. The formula may be used to find the value of either of the variables when the value of the other is known.

Example 1 Convert (a) 30 °C to the Fahrenheit scale (b) 77 °F to the Celsius scale.

Solution

(a) $F = \frac{9}{5}(30) + 32$

$ = 9(6) + 32$

$ = 54 + 32$

$ = 86$

30 °C = 86 °F

(b) $77 = \frac{9}{5}C + 32$

$45 = \frac{9}{5}C$

$5 = \frac{1}{5}C$

$25 = C$

77 °F = 25 °C

Each equation that we have considered has been an equality of an expression of the first degree and a number. Several other types of equations can be solved by elementary methods. For example, consider the equation

$$x^2 = 3^2$$

Then $x^2 = 9$

and $x = 3$ or $x = -3$ $(3)^2 = (-3)^2 = 9$

In general, any equation of the form

$$x^2 = b^2 \qquad b \neq 0$$

has the solution set $\{b, -b\}$.

Equations of these forms arise in many ways, as in the following examples.

$A = s^2$ where A is the area of a square with side of length s.

$s = 16t^2$ where s is the distance in feet that a freely falling body falls in t seconds.

$h = h_0 - 16t^2$ where h is the height in feet of a freely falling body t seconds after it was released at height h_0.

286

Example 2   Find the time required for a freely falling body to fall 2500 feet.

Solution

$$s = 16t^2$$
$$2500 = 16t^2$$
$$t^2 = \frac{2500}{16}$$
$$t = \frac{50}{4} = 12.5 \text{ seconds}$$

In Example 2, the equation had both 12.5 and -12.5 as solutions, but only 12.5 satisfied the conditions of the problem. The nature of the problem determined the replacement set.

We next use the alternative form

$$x^2 - b^2 = 0$$

of the equation $x^2 = b^2$ to introduce the *method of factorization* for solving equations. This method is based on the fact that a product of real numbers is zero if and only if at least one of the factors is zero. Recall that $x^2 = b^2$ if and only if $x = b$ or $x = -b$. Then $x^2 - b^2 = 0$ if and only if $x - b = 0$ or $x + b = 0$, that is, if and only if

$$(x - b)(x + b) = 0$$

The expression $(x - b)(x + b)$ is the *factored form* of $x^2 - b^2$. The *factorization*

This factorization is the basis for many shortcuts in arithmetic. For example,

$$58 \times 62 = (60 - 2)(60 + 2)$$
$$= 60^2 - 2^2$$
$$= 3600 - 4 = 3596.$$

Use this method to find the product 37×43.

$$x^2 - b^2 = (x - b)(x + b)$$

can be verified by multiplication.

$$
\begin{array}{r}
x \; + \; b \\
x \; - \; b \\
\hline
x^2 \; + \; bx \\
- \; bx \; - \; b^2 \\
\hline
x^2 \qquad\;\; - \; b^2
\end{array}
$$

In general, any sentence that can be expressed as a product of first-degree expressions may be solved by considering these factors. Each factor is set equal to zero and its solution obtained. The union of these solutions is the solution set of the original sentence.

Example 3   Solve (a) $x^2 - 9 = 0$ (b) $5x^2 + 3 = 23$
(c) $2x(x - 2)(x + 3) = 0$.

Solution   **(a)** $x^2 - 9 = (x - 3)(x + 3) = 0$
       If $x - 3 = 0$, then $x = 3$.
       If $x + 3 = 0$, then $x = -3$.
       The solution set is $\{3, -3\}$.

   (b) $5x^2 + 3 = 23$
       $5x^2 - 20 = 0$ Subtract 3 from both sides.
       $x^2 - 4 = 0$ Divide both sides by 5.
       $(x - 2)(x + 2) = 0$ Factor.
       If $x - 2 = 0$, then $x = 2$.
       If $x + 2 = 0$, then $x = -2$.
       The solution set is $\{2, -2\}$.

   (c) The factors of the first degree are $2x$, $x - 2$, and $x + 3$.
       If $2x = 0$, then $x = 0$.
       If $x - 2 = 0$, then $x = 2$.
       If $x + 3 = 0$, then $x = -3$.
       The solution set is $\{0, 2, -3\}$.

These methods may be extended to obtain a formula for solving any sentence that can be written in the form

$$ax^2 + bx + c = 0, \quad a \neq 0$$

that is, any **quadratic equation.** The derivation of the formula is considered in Exercise 51. The formula, called the **quadratic formula,** is usually expressed as

$$x = \frac{-b \pm \sqrt{b^2 - 4ac}}{2a}$$

In other words, the solution set of the quadratic equation $ax^2 + bx + c = 0$, $a \neq 0$, is

$$\left\{ \frac{-b + \sqrt{b^2 - 4ac}}{2a}, \frac{-b - \sqrt{b^2 - 4ac}}{2a} \right\}$$

Example 4   Use the quadratic formula to solve $2x^2 - 5x + 3 = 0$.

Solution   We compare $ax^2 + bx + c$ and $2x^2 - 5x + 3$ to obtain $a = 2$, $b = -5$, and $c = 3$. Then

$$x = \frac{-(-5) \pm \sqrt{(-5)^2 - 4(2)(3)}}{2(2)} = \frac{5 \pm \sqrt{25 - 24}}{4} = \frac{5 \pm \sqrt{1}}{4}$$

$$x = \frac{5 + 1}{4} = \frac{6}{4} = \frac{3}{2} \quad \text{or} \quad x = \frac{5 - 1}{4} = \frac{4}{4} = 1$$

The solution set is $\{1, 3/2\}$.

The Norwegian mathematician Niels Henrik Abel (1802–1829) proved the impossibility of finding solutions for fifth-degree or higher-degree equations in terms of radicals, the operations of arithmetic, and the coefficients of the original equation. Read about his life in E. T. Bell's *Men of Mathematics,* Dover Publications, 1937.

Any equation of the first degree in one variable

$$ax + b = 0, \qquad a \neq 0$$

is a *linear equation* and can be solved using arithmetic operations. Any equation of the second degree in one variable

$$ax^2 + bx + c = 0, \qquad a \neq 0$$

is a *quadratic equation* and can be solved by the quadratic formula. There exist formulas for solving equations of the third degree (cubic equations) and equations of the fourth degree (quartic equations):

$$ax^3 + bx^2 + cx + d = 0, \qquad a \neq 0$$
$$ax^4 + bx^3 + cx^2 + dx + e = 0, \qquad a \neq 0$$

There do not exist corresponding general formulas for solving equations of the fifth-degree or for higher-degree equations.

Solutions of linear and quadratic sentences in one variable are considered in the exercises. The preceding very brief introduction of the solution of quadratic equations has been included without endeavoring to be algebraically complete.

EXERCISES

Convert to the Fahrenheit scale.

1. 0 °C
2. 20 °C
3. 25 °C
4. 40 °C
5. 60 °C
6. 100 °C

Convert to the Celsius scale.

7. 32 °F
8. 50 °F
9. 86 °F
10. 95 °F
11. 122 °F
12. 212 °F

Solve the given equation for x.

13. $x^2 = 8^2$
14. $x^2 = (-11)^2$
15. $x^2 = (-25)^2$
16. $x^2 = 81$
17. $x^2 = r^2$
18. $x^2 = A$

19. Find the time required for a freely falling body to fall **(a)** 400 feet **(b)** 10,000 feet.
20. Find the height of a body that is dropped from 30,000 feet and has fallen **(a)** 20 seconds **(b)** 30 seconds.

Solve.

21. $x^2 - 25 = 0$
22. $x^2 - 49 = 0$
23. $2x^2 - 72 = 0$
24. $3x^2 = 12$

25. $2x^2 + 5 = 23$ 26. $3x^2 - 1 = 47$

27. $50 - 3x^2 = 23$ 28. $75 - 2x^2 = 3$

29. $(x - 3)(x - 5) = 0$ 30. $(x + 2)(x - 3) = 0$

31. $(x + 1)(x + 5) = 0$ 32. $(x - 6)(x + 4) = 0$

33. $(x + 2)(x - 1)(x + 3) = 0$

34. $(x + 1)(x - 2)(x + 4) = 0$

35. $2x(x + 5)(x - 2) = 0$

36. $3x(x - 1)(x - 2)(x - 5) = 0$

37. $(2x - 1)(3x - 6)(x + 5) = 0$

38. $(3x - 1)(2x + 5)(x - 7) = 0$

Use the quadratic formula and solve the given equation.

39. $x^2 - x - 12 = 0$ 40. $x^2 - 2x - 15 = 0$

41. $2x^2 + 3x - 5 = 0$ 42. $3x^2 - 10x - 13 = 0$

43. $5x^2 + 7x - 6 = 0$ 44. $2x^2 + 9x - 5 = 0$

Graph the given statement on a number line.

*45. $(x - 2)(x + 3) \leq 0$

*46. $(x + 1)(x - 3) \geq 0$

*47. $(x - 1)(x - 2) \geq 0$

*48. $(x - 1)(x - 2)(x - 3) \geq 0$

*49. $(x + 1)(x - 2)(x + 2) \leq 0$

*50. $(x + 3)(x + 1)(x - 2) \leq 0$

*51. The equation $ax^2 + bx + c = 0$, $a \neq 0$ is given. State a reason for each indicated step in this derivation of the quadratic formula.

(a) $x^2 + \dfrac{b}{a}x + \dfrac{c}{a} = 0 \times \dfrac{1}{a}$

(b) $x^2 + \dfrac{b}{a}x + \dfrac{c}{a} = 0$

(c) $\left(x^2 + \dfrac{b}{a}x + \dfrac{c}{a}\right) + \left(-\dfrac{c}{a}\right) = 0 + \left(-\dfrac{c}{a}\right)$

(d) $\left(x^2 + \dfrac{b}{a}x\right) + \left[\dfrac{c}{a} + \left(-\dfrac{c}{a}\right)\right] = 0 + \left(-\dfrac{c}{a}\right)$

(e) $\left(x^2 + \dfrac{b}{a}x\right) + 0 = 0 + \left(-\dfrac{c}{a}\right)$

(f) $\quad x^2 + \dfrac{b}{a}x = -\dfrac{c}{a}$

(g) $\quad x^2 + \dfrac{b}{a}x + \dfrac{b^2}{4a^2} = -\dfrac{c}{a} + \dfrac{b^2}{4a^2}$

(h) $\qquad\qquad\quad = \dfrac{b^2}{4a^2} + \left(-\dfrac{c}{a}\right)$

$\qquad\qquad\qquad\quad = \dfrac{b^2}{4a^2} + \left(\dfrac{-c}{a}\right)$ $\qquad$ Since $-\dfrac{c}{a} = \dfrac{-c}{a}$.

(i) $\qquad\qquad\quad = \dfrac{b^2}{4a^2} + \dfrac{-4ac}{4a^2}$

(j) $\qquad\qquad\quad = \dfrac{b^2 - 4ac}{4a^2}$

(k) $\quad \left(x + \dfrac{b}{2a}\right)^2 + \left(-\dfrac{b^2 - 4ac}{4a^2}\right)$

$\qquad\qquad\qquad\qquad = \dfrac{b^2 - 4ac}{4a^2} + \left(-\dfrac{b^2 - 4ac}{4a^2}\right)$

(l) $\qquad\qquad\qquad = 0$

(m) $\quad \left(x + \dfrac{b}{2a}\right)^2 - \dfrac{b^2 - 4ac}{4a^2} = 0$

$\qquad \left(x + \dfrac{b}{2a}\right)^2 - \left(\dfrac{\sqrt{b^2 - 4ac}}{2a}\right)^2 = 0$ $\qquad$ Properties of square root.

$\qquad x + \dfrac{b}{2a} = \dfrac{\sqrt{b^2 - 4ac}}{2a}$ or $\dfrac{-\sqrt{b^2 - 4ac}}{2a}$ $\qquad$ As in Example 3.

$\qquad x = \dfrac{-b}{2a} + \dfrac{\sqrt{b^2 - 4ac}}{2a}$ or $\dfrac{-b}{2a} - \dfrac{\sqrt{b^2 - 4ac}}{2a}$

That is, $x = \dfrac{-b \pm \sqrt{b^2 - 4ac}}{2a}$.

Explorations

In ancient Greece the equivalent of a quadratic equation was a statement about areas. Zero and negative numbers had not yet been introduced. Thus a problem that we would represent by

$$x^2 + 2x - 24 = 0$$

would have been stated in words in a form equivalent to

$$x^2 + 2x = 24$$

To solve this problem a square would be sketched and its area

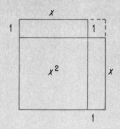

The use of the quadratic formula has been made possible during the last 400 years by the introduction of letters to represent variables. About 2300 years ago the classical Greeks represented both numbers (counting numbers) and magnitudes (other positive real numbers) by lengths of line segments. Products were represented by the areas of rectangular regions. Statements of equality were expressed in words.

considered to be x^2. A unit of length would be selected initially unrelated to the side of the square, except that for convenience the unit would be less than x. Then two rectangular regions each 1 by x would be added to the square as in the adjacent figure. Finally a square of side $x + 1$ would be *completed* by adding a square of area 1. The square of side $x + 1$ has area $24 + 1$, that is, 25. The early Greeks knew the squares of the counting numbers and thus knew that if

$$(x + 1)^2 = 25$$
then $\quad x + 1 = 5$
$$x = 4$$

The reasoning used to solve for x corresponded to the modern procedure of **completing the square.**

$$x^2 + 2x = 24$$
$$x^2 + 2x + 1 = 24 + 1$$
$$(x + 1)^2 = 25$$
$$x + 1 = 5 \text{ or } -5$$
$$x = 4 \text{ or } -6$$

However, only the positive value would have been obtained by the classical Greeks. This approach can be used for any equation of the form $x^2 + 2bx = c$, where b and c are positive numbers and a square of area b^2 is added.

Sketch and label the related figure. Find the positive solution for each equation by completing the square in the classical Greek manner.

1. $x^2 + 2x = 8$　　2. $x^2 + 2x = 15$　　3. $x^2 + 4x = 12$
4. $x^2 + 6x = 16$　　5. $x^2 + 3x = 7/4$　　6. $x^2 + 5x = 6$

Extend the reasoning just used for solving equations of the form $x^2 + 2bx = c$ to obtain a corresponding procedure for solving the following equations of the form $x^2 - 2bx = c$. For each equation in Explorations 7 through 12, solve by completing the square in the classical Greek manner.

7. $x^2 - 2x = 8$　　8. $x^2 - 2x = 15$　　9. $x^2 - 8x = 20$
10. $x^2 - 12x = 13$　　11. $x^2 - 3x = 4$　　12. $x^2 - 7x = 8$

13. Criticize the following "solution" of the given equation.

$$x^2 - 2x = 2$$
$$x(x - 2) = 2$$
$$x = 2 \quad \text{or} \quad x - 2 = 2$$
$$x = 2 \quad \text{or} \quad x = 0$$

7-6

Sentences in Two Variables

Sentences in two variables arise in many ways.

$p = 4s$ the perimeter p of a square of side s

$d = 80h$ the distance d in kilometers that a person drives in h hours at 80 kilometers per hour

$A = P + 0.09P$ the amount A of money in a savings account in which P dollars has been deposited at 9% simple interest for 1 year

For a sentence in two variables a *pair* of replacements is needed. For example, consider the sentence

$x + y = 5$

If a replacement, such as 3 for x, is made for one variable, then the resulting sentence

$3 + y = 5$

is an open sentence in the other variable. Thus a pair of re-placements is needed before an open sentence in two variables can be identified as true or false for these replacements. The two variables may be considered as an *ordered pair* such as (x, y). Then an ordered pair of numbers may be used as re-placements for the variables. For example, if $(x, y) = (3, 4)$ in the sentence $x + y = 5$, then the resulting sentence

$3 + 4 = 5$

is a false statement. If $(x, y) = (7, -2)$ in the given sentence, then the resulting sentence

$7 + (-2) = 5$

is a true statement.

By convention, for an ordered pair (x, y) the variable x is assumed to be the first of the two variables, and the first num-ber in the ordered pairs of numbers is taken as the replacement for x; the variable y is then taken as the second variable, and the second number in the ordered pair is taken as the replace-ment for y. For example, the ordered pair $(3, 2)$ implies that $x = 3$ and $y = 2$; the ordered pair $(2, 3)$ implies that $x = 2$ and $y = 3$.

The sentence $x + 2y = 8$ is true for $(2, 3)$; the sentence is false for $(3, 2)$. That is, $2 + 2(3) = 8$, but $3 + 2(2) \neq 8$. The solution set for the sentence $x + 2y = 8$ is the set of ordered pairs of numbers for which the sentence is true.

As in the case of sentences in one variable, the solution set of a sentence in two variables depends on the replacement sets of the variables used. For example, if the replacement set for x and y is the set of whole numbers, then the solution set for $x + 2y = 8$ is

$$\{(0, 4), \quad (2, 3), \quad (4, 2) \quad (6, 1), \quad (8,0)\}$$

Example 1   Solve the sentence $x + y < 3$ if the replacement set for x and y is the set of whole numbers.

Solution   When $x = 0$, y must be less than 3. Thus the ordered pairs of numbers (0, 0), (0, 1), and (0, 2) make the sentence true. Similarly, when $x = 1$, y must be less than 2; when $x = 2$, y must be less than 1. Is there a whole number y that satisfies the inequality when $x = 3$? When $x > 3$? The solution set is

$$\{(0, 0), \quad (0, 1), \quad (0, 2), \quad (1, 0), \quad (1, 1), \quad (2, 0)\}$$

Example 2   Solve the sentence $x + y \leq 3$ if x and y are counting numbers.

Solution $\{(1, 1), \quad (1, 2), \quad (2, 1)\}$

Unless we are told otherwise, we shall assume that the replacement set for the variables in a sentence is the set of real numbers. Then we graph the solution of such sentences on a plane by first drawing two perpendicular number lines to serve as the *x*-axis and the *y*-axis as in the figure.

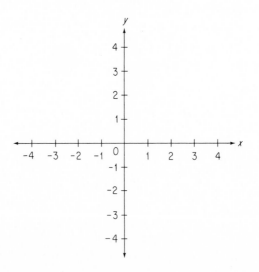

7 AN INTRODUCTION TO ALGEBRA

On a coordinate plane, two perpendicular number lines OX and OY are used as **coordinate axes.** Their intersection is at O on each number scale and this intersection is the **origin** of the **real coordinate plane.** In honor of René Descartes, such a plane is frequently referred to as the **Cartesian plane.** The numbers of each ordered pair of real numbers are the **Cartesian coordinates** of a point of the plane. Each point of a Cartesian plane can be represented by (has as its *coordinates*) an ordered pair of real numbers, and each ordered pair of real numbers can be used to identify (locate) a unique point of the plane. We follow the custom of speaking of "the point with coordinates (x, y)" as "the point (x, y)."

Let us now discuss the graph of the sentence $y = x + 2$. We may list in a **table of values** several ordered pairs of numbers that are solutions of the sentence. For $x = -1$, we have $y = -1 + 2 = 1$; for $x = 2$, we have $y = 2 + 2 = 4$. Confirm the other entries given in this table.

$y = x + 2$	x	-3	-2	-1	0	1	2	3
	y	-1	0	1	2	3	4	5

Each ordered pair of numbers (x, y) from the table can then be graphed. These points may be connected in the order of the x-coordinates and the graph of $y = x + 2$ obtained as in the following figure. The graph of $y = x + 2$ is a straight line and extends indefinitely as indicated by the arrowheads.

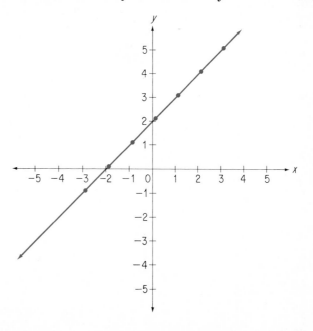

In general, any equation that can be expressed in the form

$$ax + by + c = 0$$

where a and b are not both zero, is said to be a **linear equation in x and y** and has a straight line as its graph.

Inasmuch as a straight line is determined by two points, we can graph a linear equation after locating two of its points. When $a \neq 0$ and $b \neq 0$, the most convenient points to locate are the point $(0, y)$ at which the graph crosses the y-axis and the point $(x, 0)$ at which the graph crosses the x-axis. The graph of $y = x + 2$ crosses the y-axis when $x = 0$, that is, at the point A: $(0, 2)$, where 2 is the **y-intercept** of the graph. The graph of $y = x + 2$ crosses the x-axis when $y = 0$, that is, at the point B: $(-2, 0)$, where -2 is the **x-intercept** of the graph.

Example 3   Graph $\{(x, y) \mid x - 2y = 4\}$.

Solution   In set-builder notation the given statement is read as "the set of all (x, y) such that $x - 2y = 4$." The equation may be expressed as $x - 2y - 4 = 0$ and thus has a line as its graph. For $x = 0$, $-2y = 4$ and $y = -2$. For $y = 0$, $x = 4$. The x-intercept 4 and the y-intercept -2 determine the line as shown in the graph.

Usually it is desirable to graph a third point as a check. For example, when $x = 2$ we have $2 - 2y = 4$, $-2y = 2$, and $y = -1$. Thus $(2, -1)$ may be graphed and the point should be on the line.

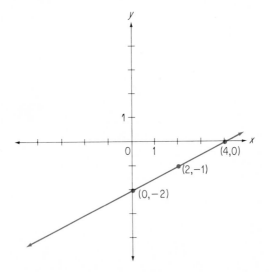

Inequalities of the first degree can be graphed, have a line as a boundary, and are called *linear inequalities*. Consider the sentence $y \leq x + 2$. Here the graph consists of all the points on the line $y = x + 2$, as well as the points in the *half-plane* below the line as indicated by the shaded portion of the graph. Note that for any value b of x the point $(b, b + 2)$ is on the line $y = x + 2$, and for any value of y less than $b + 2$ the point (b, y) is below the line. Thus the graph of $y \leq x + 2$ is the union of a line and a half-plane. A solid line is used in this graph since the points of the line are points of the graph. When the points of the line are not points of the graph a dashed line is used, as in Example 4.

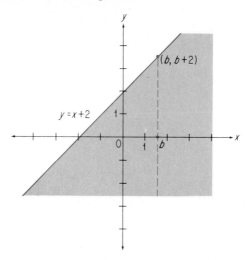

In the preceding figure, the line $y = x + 2$ separates the plane into two *half-planes*. One of these half-planes is part of the desired graph and satisfies the given inequality. One way to determine which half-plane is needed is to select a point and determine whether or not the point is part of the graph. For example, $(0, 0)$ satisfies the inequality $y < x + 2$ and therefore the point with coordinates $(0, 0)$ is in the desired half-plane. If the coordinates of any point satisfy the inequality, the point is in the desired half-plane. If the coordinates of any point, such as $(0, 3)$, do not satisfy the inequality, the point is on the opposite side of the line from the desired half-plane.

Example 4   Graph $y > x - 1$.

Solution   It is helpful to graph first the corresponding statement of equality, $y = x - 1$. The line is dashed since it is not part of the graph. This line separates the plane into two half-planes. The desired graph is the half-plane above the line. Note that the coordinates of the point $(0, 0)$ satisfy the given inequality and the point is part of the desired half-plane; the coordinates of (2, 0) do not satisfy the given inequality and the point is on the opposite side of the line from the desired half-plane.

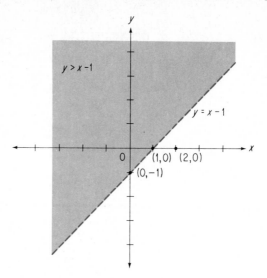

Example 5   Graph $x - 2y - 4 \geq 0$.

Solution   The given inequality can be expressed in *y-form*.

An equation or inequality that is solved for y is said to be in y-form.

$$x - 2y \geq 4 \qquad \text{Add 4 to both sides.}$$

$$-2y \geq -x + 4 \qquad \text{Subtract } x \text{ from both sides.}$$

As illustrated by $-3 < -2$ but $3 > 2$, the "sense" of an inequality is changed whenever both sides of the inequality are multiplied, or divided, by a negative number.

$$y \leq \frac{1}{2}x - 2 \qquad \text{Divide both sides by } -2.$$

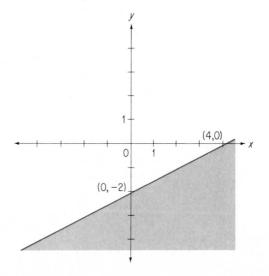

The corresponding equality $y = (1/2)x - 4$ is graphed. Then, as indicated by the last inequality, the points on and below this line constitute the desired graph. Note that $(0, 0)$ does not satisfy the inequality; $(0, -3)$ does satisfy the inequality.

The half-plane that is to be included in the graph of a linear inequality can be determined either by testing the coordinates of points or from the inequality. Note that in each case the inequality was first solved for y. This corresponds to the y-form of an equation and is the **y-form** of the inequality. The graph of the inequality is *above* the line if the y-fo. .n is $y > \ldots$; the graph is *below* the line if the y-form is $y < \ldots$.

Example 6   Solve $3x - 4y \leq 12$ for y.

Solution

$3x - 4y \leq 12$	Given.
$-4y \leq 12 - 3x$	Subtract $3x$ from both sides.
$y \geq \dfrac{3}{4}x - 3$	Divide both sides by -4.

EXERCISES   *Solve if the replacement set is the set of counting numbers.*

1. $x + y = 6$ 2. $x + y < 5$
3. $y \leq 4 - x$ 4. $y = 5 - x$

Find the solution set for each sentence for the replacement set $\{1, 2, 3, 4\}$.

5. $x + y = 4$ 6. $2x + y \leq 4$
7. $x + y \leq 3$ 8. $y = x + 1$
9. $x + y < 3$ 10. $y \leq x$

For the graph of each sentence, find (a) the x-intercept (b) the y-intercept.

11. $x + y = 8$ 12. $x - y = 3$
13. $2x - 3y = 12$ 14. $3x - 2y = 12$
15. $y = 2x + 5$ 16. $y = 3x + 5$
17. $2x - y + 8 = 0$ 18. $x + 3y - 6 = 0$
19. $5x + 4y = 10$ *20. $ax + by + c = 0$

Write each sentence in y-form, that is, solve for y.

21. $2x + y = 9$ 22. $4x - 2y = 12$ 23. $2x + y \geq 8$
24. $x + 3y \leq 6$ 25. $2x - y \leq 8$ 26. $3x - 2y \geq 12$

Graph each linear equation.

27. $y = x + 3$ 28. $y = x - 1$ 29. $x + y = 2$
30. $x - y = 5$ 31. $2x + 3y + 6 = 0$ 32. $3x + 2y - 6 = 0$
33. $x = 3$ 34. $y = -2$

For Exercises 35 through 40 select from the inequalities in (a) *through* (i) *the inequality that best describes the given graph.*

(a) $x + y - 1 > 0$ (b) $x + y - 1 \geq 0$ (c) $x + y - 1 \leq 0$
(d) $x + y - 1 < 0$ (e) $x + y + 1 \leq 0$ (f) $x - y + 1 \leq 0$
(g) $x - y + 1 \geq 0$ (h) $x - y - 1 \leq 0$ (i) $x - y - 1 \geq 0$

35.
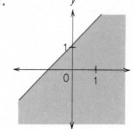

36.
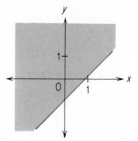

37.
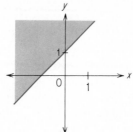

38.
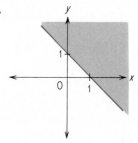

39.
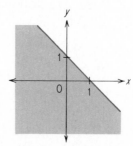

40.
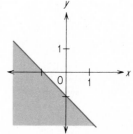

Graph each linear inequality.

41. $y \leq x + 2$ 42. $y \geq x - 2$
43. $y > 2x - 1$ 44. $y < 3x + 1$
45. $2x + y > 6$ 46. $3x + 2y > 6$

47. $x + 2y - 4 \leq 0$

48. $2x + y \leq 4$

49. $x - y \leq 2$

50. $2x - y \geq 3$

51. $3x - 2y + 6 \geq 0$

52. $3x - 4y + 12 \geq 0$

Explorations

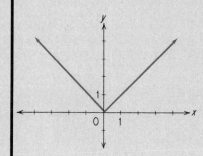

The overhead projector is an effective device for showing graphs on a plane, especially through the use of translations and rotations. For example, prepare a set of coordinate axes and grid lines. On a separate sheet of acetate draw the graph of $y = |x|$.

$$y = |x| = x \qquad \text{for } x \geq 0$$
$$y = |x| = -x \qquad \text{for } x < 0$$

1. Demonstrate each of the following graphs by appropriate translations of the graph of $y = |x|$.
    (a) $y = |x| + 1$. Shift graph 1 unit up.

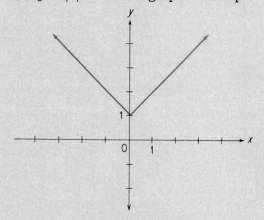

(b) $y = |x| - 1$. Shift graph 1 unit down.

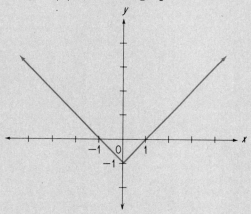

(c) $y = |x + 1|$. Shift graph 1 unit to the left.

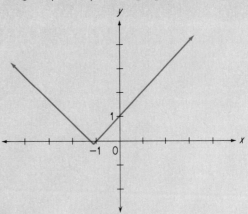

(d) $y = |x - 1|$. Shift graph 1 unit to the right.

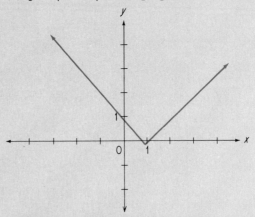

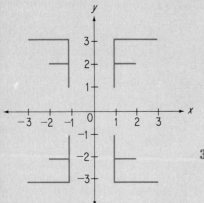

2. The graph of $y = -|x|$ may be obtained by a rotation of the graph of $y = |x|$ through 180° about the x-axis, that is, by a reflection in the x-axis. Demonstrate the use of rotations and translations of the graph of $y = |x|$ to obtain the graphs of each of these equations.

   (a) $y = -|x|$ **(b)** $y = -|x| + 1$ **(c)** $y = -|x| - 1$

   (d) $y = |x| + 2$ **(e)** $y = |x| - 2$ **(f)** $y = 2 - |x|$

   (g) $y = |x - 1|$ **(h)** $y = -|x - 1|$ **(i)** $y = -|x - 2|$

   (j) $y = -|x + 1|$ **(k)** $y = 2 - |x - 1|$ **(l)** $y = 1 - |x + 1|$

3. In the adjacent figure the letter **F** was drawn in the first quadrant and reflections in the coordinate axes were used to obtain its images in the other quadrants. Copy the figure, assign coordinates to each of the vertices in the first

quadrant, and label each of the vertices in the other quadrants with their coordinates.

4. Capital letters of the English language are drawn in the first quadrant with images in the other quadrants from reflections in the coordinate axes. Which letters can be formed so that the figures appear the same in **(a)** the first and second quadrants **(b)** the first and fourth quadrants **(c)** the first and third quadrants?

5. Repeat Exploration 4 for decimal digits and write a five-digit numeral that would appear the same in all four quadrants.

6. Try to write the word **SEVEN** as it would appear if held up to a mirror. Then use a mirror to check your results. Try other positions of the mirror as needed until you can describe the mirror images in terms of a line reflection.

7-7
Linear Systems

Many applications make use of compound sentences involving two variables. Consider, for example, the compound sentence

$$x + y - 3 = 0 \quad \text{and} \quad x - y - 1 = 0$$

Frequently such sentences are written in the form

$$\begin{cases} x + y - 3 = 0 \\ x - y - 1 = 0 \end{cases}$$

(with or without the brace) and are referred to as a **system of linear equations** or as a set of **simultaneous linear equations**. To solve a set of two simultaneous equations in two variables, we find the set of ordered pairs that are solutions of both equations. Often a graphical approach is useful. For the given system of equations, we have the following graph.

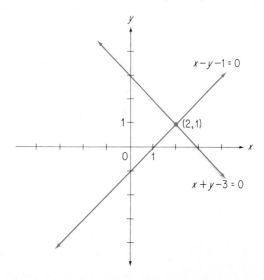

The point located at (2, 1) is on both lines and (2, 1) is the solution of the given system of equations. We may express the solution of the given system in set-builder notation.

$$\{(x, y) \mid x + y - 3 = 0\} \cap \{(x, y)\} \mid x - y - 1 = 0\} = \{(2, 1)\}$$

Consider next the compound sentence

$$x + y - 3 = 0 \quad \text{or} \quad x - y - 1 = 0$$

The word *or* indicates that we are to find the set of ordered pairs of numbers that are solutions of either one or of both of the given equations.

$$\{(x, y) \mid x + y - 3 = 0\} \cup \{(x, y) \mid x - y - 1 = 0\}$$

The graph of the solution set consists of all the points that are on at least one of the two lines; that is, the graph is the union of the points of the two lines shown in the preceding figure.

Example 1 Graph **(a)** $x - y - 1 = 0$ or $x - y + 2 = 0$
(b) $(x - y - 1)(x - y + 2) = 0$.

Solution **(a)** The graph consists of the union of the points of the two lines.

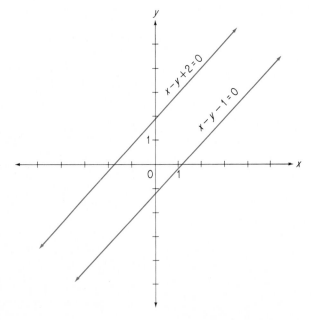

Notice that the two lines in the solution for Example 1(a) are parallel. The solution set of the sentence "$x - y - 1 = 0$ *and* $x - y + 2 = 0$" is the empty set.

(b) Since any product $a \times b = 0$ implies that $a = 0$ or $b = 0$, the given sentence can be written as in part (a) and has the same graph.

Systems of inequalities can also be solved graphically as the union or intersection of half-planes. Consider the system

$$\begin{cases} x + y - 3 > 0 \\ x - y - 1 > 0 \end{cases}$$

The corresponding statements of equality have been graphed earlier in this section. The graphs of the inequalities are shown in the next two figures.

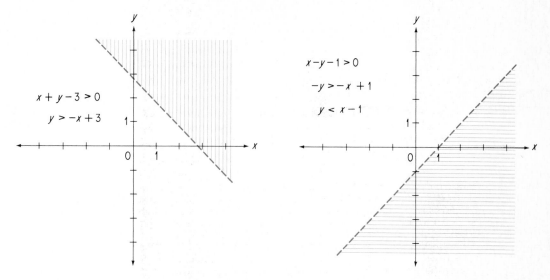

$x + y - 3 > 0$

$y > -x + 3$

$x - y - 1 > 0$

$-y > -x + 1$

$y < x - 1$

Note that the corresponding lines are dashed since the points of the lines are *not* included in the graphs. Also one graph has vertical shading and the other graph has horizontal shading. The graph of the system consists of the intersection of these two graphs, that is, the points of the region that is shaded both vertically *and* horizontally when the graphs of the two inequalities are drawn on the same coordinate plane.

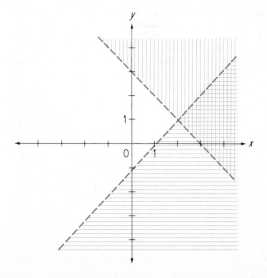

305

We can use the previous figure to determine the graph of the compound sentence $x + y - 3 > 0$ or $x - y - 1 > 0$, namely, the union of the points of the two shaded regions. The unshaded region (including adjacent parts of the dotted lines) represents the graph of the solution set of the system

$$\begin{cases} x + y - 3 \leq 0 \\ x - y - 1 \leq 0 \end{cases}$$

Example 2 Graph each set of points.

(a) $\begin{cases} x - y + 2 \leq 0 \\ 2x + y - 4 \geq 0 \end{cases}$

(b) $\{(x, y) \mid x - y + 2 \leq 0\} \cup \{(x, y) \mid 2x + y - 4 \geq 0\}$

Solution (a) The graph of $x - y + 2 \leq 0$ has horizontal shading in the figure; the graph of $2x + y - 4 \geq 0$ has vertical shading. The graph of the solution set consists of the points in the region with both horizontal and vertical shading and includes the points of the two rays that serve as the boundary of this region.

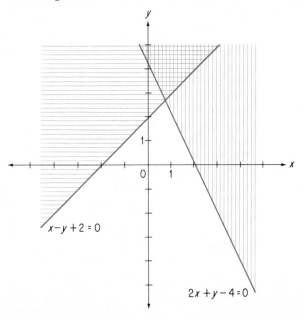

(b) Consider the graph for part (a). The solution set in part (b) consists of all the points in the regions that are shaded in any way, that is, all points except those in the unshaded region.

Many word problems that involve two unknown numbers can be solved graphically on a coordinate plane.

Example 3 Use the graph of a system of equations to find two numbers such that the sum of the numbers is 12 and the first number is twice the second number.

Solution Let the numbers be x and y.

$$\begin{cases} x + y = 12 \\ x = 2y \end{cases}$$

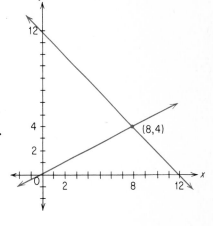

The graphs intersect at (8, 4).
The numbers are 8 and 4.

EXERCISES *Graph.*

1. $\begin{cases} x + y - 6 = 0 \\ x - y + 4 = 0 \end{cases}$ 　　　　 2. $\begin{cases} x + y - 5 = 0 \\ x - y - 3 = 0 \end{cases}$

3. $\begin{cases} x + y = 2 \\ x + 2y = 4 \end{cases}$ 　　　　 4. $\begin{cases} 2x - y = 8 \\ x + 2y = -4 \end{cases}$

5. $\begin{cases} 2x - y - 2 = 0 \\ 4x - 2y + 4 = 0 \end{cases}$ 　　 6. $\begin{cases} x + 2y - 4 = 0 \\ 2x + 4y - 8 = 0 \end{cases}$

7. $(x + y - 3)(x + y + 4) = 0$

8. $(x - y + 2)(2x + 3y + 6) = 0$

9. $(x - 2y + 6)(x + 2y - 6) = 0$

10. $(3x + 2y - 6)(x - 2y - 4) = 0$

11. $(x - 3y + 3)(2x + y - 2) = 0$

12. $(2x - 4y + 4)(x - 2y + 2) = 0$

Use the graph of a system of equations to find two numbers that satisfy the given conditions.

13. The sum of the numbers is 19. A difference of the numbers is 7.

14. The sum of the numbers is 21. A difference of the numbers is 13.

15. The sum of the numbers is 25. Twice the larger number is equal to three times the smaller.

16. The sum of the numbers is 23. The larger number is 5 more than twice the smaller.

17. The sum of the larger number and twice the smaller is 12. A difference of the numbers is 3.

18. The sum of the larger number and three times the smaller is 12. The larger is 2 less than four times the smaller.

Graph.

19. $\begin{cases} x - y + 3 > 0 \\ x + y - 3 > 0 \end{cases}$ 20. $\begin{cases} 2x + y - 6 < 0 \\ x - 2y + 6 < 0 \end{cases}$

21. $\begin{cases} 3x - 2y - 6 \le 0 \\ 2x + 3y + 6 \ge 0 \end{cases}$ 22. $\begin{cases} x + 2y - 6 \le 0 \\ 2x - y + 4 \ge 0 \end{cases}$

23. $\{(x, y) \mid 3x - y + 6 \le 0\} \cup \{(x, y) \mid 3x + 4y + 12 \ge 0\}$

24. $\{(x, y) \mid x + 4y - 8 \ge 0\} \cup \{(x, y) \mid 4x - 2y + 8 \ge 0\}$

25. $xy \ge 0$ 26. $(x - 1)(y + 1) \ge 0$

27. $(x + 2)(y - 3) \le 0$ 28. $x(x - y) \le 0$

29. $(2x - 3y + 6)(x + 2y - 4) \le 0$

30. $(x + 2y - 4)(x - y + 2) \le 0$

Explorations

1. The equation $F = 2C + 30$ can be used to approximate the relationship between temperatures in degrees Fahrenheit and degrees Celsius. The following graph shows this equation as well as the actual relationship.

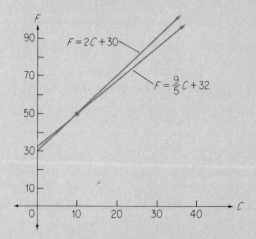

7 AN INTRODUCTION TO ALGEBRA

(a) What is the point of intersection of the two graphs?

(b) For what value of C do both equations give the same value for F?

(c) Compare the values for F from the two formulas when $C = 0$.

(d) Compare the values for F from the two formulas when $C = 30$.

(e) For what values of C do the two formulas give values for F that differ by at most 2?

(f) Extend the graphs for negative values of C.

2. In 1980 Mary was 5 years older than her brother John. Mary will be 16 in 1990. Use a coordinate plane with x the number of years after 1980 and y the person's age. Remember that each person gets one year older each year and graph the line that represents Mary's age. Graph the line that represents John's age.

(a) How old was Mary in 1980?

(b) How old was John in 1980?

(c) Find and explain a method for determining when Mary is twice as old as John.

Chapter 7 Review

Solutions to the following exercises may be found within the text of Chapter 7. Try to complete each exercise without referring to the text.

Section 7-1   Sentences and Statements

1. Solve each statement.
   (a) $x + 1 > 4$ for whole numbers x.
   (b) $n + 2 = 2 + n$ for real numbers n.
   (c) $x + 2 < x$ for integers x.

2. Graph the given sentence.
   (a) $x + 3 = 5$ for integers x.
   (b) $x + 3 \le 5$ for whole numbers x.
   (c) $x + 3 > 5$ for real numbers x.

3. Graph the given sentence for real numbers x.
   (a) $x + 3 \not< 5$ \hfill (b) $-1 \le x \le 3$

Section 7-2   Compound Sentences

4. Solve for integers x.
   $$x \ge -2 \quad \text{and} \quad x + 1 \le 4$$

5. Solve for real numbers x.
   $$x + 3 < 5 \quad \text{and} \quad 3 < 1$$

6. Solve (a) $5 > 1$ or $x + 2 < 5$
   (b) $x + 2 \ne 2 + x$ or $x + 2 < x$.

7. Graph (a) $x \le -1$ or $x \ge 2$ (b) $x = -2$ or $x \ge 1$.

8. Solve $x + 3 = 7$ and explain each step.
9. Solve $2x - 3 = 7$ and explain each step.
10. Solve $\frac{1}{2}x - 2 > 5$ and explain each step.
11. Solve $-2x + 3 < 7$ and explain each step.

Section 7-4   *Problem Solving*

12. The sum of two numbers is 20; their difference is 6. Find the numbers.
13. Doris' class is renting a bus for a short field trip. If the members of the class pay $1 each, there is a surplus of $2. If the members pay 90¢ each, there is a shortage of $1. How many members are in the class and what is the charge for renting the bus?
14. Ginny and Bill drive small cars. Ginny averages 40 miles per gallon with her car; Bill averages 30 miles per gallon with his. They attend the same college and live the same distance from campus. If Bill needs one gallon more fuel than Ginny for five round trips to campus, how far does each live from campus?

Section 7-5   *Equations in One Variable*

15. Convert   (a) 30 °C to the Fahrenheit scale   (b) 77 °F to the Celsius scale.
16. Find the time required for a freely falling body to fall 2500 feet.
17. Solve   (a) $5x^2 + 3 = 13$ (b) $2x(x - 2)(x + 3) = 0$
18. Use the quadratic formula and solve $2x^2 - 5x + 3 = 0$.

Section 7-6   *Sentences in Two Variables*

19. Find the solution set for the equation $x + 2y = 8$ if the replacement set for x and y is the set of whole numbers.
20. Solve the sentence $x + y < 3$ if the replacement set for x and y is the set of whole numbers.
21. Graph $\{(x, y) \mid x - 2y = 4\}$.
22. Graph $x - 2y - 4 \geq 0$.
23. Solve $3x - 4y \leq 12$ for y.

Section 7-7   *Linear Systems*

24. Graph.
    (a) $x - y - 1 = 0$ or $x - y + 2 = 0$
    (b) $(x - y - 1)(x - y + 2) = 0$

25. Graph each set of points.

(a) $\begin{cases} x - y + 2 \le 0 \\ 2x + y - 4 \ge 0 \end{cases}$

(b) $\{(x, y) \mid x - y + 2 \le 0\} \cup \{(x, y) \mid 2x + y - 4 \ge 0\}$

26. Use the graph of a system of equations to find two numbers such that the sum of the numbers is 12 and the first number is twice the second number.

Chapter 7 Test

1. Solve for whole numbers x.

   (a) $x + 7 < 11$ (b) $11 - x > 7$

2. Describe the solution set of the given sentence for real numbers x.

   (a) $x + 2 < 7$ (b) $x + 7 \ge 5$

Graph for real numbers x.

3. $x + 3 \le -1$ 4. $x - 2 \not< 3$

5. $-1 \le x \le 2$ 6. $x = 2$ or $x \ge 3$

7. $x + 3 < 4$ and $x \ge 0$ 8. $x \le 1$ or $x \ge 2$

9. $x + 3 < 4$ or $x \ge 0$ 10. $x < x + 2$ and $x + 2 < 5$

Solve for x.

11. $2x + 5 = 13$ 12. $7 - 4x = 11$

13. $\dfrac{1}{2}x - 3 > 7$ 14. $3 - \dfrac{2}{3}x < 1$

15. $5x + 6 = 4x + 5$ 16. $11 - 2x^2 = 3$

17. The sum of two numbers is 25; their difference is 7. Find the numbers.

18. The mother of one of Jane's students sent a box of cookies for the members of the class. If 2 cookies were given to each student, there was a surplus of 10 cookies. If 3 cookies were distributed to each student, there was a shortage of 5 cookies. How many students were there in the class?

19. For the graph of $3x - 2y + 18 = 0$ find (a) the x-intercept   (b) the y-intercept.

20. Graph (a) $y = x - 2$ (b) $2x - y = 4$.

Graph the given sentence or system.

21. $2x - 3y \le 6$ 22. $(x + y - 2)(x - y + 3) = 0$

23. $\begin{cases} x + y - 5 = 0 \\ x - y + 1 = 0 \end{cases}$ 24. $\begin{cases} x + y - 3 \ge 0 \\ x - y + 2 \le 0 \end{cases}$

25. Use the graph of a system of equations to find two numbers such that the first number is three times the second and the sum of the first number and twice the second number is 10.

Elements of Geometry

8

The earliest geometry was concerned with space figures *(solids)* that could be recognized by their shapes, reproduced from clay or other materials, handled, admired, and used. Some of the very early accounts of religious rituals included detailed descriptions of the altars that were to be used. Many common geometric figures may be recognized in the shapes of temples, houses, building blocks, pottery, and other objects. In our three-dimensional world, experiences with geometry naturally started with solids. However, our descriptions of solid figures make use of the names of *plane figures*.

Look at the plane figures that occur on solid figures around you.

The annual flooding of the Nile River in Egypt made it necessary to relocate the boundaries of fields each year. This early surveying provided the basis for the word *geometry* (geometry) from the Greek words for *earth measure*. About 300 B.C. Euclid wrote these definitions.

A *point* is that which has no part.

A *line* is breadthless length.

A *straight line* is a line which lies evenly with the points on itself.

A *surface* is that which has length and breadth only.

A *plane surface* is a surface which lies evenly with the straight lines on itself.

These descriptions no longer provide satisfactory definitions since we now recognize that some *undefined terms* are needed to provide words for use in defining other terms.

We take **point** and **line** (straight line) as undefined terms. We assume *(postulate)* these statements.

Two points determine one and only one line.

A line and a point that is not on the line determine one and only one plane.

If two points of a line are on a plane, then every point of the line is on the plane.

A line is a set of points. Two lines that have exactly one point in common are **intersecting lines**. A plane is also a set of points. We think of a plane as a *flat surface*.

Any line may be taken as a real number line. Then each point of the line has a unique real number as its *coordinate* and each real number has a unique point of the line as its *graph*. Properties of real numbers may be used as in Section 7-1 to define half-line, opposite half-lines, ray, opposite rays, endpoint of a ray, line segment, and endpoints of a line segment. In the present section we use paper folding to explore a few geometric figures. Lines are represented by creases (folds) in the paper. The word *line* is used for both the abstract line and the crease, just as the word *number* is used for both the abstract number and its numeral.

Almost any kind of paper may be used for paper folding; however, the creases are particularly visible when waxed paper is used. A point is represented by the intersection of two lines, often by short creases "pinched" into the paper. Points and lines may be labeled for reference as on sketches and other representations of geometric figures.

Shown in the first three figures are points *A* and *B* on a piece of paper, a fold through *A* and *B*, and a line *AB*. Since one and only one fold can be made through two given points, we have demonstrated that

There is one and only one line through two given points.

Points are usually named by capital letters. A line may be named by any two of its points or by a single lower case letter. For example, the line *AB* containing the point *C* also may be named *AC* or *BC*, or by a letter such as *t* or *m*. The *line AB* is often written as $\overleftrightarrow{AB}$, the *line segment AB* as $\overline{AB}$, and the *ray AB* as $\overrightarrow{AB}$.

Let *C* be a point of the ray opposite to ray *AB*. Then *A*, *B*, and *C* are points of the same line, that is, **collinear points**. If ray *AB* is folded onto its opposite ray *AC*, the resulting crease can be formed in one and only one position. On this crease, label the opposite rays with endpoint *A* as *AD* and *AE*. The lines *AB* and *AD* appear to have a special relationship to each other. Before describing this relationship we define angles and right angles.

Any two rays *PQ* and *PR* with a common endpoint form an **angle.** The common endpoint *P* is the **vertex** of the angle; the rays *PQ* and *PR* are the **sides** of the angle. The angle may be named as ∠ *QPR* or as ∠ *RPQ* with the letter for the vertex as the middle letter, or if there is only one angle with vertex *P*, the angle may be named ∠ *P*.

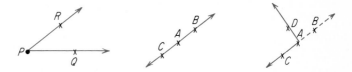

An angle with sides that are opposite rays, such as ∠ *CAB*, is often referred to as a **straight angle.** In the paper folding of ray *AB* onto its opposite ray *AC*, line *AD* was formed and ∠ *DAB* was folded onto (**superimposed upon**) ∠ *DAC*. We assume that if one figure may be superimposed upon another figure, the two figures are **congruent** (have the same measures). Then

∠ *DAB* ≅ ∠ *DAC*   (angle *DAB* is congruent to angle *DAC*)

and each angle is one-half of a straight angle; that is, each angle is a **right angle.**

Any two intersecting lines that form right angles are **perpendicular lines.** Thus the special relationship between lines *AB* and *AD* in the paper-folding example may be described by identifying the lines as perpendicular at the point *A*. The procedure used demonstrates the statement

> On a plane there is one and only one line that is perpendicular to a given line at a given point of the line.

Statements in geometry may be **assumptions** (postulates and definitions) or **theorems** (statements proved from assumed statements). Whether a particular statement is an assumption or a theorem often depends upon the assumptions that have been selected.

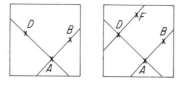

Let us return to the paper-folding example with perpendicular lines *AB* and *AD*. A line *DF* perpendicular to the line

DA at *D* may be obtained by folding the ray *DA* onto its opposite ray. The lines *DF* and *AB* do not intersect on the paper and do not appear to intersect even if extended indefinitely. Since the lines are on the same plane, they are coplanar lines. Also as coplanar lines that do not intersect, lines *AB* and *DF* are parallel lines, which is written *AB* ∥ *DF* (*AB* is parallel to *DF*). We have used paper folding to demonstrate the following theorem.

> Any two coplanar lines that are perpendicular to the same line are parallel lines.

Two line segments are parallel if they are on parallel lines; two line segments are perpendicular if they are on perpendicular lines.

In preparation for the exercises on the development of common figures and the demonstrations of theorems by paper folding, we need to define a half-plane and provide another illustration of the use of paper folding. The definition of a half-plane is very similar to that of a half-line. A line may be determined by two points *A* and *B*. As shown in the adjacent figure, the point *A* separates the line into three parts,

Your "line of vision" as you look in any specified direction represents a half-line.

the point *A*

the half-line that contains *B* and has endpoint *A*

the opposite half-line

The point *A* is often called the endpoint of each half-line even though it is not a point of either half-line.

A plane may be determined by a line *t* and a point *D* that is not a point of the line *t*. As shown in the figure, the line *t* separates the plane into three parts,

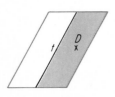

When the horizon appears to be a line, the sky represents a half-plane and the scenery is on the opposite half-plane.

the line *t*

the half-plane that contains *D* and has edge *t*

the opposite half-plane

8 ELEMENTS OF GEOMETRY

The line *t* is often called the **edge** of each half-plane even though it is not a part of either half-plane.

Triangle *ABC* Interior points of triangle *ABC*

Several definitions are included in the exercises. A few others are desirable for effective communication.

Label three noncollinear points *A*, *B*, and *C* on a piece of paper. Make folds for the lines *AB, BC,* and *AC.* The union of the line segments *AB, BC,* and *AC* is the **triangle *ABC*,** written △*ABC*. The three line segments are the **sides** of the triangle. The points *A, B,* and *C* are the **vertices** (singular, *vertex*) of the triangle. The **interior points** of the triangle are the points of intersection of

> the half-plane that has edge *AB* and contains *C*
>
> the half-plane that has edge *BC* and contains *A*
>
> the half-plane that has edge *AC* and contains *B*

Triangular region   Exterior points of a triangle

The union of the triangle and its interior points is a **triangular region.** The points that are on the plane of the triangle but are not points of the triangular region are **exterior points** of the triangle.

Any set of points on a plane may be called a **plane figure.** Any union of line segments such as *AB, BC, CD, DE,* and *EF* is a **broken line *ABCDEF*.** A broken line such as *PQRSTUW* for which the line segments have distinct endpoints and no interior point of any line segment is also a point of another line

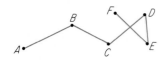

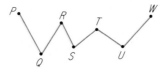

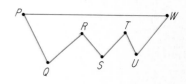

As an example of a broken line consider the path of a chess piece in a game or the path of a child's foot in hopscotch.

segment is called a **simple broken line.** A simple broken line such as *PQRSTUWP* with the same point as the first and last endpoints of its segments is a **simple closed broken line,** that is, a **simple closed union of line segments.**

As examples of polygons consider the figures formed by the exterior walls of most buildings.

Any plane figure that is a simple, closed, connected, union of line segments is a **polygon**. The line segments are **sides** of the polygon; the endpoints of the line segments are **vertices** of the polygon. A polygon is a **convex polygon** if every line segment PQ with points of the polygon as endpoints is either a subset of a side of the polygon or has only the points P and Q in common with polygon. A polygon that is not a convex polygon is a **concave polygon**.

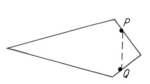
Convex polygon

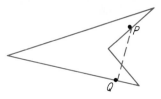
Concave polygon

If P and Q are points of a convex polygon and the line segment PQ is not a subset of a side of the polygon, then the points of the open line segment PQ are **interior points** of the convex polygon. The union of the points of a convex polygon and its interior points is a **polygonal region**. The points of the plane of a convex polygon that are not points of the polygonal region are **exterior points** of the polygon.

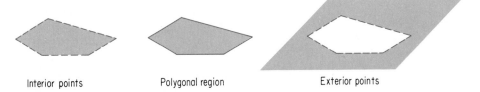

Interior points   Polygonal region   Exterior points

Fold a sheet of paper to represent each type of polygon. Note that a polygonal region must have at least three sides.

Plane polygons are classified according to the number of sides that each has. Most of the common names are shown in the next array.

NUMBER OF SIDES	NAME OF POLYGON
3	triangle
4	quadrilateral
5	pentagon
6	hexagon
7	heptagon
8	octagon
9	nonagon
10	decagon
12	dodecagon
n	n-gon

8 ELEMENTS OF GEOMETRY

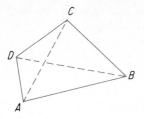

Any quadrilateral may be named by listing its vertices as if you were walking around the figure.

This square has two names starting with S. They are SENW and SWNE. You can start at any one of the four vertices and proceed from that vertex in either of two directions. State the other six ways of using its vertices to name the given square.

The quadrilateral has the points *A*, *B*, *C*, and *D* as vertices. Any two vertices, such as *A* and *B*, that are endpoints of a common side are **consecutive vertices**. The quadrilateral has the line segments *AB*, *BC*, *CD*, and *DA* as sides and the line segments *AC* and *BD* as **diagonals**. Any two sides, such as *AB* and *BC*, that have a common endpoint are **adjacent sides**. Any two sides, such as *AB* and *CD*, that are not adjacent sides are **opposite sides**. The quadrilateral may be named by listing the points *A*, *B*, *C*, and *D* in any order of consecutive vertices. For example, *CBAD* and *CDAB* are names for the given quadrilateral but *CADB* and *CBDA* are not names for the given quadrilateral. The four vertices of the quadrilateral determine six line segments: the four sides and the two diagonals *AC* and *BD*. In the name of any quadrilateral, consecutive letters must identify a side, and not a diagonal, of the quadrilateral.

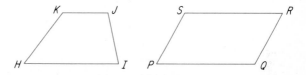

Parallelism and perpendicularity (right angles) may be used to classify quadrilaterals. A quadrilateral is a **trapezoid** if at least one pair of opposite sides are parallel; a quadrilateral is a **parallelogram** if both pairs of opposite sides are parallel. In the figures *HIJK* and *PQRS* both appear to be trapezoids; only *PQRS* appears to be a parallelogram. A parallelogram with at least one (actually all four) of its pairs of adjacent sides perpendicular is a **rectangle**. A parallelogram with at least one (actually all four) of its pairs of adjacent sides congruent is a **rhombus**. A parallelogram that is both a rhombus and a rectangle is a **square**.

Rectangle   Rhombus   Square

EXERCISES *Use paper folding to illustrate the given theorem. In many cases steps, labels, and definitions are suggested in the exercises.*

1. *Theorem* There are infinitely many points on any given line.

    To illustrate this theorem it is only necessary to show that for any given number of points on a line, another point can be added.

(a) Represent a line t containing points A and B.

(b) Represent additional points C, D, E, F, G, and H on line t.

2. *Theorem* There are infinitely many lines through any given point.

   To illustrate this theorem it is only necessary to show that for any given number of lines through a point, another line can be added.

   (a) Represent lines a and b through a point P.

   (b) Represent additional lines c, d, e, f, g, and h through the point P.

3. *Theorems* Any line segment has one and only one midpoint.

   Any line segment has one and only one perpendicular bisector.

   The point M of a line segment AB such that the line segments AM and BM are congruent is the **midpoint** of the line segment AB. The line t that is perpendicular to line AB at M is the **perpendicular bisector** of segment AB.

   (a) Represent a line containing points A and B.

   (b) Fold the paper so that the point B is superimposed upon the point A.

   Label the line obtained t and its intersection with AB as M.

   (c) Explain why M is the one and only midpoint of the line segment AB.

   (d) Explain why t is the one and only perpendicular bisector of AB.

4. *Theorem* Any two lines that are respectively perpendicular to intersecting lines are themselves intersecting lines.

   (a) Represent an angle KLM such that the lines KL and LM are intersecting lines.

   (b) Represent the line k that is perpendicular to KL at K and the line m that is perpendicular to ML at M.

5. *Theorem* The base angles of an isosceles triangle are congruent.

   Any triangle with at least two sides congruent is an **isosceles triangle.** The third side is the **base** of the isosceles triangle. The endpoints of the base are the vertices of the **base angles** of the isosceles triangle.

   (a) Represent a line segment RS with perpendicular bisector p and midpoint M.

   (b) Select a point T different from M on the line p and make folds for the line segments *TR* and *TS*.

(c) Show that the line segments *RT* and *ST* are congru-
ment and thus that the line segment *RS* is the base of
an isosceles triangle *RST* with base angles *RST* and
SRT.

(d) Show that the angles *RST* and *SRT* are congruent.

6. *Theorems* There are infinitely many isosceles triangles
with a given line segment *AB* as a base.

There are infinitely many scalene triangles
with a given line segment *AB* as a side.

There are infinitely many right triangles with a
given line segment *AB* as a side and a right
angle at *A*.

Any triangle that is not an isosceles triangle is a **scalene triangle.** Any triangle *ABC* with perpendicular sides
AB and *AC* is a *right triangle* with the right angle at *A* and
the side *BC* as *hypotenuse*.

7. *Theorem* There is one and only one line that is perpendicular to a given line and contains a given
point that is not a point of the given line.

(a) Represent a line *AB* and a point *C* that is not a point
of *AB*.

(b) Represent a line *t* that contains *C* and is perpendicular
to *AB*.

8. *Theorem* The sum of the measures of the angles of any
triangle is equal to the measure of a straight
angle.

(a) Represent a triangle *PQR* with the vertices labeled so
that there is a line *PS* perpendicular to the line *QR* and
with *S* an interior point of the line segment *QR*.

(b) After cutting out the triangular region *PQR*, superim-
pose all three points *P*, *Q*, and *R* on the point *S*.

9. *Theorem* Any angle *ABC* such that the points *A*, *B*, and
C are not collinear points has a unique angle
bisector.

(a) Represent an angle *ABC* such that the points *A*, *B*,
and *C* are not collinear points.

(b) Represent points *D*, *E*, *F*, *G*, and *H* that are on the
intersection of the half-plane that has edge *BA* and
contains *C* with the half-plane that has edge *BC* and
contains *A*, that is, points that are **interior points** of
∠ *ABC*.

(c) Represent the ray *BK* where *K* is an interior point of
∠ *ABC* and the ray *BK* is on the line obtained when
the ray *BA* is superimposed on the ray *BC*.

(d) Explain why angles *ABK* and *CBK* are congruent, that is, ray *BK* is the angle bisector of ∠ *ABC*.

10. *Theorem*   The angle bisectors of any triangle are concurrent.

     Three or more lines (or other figures) that contain a given point are concurrent at that point.

11. *Theorem*   The perpendicular bisectors of the sides of any triangle meet at a single point, that is, are concurrent.

12. *Theorem*   The medians of any triangle are concurrent.

     A line segment with a vertex of a triangle as one endpoint and the midpoint of the *opposite side* (the side determined by the other two vertices) of the triangle as the other endpoint is a median of the triangle.

13. *Theorem*   The altitudes of any triangle are on concurrent lines.

     An altitude of a triangle is a line segment that has a vertex of the triangle as an endpoint and is perpendicular to the opposite side of the triangle. The intersection of the altitude with the line determined by the opposite side of the triangle is the other endpoint, often called the *foot,* of the altitude. The foot may be an interior point, an endpoint, or an exterior point of the opposite side.

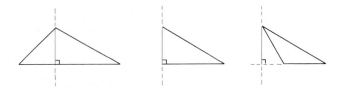

*14. *Theorem*   There is one and only one line that is parallel to a given line *m* and contains a given point that is not on the line *m*.

*15. *Theorem*   The midpoint of the hypotenuse of a right triangle is equidistant from the vertices of the triangle.

*16. *Theorem*   There are infinitely many right triangles with a given line segment *AB* as hypotenuse.

*17. *Theorem*   For a given line segment *AB* with midpoint *M*, it is possible to use paper folding to construct as many points as desired of the circle with center *M* and radius *MA*.

Use paper folding for each exercise.

See Exercise 13. 18. (a) Make folds for the sides of a triangle *ABC* and the altitude *AD*.
   (b) Fold the point *A* onto the point *D* and label the crease *EF*, where *E* is on the line *AB* and *F* is on *AC*.
   (c) Identify the type of quadrilateral represented by *BCFE*.

19. (a) Repeat Exercise 18 for an isosceles triangle *ABC* with *A* on the perpendicular bisector of the side *BC*.
   (b) Explain why line segments *BE* and *CF* are congruent, that is, why the trapezoid *BCFE* is an isosceles trapezoid.

20. For any given triangle *ABC* describe a construction by paper folding for a parallelogram *ABCD* and a different parallelogram *ABFC*.

21. (a) Make folds for any line segment *AB* with line *AM* perpendicular to *AB* at *A* and line *BK* perpendicular to *AB* at *B*.
   (b) Make a fold for the bisector of ∠ *BAM* and label the intersection *C* of this angle bisector and line *BK*.
   (c) Make a fold for the line that is perpendicular to line *BC* at *C* and label its intersection *D* with line *AM*.
   (d) Explain why quadrilateral *ABCD* is a rectangle.
   (e) Make creases for the angle bisectors and show that the sides of the rectangle are all congruent; that is, the rectangle is a *square*.

Use paper folding to demonstrate each theorem.

22. For any rectangle:
   (a) The opposite sides of the rectangle are congruent.
   (b) The diagonals separate the rectangular region into two pairs of congruent triangular regions.
   (c) The triangles are isosceles triangles.
   (d) The diagonals of a rectangle are congruent line segments and bisect each other.

23. The diagonals of any parallelogram bisect each other.

Conjecture a general statement for each specified situation.

24. A single point on a line separates the line into two nonintersecting *(disjoint)* parts, besides the point itself. Into how many disjoint parts, besides the points themselves, do *n* distinct points of a line separate the line?

25. A single point of a circle does not separate a circle; thus there remains one part beside the point. Into how many disjoint parts, besides the points themselves, do n distinct points of a circle separate the circle?

In 1852 Francis Guthrie mentioned in a letter to his brother Frederick that it seemed that every map drawn on a sheet of paper could be colored with only four colors in such a way that countries sharing a common boundary have different colors. For 124 years the problem provided a challenge for both amateur and professional mathematicians. In 1976 Kenneth Appel and Wolfgang Haken proved Francis Guthrie's conjecture by making extensive use of computers. (See their article cited in Exploration 17.) Searches continue for a proof that does not depend on computers.

26. Consider a map composed of a single central region and n neighboring regions as in the figure, where $n = 4$. If $n = 1$, the map can be colored with two colors. In general, regions that have a common edge (boundary) should have different colors. For this special type of map, how many colors are needed (what is the smallest possible number of colors) for the central region and its n neighbors?

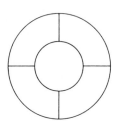

Explorations

1. Four lines are to be drawn on a sheet of paper. Can the lines be drawn so that there are no points of intersection (even if the drawings were extended)? One point of intersection? Two points? Three points? Four points? Five points? Six points? Only one of these seven cases is impossible. Two cases can occur in different appearing ways. Sketch figures for all possible situations.

2. For any given triangle two line segments are to be drawn with points of the triangle as endpoints. Three or four non-overlapping polygons are formed and the sum of the numbers of sides of these polygons determined. Here are two examples.

 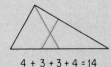

3 + 4 + 3 = 10   4 + 3 + 3 + 4 = 14

Sketch such figures for as many different sums as possible and list those sums from smallest to largest.

3. The early Greeks considered sets of points that formed geometric figures. Find a formula for T_n, the nth triangular number, which is the sum of the integers 1 through n.

T_1 T_2 T_3 T_4 ...

Hint: Compare R_n and T_n (see Exploration 3).

*4. One line separates a plane into two regions. Two lines separate the plane into at most four regions. Three lines separate a plane into at most seven regions. Find a formula for the greatest possible number R_n of regions into which n lines can separate a plane.

5. Make a list of places in which you encounter mirror images in daily life. This list may range from such activities as simply combing your hair while looking in a mirror, to seeing the word "ambulance" printed on the front of an ambulance so that it reads correctly when you see it in your rearview mirror.

Games may be used to present many mathematical concepts very effectively. Frequently slight modifications of the rules for a particular game lead to another very different game. The following games are described in terms of points and line segments on a piece of paper. However, other procedures such as pegs and elastics on a peg board may be used instead.

6. Make several dots for points on a piece of paper as in the adjacent figure. Two players take turns drawing line segments with the given dots as endpoints. No line segment may cross another or contain more than two of the given points. The last person able to play is the winner.

7. Proceed as in Exploration 6, but with the added restriction that no point may be the endpoint of more than two line segments.

8. Proceed as in Exploration 6, but with the added restriction that no point may be the endpoint of more than three line segments.

9. In Exploration 6, can you win if you begin with three dots and your opponent has the first move?

When doing paper folding, designs and patterns may be formed by folding a given point P to each of many points of a given line, circle, or other figure. Frequently the creases represent the lines that are tangent to a familiar curve. Explore the possibilities of such an approach using the indicated figure and point. A point S is an interior point of a circle with center O and radius r if $OS < r$; a point T is an exterior point if $OT > r$.

10. A circle and its center O.

11. A circle and a point Q that is a point of the circle.

12. A line m and a point P that is not a point of m.

13. A line m and a point P that is a point of m.

14. A circle and a point S that is an interior point but not the center of the circle.

15. A circle and a point T that is an exterior point of the circle.

16. Read and prepare a report on *Flatland: A Romance in Many Dimensions* by E. A. Abbott, fifth revised edition, Barnes and Noble, 1963.

17. Prepare a report on the *four-color problem*. For example, see "Snarks, Boojums, and Other Conjectures Related to the Four-Color-Map Theorem" by Martin Gardner on pages 126 through 130 of the April 1976 issue of *Scientific American* and "The Solution of the Four-Color-Map Problem" by Kenneth Appel and Wolfgang Haken on pages 108 through 121 of the October 1977 issue of *Scientific American*.

8-2

Measures of Plane Figures

The basic concepts of measurement are the same in all cases. Consider the measurement of the line segment PQ in the next figure. A *unit of linear measure* is needed. As usual any convenient unit of measure can be used. The selected unit is repeated as indicated on the scale *(ruler)*.

8 ELEMENTS OF GEOMETRY

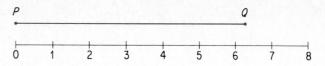

In geometry, particularly in applied situations, we often are concerned with the size of a figure. Thus we frequently measure distances, areas, volumes, and amounts of rotation.

To the nearest unit, $PQ = 6$ and we write

$$PQ \approx 6$$

(Recall that as before, the symbol $\approx$ is read "is approximately equal to.") Some people mentally subdivide the unit of length on a scale and *estimate* measures to smaller units. For example, different people might give the following estimates.

$PQ = 6$ to the nearest half unit.

$PQ \approx 6\frac{1}{4}$, that is, 6 and $\frac{1}{4}$ to the nearest quarter unit.

$PQ \approx 6.2$, that is, 6 and $\frac{2}{10}$ to the nearest tenth of a unit.

Whatever subdivisions are used, the measure is based on someone's "reading the scale" and thus is approximate. All linear measurements are approximate since they are based on estimation from reading a scale. Indeed, all measurements are approximate. The *length* of a line segment, such as $\sqrt{2}$ for a diagonal of a unit square, may be an irrational number but any *measurement* of that line segment must be a rational number of the units used in making the measurement. Even though a measurement is made with a very precise instrument, there is always a final estimation and thus an approximation. The smallest unit that is actually used is the **unit of precision**. Suppose that the selected unit that is marked on the scale is one centimeter. Then in a measurement to the nearest unit, the unit of precision would be one centimeter; in a measurement to the nearest tenth of a centimeter, one millimeter would be the unit of precision.

Example 1 State the unit of precision for the given measurement.

(a) 5.2 kilometers   (b) $3\frac{1}{8}$ units

(c) 5000 to the nearest hundred people

Solution (a) One-tenth kilometer   (b) One-eighth unit
(c) 100 people

   A *count* of individual items is a special type of measurement. Whenever a count is actually made, the measurement is

In applications of mathematics the distinction between exact and approximate measurements can require careful consideration. For example, the measurement involved in buying *two 1-quart packages* of ice cream is exact but the measurement involved in buying *two quarts* of ice cream that is hand packed into cartons is an approximate measurement since the measurement of each quart is approximate.

exact and the greatest possible error is zero. For example, the number of registered students who are in your class and eventually obtain credit for the course is an exact number. The number of registered students who have studied from one of the editions of this book and obtained college credit for the course before January 1, 1984 is an exact number, although we cannot readily count the students to determine that number. It should be possible to identify any given measurement as either an **exact measurement** based on counting or as an **approximate measurement** with a positive greatest common error and therefore a positive unit of precision.

Example 2 Identify the given measurement as exact or approximate.
(**a**) June lives *2 kilometers* from school.
(**b**) Pedro's next class is *three doors* down the corridor from this class.
(**c**) Susan bought *two new tires.*
(**d**) Bob bought *1 kilogram* of grapes.

Solution (**a**) Approximate (**b**) Exact (**c**) Exact (**d**) Approximate

The unit of area measure is a *unit square,* that is, a square with side 1 linear unit. The common units of angular measure are defined so that

$$1 \text{ revolution} = 360 \text{ degrees} = 4 \text{ right angles}$$
$$= 2 \text{ straight angles} = 2\pi \text{ radians}$$

where a **radian** is the SI unit of angle measure.

An angle has been defined as the union of two rays with a common endpoint. To measure an angle it is customary to select one ray as the **initial side,** the other ray as the **terminal side,** and a direction of rotation (clockwise, $-$, or counterclockwise, $+$) about the vertex of the angle. The measure of the angle is the amount of rotation needed to rotate a ray from the position of the initial side to the position of the terminal side. When an angle is represented but an initial side is not indicated, the measurement is usually taken as a number n of degrees, $0 \leq n \leq 180$. Thus, in the first figure, $\angle ABC = 60°$ unless curved arrows are added to indicate the initial side and a direction of rotation that requires a different angle measurement.

8 ELEMENTS OF GEOMETRY

An angle ABC is

an **acute angle** if $0° < \angle ABC < 90°$

a **right angle** if $\angle ABC = 90°$

an **obtuse angle** if $90° < \angle ABC < 180°$

a **straight angle** if $\angle ABC = 180°$

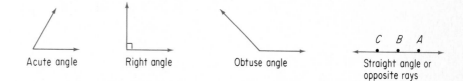

| Acute angle | Right angle | Obtuse angle | Straight angle or opposite rays |

Measurements enable us to extend our paper-folding definitions of the congruence of line segments and the congruence of angles. Any two line segments with equal measurements are **congruent line segments**; any two angles with equal measurements are **congruent angles**.

Any polygon or other simple closed plane curve is the common boundary of a bounded region (its **interior**) and an unbounded region (its **exterior**). The **area** $\mathcal{A}$ of the polygon is the number of unit square regions that is "equivalent" to the interior of the polygon. It is customary to speak of the "area of the polygon" rather than to specify the "area of the region bounded by the polygon." Here are some common formulas for areas of polygons.

Square $\qquad \mathcal{A} = s^2$

Rectangle $\qquad \mathcal{A} = bh$

Parallelogram $\qquad \mathcal{A} = bh$

Right triangle $\qquad \mathcal{A} = \dfrac{1}{2}bh$

Any triangle $\qquad \mathcal{A} = \dfrac{1}{2}bh$

Trapezoid $\qquad a = \dfrac{1}{2}h(b_1 + b_2)$

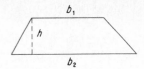

$s = 3$
$a = 9$

$b = 4, \quad h = 2$
$a = 8$

The area of any square with a side of integral length can be found by counting the number of unit squares in the square region. For example, a square with side 3 can be seen to "contain" 9 unit squares. Similarly, the area of any rectangle with sides of integral lengths can be obtained by counting. We assume here without formal proof that the given formulas hold for all squares and rectangles.

Any parallelogram is equivalent to a rectangle with the same base and height, as shown in the figures.

   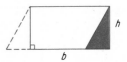

Any right triangle is half of a rectangle; any triangle is half of a parallelogram.

Any trapezoid is equivalent to a sum of two triangles.

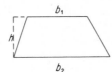 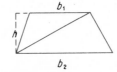 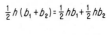

$\frac{1}{2}h(b_1 + b_2) = \frac{1}{2}hb_1 + \frac{1}{2}hb_2$

These visualizations of areas of parallelograms, triangles, and trapezoids may be demonstrated by constructions and also by cutting out and reassembling figures. Formal proofs are left for advanced courses. The word *area* is used for both the area measure (number) and the area measurement (number of units). However, the units should be specified in the answer to a problem whenever units have been designated.

Example 3 Find the area of the region shown at the top of the next page.

8 ELEMENTS OF GEOMETRY

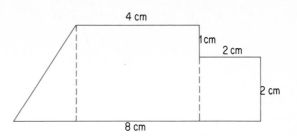

We subdivide the region into triangular and rectangular regions, as shown in the next figure. The triangular region has area $(1/2)(2)(3)$, that is, 3; the rectangular regions have areas 3×4 and 2×2, that is, 12 and 4. Thus, in square centimeters, the area of the given region is 19 cm^2.

Any circle of radius r has circumference C and area $Ⅽ$ as given by these formulas:

The value, π, of the ratio of the circumference to the diameter of a circle has had utilitarian and theoretical significance for thousands of years. The very early Babylonians, Hindus, and Chinese used the value 3. This same value is used in the Old Testament in I Kings vii:23, where a molten sea is described with diameter 10 cubits and circumference 30 cubits.

$$C = 2\pi r$$
$$Ⅽ = \pi r^2$$

We accept these formulas here without proof. The first formula serves as a definition of π and the second may be explained as in Exploration 2. The length of a simple closed curve is called the *circumference* if the curve is a circle and the *perimeter* if the curve is a polygon. Often the words *perimeter* and *circumference* are used for other curves.

Example 4   Find the length and the area of the given curve formed by line segments and semicircles. The line segments intersect at right angles.

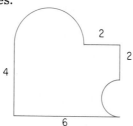

Solution   The semicircles have radii 1 and 2 and thus have lengths π and 2π. The length of the curve is

$$4 + 6 + \pi + 2 + 2 + 2\pi, \quad \text{that is,} \quad 14 + 3\pi \text{ units.}$$

To find the area we may think of a 4×6 rectangle with one semicircular region added and the other subtracted. Then the area is $24 + 2\pi - \pi/2$, that is, $24 + (3/2)\pi$ square units.

EXERCISES   *Without measuring, choose the length that is your best estimate for the length of the given line segment.*

1.   (a) 2 inches,   3 inches, $2\frac{1}{2}$ inches, $1\frac{1}{2}$ inches
    (b) 4 cm,   5 cm,   6 cm,   7 cm,   8 cm

2.   (a) $3\frac{1}{2}$ inches, $3\frac{3}{4}$ inches,   4 inches, $4\frac{1}{4}$ inches, $4\frac{1}{2}$ inches

    (b) 9 cm,   9.5 cm,   10 cm,   10.5 cm,   11 cm

State the unit of precision for the given measurement.

3. (a) 3 kilometers   (b) 1.02 centimeters

4. (a) $2\dfrac{1}{4}$ meters   (b) 2.25 meters

5. (a) $3\dfrac{1}{2}$ liters   (b) 3.5 liters

6. (a) $22\dfrac{1}{2}$ degrees   (b) $22\dfrac{2}{4}$ degrees

Identify the indicated measurement as exact or approximate.

7.   (a) Doris has *three brothers*.
    (b) The oldest is *1.5 meters* tall.
8.   (a) Art drove *500 miles* today.
    (b) He visited *seven customers*.
9.   (a) Carol drove for *eight hours*.
    (b) She made only *three stops*.
10.   (a) Sue painted *two rooms* last week.
    (b) She used *four and one-half gallons* of paint.

If possible, sketch and label the indicated figure. If not possible, explain why it is not possible.

11.   An acute angle *ABC*.
12.   Two acute angles *CDE* and *FGH* that are not congruent.
13.   A right angle *IJK*.

14. Two right angles *LMN* and *OPQ* that are not congruent.
15. An obtuse angle *RST*.
16. Two obtuse angles *UVW* and *XYZ* that are not congruent.

Find the area of the given region.

17.

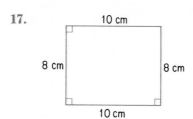

18.

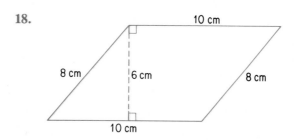

19.

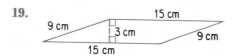

20.

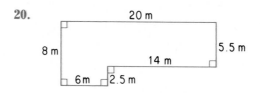

21.

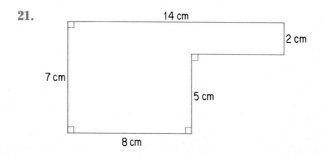

22.

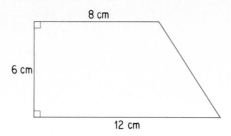

8 cm

6 cm

12 cm

23.

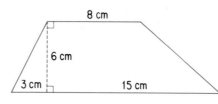

8 cm

6 cm

3 cm   15 cm

24.

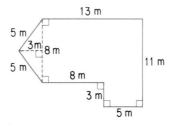

13 m

5 m

3 m 8 m

5 m   11 m

8 m

3 m

5 m

About 240 B.C. Archimedes found the perimeter of a polygon with its vertices at 96 points equally spaced around a circle and also the perimeter of a polygon with its sides perpendicular to the radii to those points. He used the fact that the circumference of the circle would be greater than the perimeter of the first polygon (the inscribed polygon) and less than the perimeter of the second polygon (the circumscribed polygon) to show that

$$3\frac{10}{71} < \pi < 3\frac{1}{7}$$

and thus that π is approximately 3.14.

Assume that the radius is exact, use π as 3.14, and find the approximate area of any circle with the given radius.

25.	3 cm	**26.**	10 cm	**27.**	9 mm
28.	5 mm	**29.**	6.1 km	**30.**	0.5 km

The curves in Exercises 31 and 32 are formed by line segments and circular arcs. For the given curve find **(a)** *the length* **(b)** *the area.*

31.

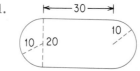

30

10 20   10

32.

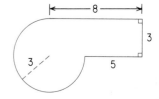

8

3

3   5

33. A border 2 centimeters wide is part of a rectangular sheet of paper 20 centimeters by 30 centimeters. Find the area of the border.

34. A walk 1 meter wide is to be built outside and as a boundary of a rectangular garden 15 meters by 20 meters. Find the area of the walk if the outside "corners" of the walk are (**a**) square (**b**) arcs of a circle.

*35. Suppose that the figure in Exercise 31 is the inside rail of a small track and the dimensions of the track are in meters. The surface of the track is 4 meters wide. Find the area of the surface of the track.

*36. Suppose that the dimensions of the figure in Exercise 32 are in meters and a border 1 meter wide (rounded corners where appropriate) is constructed around the outside of the region. Find the approximate area of the border. Explain why your answer is approximate rather than exact.

Explorations

1. Suppose that you have two pieces of paper and a pencil. Explain how you can use just these tools to make a ruler with multiples and subdivisions of a selected unit.

Many algebraic formulas were originally expressed in words as statements about areas of geometric figures. Explain a possible use of each figure as a basis for the given algebraic formula.

2.

$$a(b+c+d) = ab + ac + ad$$

3.

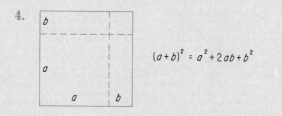

$$(a+b)(c+d) = ac + ad + bc + bd$$

4.

$$(a+b)^2 = a^2 + 2ab + b^2$$

5.

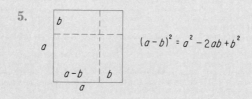

$$(a-b)^2 = a^2 - 2ab + b^2$$

6.

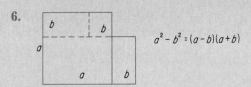

$$a^2 - b^2 = (a - b)(a + b)$$

7. Find and report on a reference that explains the historical basis for our division of a circle into 360 congruent parts to obtain measures of angles.

8. Locate a number of circular objects. Measure the circumference C and the diameter d of each object. Then compute the ratio C/d for each object and see how closely you approximate the value of π. (Note that since $C = 2\pi r$, then $C = \pi(2r) = \pi d$, and $\pi = C/d$.)

Four thousand decimal places of π may be found in Philip J. Davis's The Lore of Large Numbers, *now published by the Mathematical Association of America. Here are a few.*

$\pi = 3.14159\ 26535\ 89793\ 23846\ 26433$
$83279\ 50288\ 41971\ 69399\ 37510$
$58209\ 74944\ 59230\ 78164\ 06286$
$20899\ 86280\ 34825\ 34211\ 70679$
$82148\ 08651\ 32823\ 06647\ 09384$
$46095\ 50582\ 23172\ 53594\ 08128$
$48111\ 74502\ 84102\ 70193\ 85211$
$05559\ 64462\ 29489\ 54930\ 38196$

9. Mnemonic devices are often used by students as an aid to memorization. A very famous one that gives 13 digits for π is this: "See, I have a rhyme assisting my feeble brain, its tasks ofttimes resisting." Replace each word by the number of letters in that word. Thus "See" = 3, "I" = 1, "have" = 4, and so on. This gives $\pi = 3.141592653589$ correct to 12 places to the right of the decimal point. See what other mnemonic devices you can find, or invent, that are helpful in memorizing important mathematical facts.

10. Write a report on the history of the number π. For example, see "A Chronology of π" on pages 85 through 90 of *An Introduction to the History of Mathematics*, fifth edition, by Howard Eves, Saunders College Publishing, 1983.

8-3
The Golden Ratio

Evidence of the golden ratio is found in the shape of the Great Pyramid at Gizeh, Egypt; in the shape of the Parthenon at Athens, Greece; in the art of Leonardo da Vinci, Albrecht Dürer, and Georges Seurat; in the modern architecture of Le Corbusier; in numberous common rectangular shapes such as 5 × 8 cards, playing cards, and pages of books; and even in most human bodies (where the navel divides the height in approximately the golden ratio).

For more than two thousand years special attention has been given to the measures of plane figures that are pleasing to the eye. Each of the following rectangles is labeled with the approximate decimal value of the ratio of its length to its width. Select the rectangle that you find most pleasing to look at.

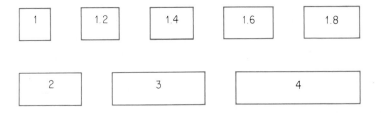

Rectangular shapes with sides having ratio approximately 16/10 have been considered the most pleasing to many people for at least 2500 years. The famous astronomer Johannes

The Parthenon

Kepler (1571–1630) credited Pythagoras (ca.540 B.C.) with two major achievements: the proof of the Pythagorean theorem, and the construction of a point that divides a line segment so that the ratio of the lengths of the two segments is the **golden ratio**, G. The decimal value of G is approximately 1.6. The early Greeks felt that the ideal shape for a rectangle was obtained when the ratio of its length to its width was G. These **golden rectangles** appear throughout Greek, Renaissance, and modern art and architecture.

If the sides $\overline{AB}$ and $\overline{BC}$ of a golden rectangle $ABCD$ are used to form a single line segment AC', then the point B divides $\overline{AC'}$ such that

$$\frac{AC'}{AB} = \frac{AB}{BC'} = G$$

We take $BC = BC' = 1$ and $AB = G = x$. Then $AC' = x + 1$ and

$$\frac{x + 1}{x} = \frac{x}{1}$$

$$x + 1 = x^2$$

$$x^2 - x - 1 = 0$$

$$x = \frac{1 \pm \sqrt{5}}{2}$$

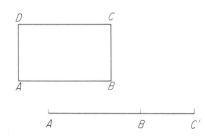

The values $(1 + \sqrt{5})/2$ and $(1 - \sqrt{5})/2$ may be obtained using the quadratic formula as in Section 7-5. Since $\sqrt{5} > 1$, the value $(1 - \sqrt{5})/2$ is negative and cannot be the length of a line segment.

Since $x > 0$, $x = (1 + \sqrt{5})/2 \approx 1.618034$. Thus $G \approx 1.618$.

Any two positive numbers p and q may be used to form two ratios, p/q and q/p. The reciprocal of G is often denoted by r, $1/G = r$. Since $1 < G$, we know that $r < 1$ and we shall find it convenient to use $AC' = 1$ and $AB = r$ to derive a value for r without using the value of G. For the golden rectangle $ABCD$ and the line segment ABC' we have the following relations.

$$r = \frac{1}{G} = \frac{AB}{AC'} = \frac{BC'}{AB}$$

$$\frac{r}{1} = \frac{1-r}{r}$$

$$r^2 = 1 - r$$

$$r^2 + r - 1 = 0$$

$$r = \frac{-1 \pm \sqrt{5}}{2}$$

Since $r > 0$, $r = (-1 + \sqrt{5})/2 \approx 0.618034$. The similarity of the decimal fraction portions of G and r may seem unexpected. However, since

$$G - r = \frac{1 + \sqrt{5}}{2} - \frac{-1 + \sqrt{5}}{2} = \frac{2}{2} = 1$$

their decimal portions are identical.

Since $r = 1/G$ and $G = 1 + r$

$$G = 1 + \frac{1}{G}$$

Suppose that we repeatedly substitute the expression $1 + (1/G)$ for G in the preceding equation. Then we have

$$1 + \frac{1}{G}, \quad 1 + \cfrac{1}{1 + \cfrac{1}{G}}, \quad 1 + \cfrac{1}{1 + \cfrac{1}{1 + \cfrac{1}{G}}}, \quad 1 + \cfrac{1}{1 + \cfrac{1}{1 + \cfrac{1}{1 + \cfrac{1}{G}}}}, \quad \ldots$$

Each of these expressions is equal to G. Accordingly, the sequence of numbers

$$1 + 1, \quad 1 + \cfrac{1}{1 + 1}, \quad 1 + \cfrac{1}{1 + \cfrac{1}{1 + 1}}, \quad 1 + \cfrac{1}{1 + \cfrac{1}{1 + \cfrac{1}{1 + 1}}}, \quad \ldots$$

approaches G. We may think of the last sequence as

$$1 + \frac{1}{1} = \frac{2}{1}, \quad 1 + \frac{1}{2} = \frac{3}{2}, \quad 1 + \frac{1}{\frac{3}{2}} = \frac{5}{3}, \quad 1 + \frac{1}{\frac{5}{3}} = \frac{8}{5}, \quad \ldots$$

There is a pattern in the last sequence. The next five terms are

$$\frac{13}{8}, \quad \frac{21}{13}, \quad \frac{34}{21}, \quad \frac{55}{34}, \quad \frac{89}{55}$$

Try to find the pattern. The denominator of each fraction after the first is the numerator of the preceding fraction; the numerator of each fraction is the sum of the numerator and the denominator of the preceding fraction. The sequence of numbers used to form the fractions is

$$1, \quad 1, \quad 2, \quad 3, \quad 5, \quad 8, \quad 13, \quad 21, \quad 34, \quad 55, \quad 89, \quad \ldots$$

where each number after the second is the sum of its two immediate predecessors. Such a sequence is called a **Fibonacci sequence** and may be used to approximate the value of G (Exercise 3). In 1202, Leonardo of Pisa, also known as Fibonacci, published a book that included a discussion of this sequence. Fibonacci originally suggested the sequence as the number of pairs of rabbits that an owner would have each month if a single pair one month old was acquired in the first month and if starting with the second month after birth each pair produced another pair each month (Exercise 4). Fibonacci sequences arise in numerous situations including architecture, art, astronomy, biology, botany, and music. Each Fibonacci sequence has the general form

$$a, \quad b, \quad a + b, \quad a + 2b, \quad 2a + 3b, \quad \ldots$$

EXERCISES

1. A starfish has approximately the shape of a regular pentagon. List the line segments of the complete regular pentagon that appear to be the same length as (a) $\overline{AB}$ (b) $\overline{AC}$ (c) $\overline{AG}$ (d) $\overline{GH}$.

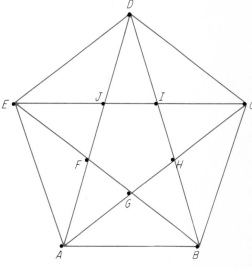

2. On the figure for Exercise 1 or a larger copy of that figure measure at least one line segment from each of the sets considered in Exercise 1. Then identify the three ratios of these lengths that are approximately equal to G. Describe one of these ratios for the picture of a starfish.

3. The sequence of ratios of the terms of the Fibonacci sequence

$$1, \quad 1, \quad 2, \quad 3, \quad 5, \quad 8, \quad 13, \quad 21, \quad 34, \quad 55, \quad 89, \quad 144, \quad 233, \quad 377, \quad \ldots$$

to their immediate predecessors is

$$\frac{1}{1}, \frac{2}{1}, \frac{3}{2}, \frac{5}{3}, \frac{8}{5}, \frac{13}{8}, \frac{21}{13}, \frac{34}{21}, \frac{55}{34}, \frac{89}{55}, \frac{144}{89}, \frac{233}{144}, \frac{377}{233}, \ldots$$

Express each of these ratios in decimal form and compare the value of G with the "trend" of these values as one continues along the sequence. The use of a calculator is recommended.

4. Assume that a friend acquires a one-month-old pair of rabbits and that starting with the second month after birth each pair of rabbits produces another pair each month. Start counting from the acquisition of the first pair of rabbits and state the number of pairs of rabbits that your friend will have at the beginning of the

    (a) first month   (b) second month

    (c) third month   (d) twelfth month

In the figure the height of the pyramid is h, the altitude of each triangular face of the pyramid is a, and each side of the base is 2b. Find each of these areas in terms of a, b, and h.

The early Greek traveler Herodotus is reported to have observed that the area of a square with edge equal to the height of the Great Pyramid at Gizeh would be equal to the area of any one of the triangular faces of the pyramid.

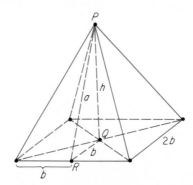

5. (a) The area of the base of the pyramid.
   (b) The area of a square with side h.

6. (a) The area of a triangular face of the pyramid.
   (b) The total area of the four triangular faces.

For the previous figure use Herodotus' assumption that $h^2 = ab$ and the Pythagorean theorem $a^2 = b^2 + h^2$ for right triangle PQR.

7. Show that $(b + a)/a = a/b$ and thus that $a/b = G$.

8. Show that the ratio of the total area of the four triangular faces of the pyramid to the area of the base of the pyramid is G.

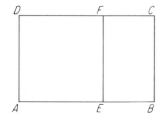

9. A square region *AEFD* is marked off from the region of a golden rectangle *ABCD* as in the figure. Let $AB = G$ and $BC = 1$. Find the lengths of the line segments *AE* and *EB*.

10. Use the areas of the square and the rectangles and the relationship $G = 1 + (1/G)$ to show that rectangle *BCFE* in the figure for Exercise 9 is a golden rectangle.

11. Assume that you are given the figure for Exercise 9. Let *M* be the midpoint of $\overline{AE}$. Find the length of (a) $\overline{ME}$ (b) $\overline{MF}$ (c) $\overline{MB}$.

12. Use the information obtained in Exercise 11 and describe a construction for a golden rectangle *ABCD* from a given square *AEFD*.

13. As in Exercise 12 construct a golden rectangle 10 cm wide. Mark off a 10-cm square to obtain a second golden rectangle. Mark off a square from the second golden rectangle to obtain a third golden rectangle. Finally, mark off a square from the third golden rectangle to obtain a fourth golden rectangle.

14. Make a scale drawing, approximately twice as long and twice as wide, of the figure for Exercise 9. Mark off a square region (a) from *BCFE* to obtain a golden rectangle *CFGH* (b) from *CFGH* to obtain a golden rectangle *FGIJ* (c) from *FGIJ* to obtain a golden rectangle *GIKL* (d) from *GIKL* to obtain a golden rectangle *IKMN*.

The Bernoulli brothers Jacob (also known as James or Jacques, 1654–1705) and Johann (also known as John, 1667–1748) both left other careers and became talented mathematicians with many interests and discoveries. Jacob found many properties of the spiral and requested that it be engraved on his tombstone with a Latin inscription meaning: "Though changed, I arise again the same."

15. In the figure for Exercise 14 note that the points *B*, *F*, *I*, and *M* are on the diagonal $\overline{BF}$; also the points *A*, *C*, *G*, *K*, *O*, and $P = \overline{MN} \cap \overline{AC}$ are on the diagonal $\overline{AC}$. Sketch a "smooth" curve *OPMKIGFCBA* using the points on the diagonals. This curve was named a *logarithmic spiral* in the seventeenth century by Jacob Bernoulli. Logarithmic spirals are frequently found in nature. The three examples shown on page 342 are the distribution of seeds in the head of a sunflower, the distribution of scales on a pineapple, and the shell of a chambered nautilus.

Explorations

1. The numbers 5 and 8 are consecutive numbers in the Fibonacci sequence 1, 1, 2, 3, 5, 8, 13, ... As indicated in the figure, cut a square region with 8 units on each side into four pieces and reassemble to form a rectangle 5 by 13 units. The square has area 64 square units; the rectangle has area 65 square units. How did the extra unit of area arise?

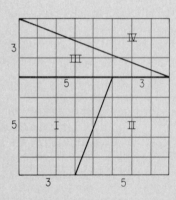
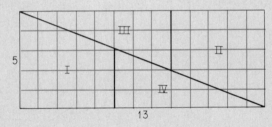

2. Repeat Exploration 1 for two other consecutive numbers from the given Fibonacci sequence.

3. For the second, third, fourth, fifth, sixth, and seventh terms of the Fibonacci sequence 1, 2, 3, 5, 8, 13, 21, 34, . . . compute the square of the indicated term minus the product of its immediate neighbors.
   (a) State the pattern formed by these differences.
   (b) Discuss two results that could arise in Exploration 2.

4. The sums in the column n are obtained by adding the ele-

See the explorations of Section 9-7 for a discussion of Pascal's triangle and probability.

ments on "diagonals" of **Pascal's triangle** as in the figure. Extend Pascal's triangle, find the first ten numbers n, and compare the sequence of numbers n with a Fibonacci sequence.

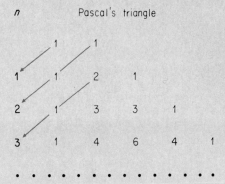

n Pascal's triangle

*5. Use the numbers $F(n)$ of the Fibonacci sequence as in the table. Then select values for n and explore the patterns formed by the values of the given expression.

n	1	2	3	4	5	6	7	8	9	...
$F(n)$	1	1	2	3	5	8	13	21	34	...

(a) $[F(n)]^2 - [F(n-1)][F(n+1)]$

(b) $1 + F(1) + F(2) + F(3) + \cdots + F(n)$

(c) $[F(n)]^2 + [F(n+1)]^2$

8-4
Figures on a Coordinate Plane

The application of both algebra and geometry to the study of a single mathematical topic is illustrated by the many uses of coordinates.

Any two points R and S determine a line segment RS and a line RS. The line RS may be taken as a coordinate line.

Let r be the coordinate of point R and let s be the coordinate of point S. If $s \geq r$, the *length* of the line segment RS is $s - r$; if $s < r$, the length is $r - s$. Thus in all cases the length of the line segment RS is $|s - r|$, the *absolute value* of $s - r$. If $R \neq S$, the line segment has positive length. The **directed distance** from R to S is $s - r$; the directed distance from S to R is $r - s$.

The *midpoint M* of any line segment *RS* on a number line may be found by starting at *R* with coordinate *r* and going halfway to *S* with coordinate *s*, that is, going $(\frac{1}{2})(s - r)$ units from *R* toward *S*. Thus *M* has coordinate

$$r + \frac{1}{2}(s - r)$$

Since

$$r + \frac{1}{2}(s - r) = r + \frac{1}{2}s - \frac{1}{2}r = \frac{1}{2}r + \frac{1}{2}s = \frac{1}{2}(r + s)$$

the coordinate of *M* is $(\frac{1}{2})(r + s)$. If *R* has coordinate -1 and *S* has coordinate 5, then the directed distance from *R* to *S* is 6,

$$s - r = 5 - (-1) = 6$$

and the coordinate of the midpoint *M* is $(\frac{1}{2})[(-1) + 5]$, that is, 2.

If *R* has coordinate 7 and *S* has coordinate 3, then the directed distance from *R* to *S* is -4 and the coordinate of the midpoint *M* is $(\frac{1}{2})(7 + 3)$, that is, 5.

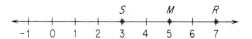

These procedures for finding the length and midpoint of any line segment on a coordinate line may be used on a coordinate plane for any line segments on lines that are parallel to a coordinate axis.

Example 1 Find the length and the midpoint of the line segment *AB* for *A*: (1, 5) and *B*: (7, 5).

Solution Graph the line segment *AB*. On the line $y = 5$, the line segment *AB* has length $7 - 1$, that is, 6. The points *A*, *B*, *C*: (1, 0), and *D*: (7, 0) are the vertices of a rectangle *BACD* with $AB = CD = 6$. The *y*-coordinate of the midpoint *M* of the line segment *AB* must be 5, since every point of the line *AB* has *y*-coordinate 5. The midpoint *E* of the line segment *CD* has *x*-coordinate $(1 + 7)/2$, that is, 4. Since the three parallel lines *AC*, $x = 4$, and *BD* intercept (cut off) congruent line segments

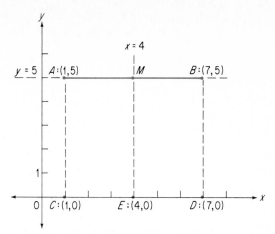

(CE and *ED)* on the *x*-axis, these parallel lines intercept congruent line segments *(AM* and *MB)* on the line $y = 5$. Therefore, *M* has coordinates (4, 5).

Any point on a coordinate plane may be identified by its coordinates. Thus in Example 1 the point *M* may also be identified as the point (4, 5). In general, any two points *A* and *C* that are on a line $y = y_1$ parallel to the *x*-axis are endpoints of a line segment with midpoint *Q*.

$$A: \quad (x_1, y_1) \qquad C: \quad (x_2, y_1) \qquad Q: \quad \left(\frac{x_1 + x_2}{2}, y_1 \right)$$

Similarly, any two points *A* and *D* that are on a line $x = x_1$ parallel to the *y*-axis are endpoints of a line segment with midpoint *P*.

$$A: \quad (x_1, y_1) \qquad D: \quad (x_1, y_2) \qquad P: \quad \left(x_1, \frac{y_1 + y_2}{2} \right)$$

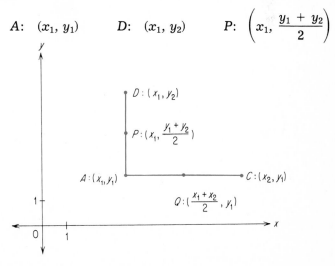

Example 2   Find the coordinates of the midpoints of the line segments with endpoints   **(a)** $(1, 0)$ and $(-3, 0)$ **(b)** $(2, 1)$ and $(2, 7)$.

Solution   **(a)** $\left(\dfrac{1 + (-3)}{2}, 0\right) = (-1, 0)$ **(b)** $\left(2, \dfrac{1 + 7}{2}\right) = (2, 4)$

     An extension of the procedure in Example 1 may be used to obtain a midpoint formula for any line segment with endpoints $A: (x_1, y_1)$ and $B: (x_2, y_2)$ that are not on a line parallel to a coordinate axis. Graph the points A and B. Draw the line AB, the lines $x = x_1$ and $y = y_1$ through A, and the lines $x = x_2$ and $y = y_2$ through B. The lines $x = x_2$ and $y = y_1$ intersect at a point $C: (x_2, y_1)$. On the line $y = y_1$ the line segment AC has midpoint $Q: ((x_1 + x_2)/2, y_1)$. The parallel lines

$$x = x_1 \qquad x = \frac{x_1 + x_2}{2} \qquad x = x_2$$

intercept congruent segments on the line AC and therefore on the line AB, that is, the line $x = (x_1 + x_2)/2$ contains the midpoint M of the line segment AB.

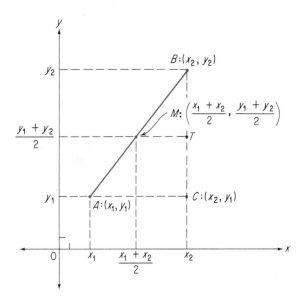

     On the line $x = x_2$, the line segment CB has midpoint $T: (x_2, (y_1 + y_2)/2)$. The parallel lines

$$y = y_1 \qquad y = \frac{y_1 + y_2}{2} \qquad y = y_2$$

8 ELEMENTS OF GEOMETRY

intercept congruent segments on the line CB and therefore on the line AB, that is, the line $y = (y_1 + y_2)/2$ contains the midpoint M of the line segment AB.

> Any line segment AB with endpoints A: (x_1, y_1) and B: (x_2, y_2) has midpoint M: (x, y) where
>
> $$(x, y) = \left(\frac{x_1 + x_2}{2}, \frac{y_1 + y_2}{2} \right)$$

This is the **midpoint formula** for points on a coordinate plane. The formula holds whether or not the line segment is parallel to a coordinate axis.

Example 3   Find the midpoint M of the line segment AB with the given endpoints.
(a) A: $(6, 1)$ and B: $(-2, 5)$
(b) A: $(-3, 4)$ and B: $(1, -2)$
(c) A: $(5, 3)$ and B: $(-1, 3)$

Solution   (a) $\left(\dfrac{6 + (-2)}{2}, \dfrac{1 + 5}{2} \right) = (2, 3)$

(b) $\left(\dfrac{-3 + 1}{2}, \dfrac{4 + (-2)}{2} \right) = (-1, 1)$

(c) $\left(\dfrac{5 + (-1)}{2}, \dfrac{3 + 3}{2} \right) = (2, 3)$

Example 4   For A: $(2, 4)$ and M: $(3, 2)$, find the coordinates of B: (x, y) such that M is the midpoint of the line segment AB.

Solution $\left(\dfrac{2 + x}{2}, \dfrac{4 + y}{2} \right) = (3, 2)$

$\dfrac{2 + x}{2} = 3, \qquad x = 4$

$\dfrac{4 + y}{2} = 2, \qquad y = 0$

The coordinates of B are $(4, 0)$.

For any two points A: (x_1, y_1) and B: (x_2, y_2) on a coordinate plane, there are the following three possibilities for the line segment AB.

The absolute value notation, $|y_2 - y_1|$ and $|x_2 - x_1|$, is used to give the lengths as positive numbers.

If $x_1 = x_2$, the line segment is parallel to the y-axis and has length $|y_2 - y_1|$.

If $y_1 = y_2$, the line segment is parallel to the x-axis and has length $|x_2 - x_1|$.

If $x_1 \neq x_2$ and $y_1 \neq y_2$, the line segment is not parallel to either axis, there is a point $C: (x_2, y_1)$ such that triangle ABC is a right triangle, and by the Pythagorean theorem

$$(AB)^2 = (AC)^2 + (BC)^2$$

$$(AB)^2 = (x_2 - x_1)^2 + (y_2 - y_1)^2$$

$$\boxed{AB = \sqrt{(x_2 - x_1)^2 + (y_2 - y_1)^2}}$$

This is the **distance formula** on a plane. The formula holds whether or not the line segment is parallel to a coordinate axis.

Graphical treatments of data are very common, used for many purposes, and generated by many types of machines. For example, electrocardiographs are used to record the action of a human heart. The following graph shows a typical electrocardiogram taken from a healthy person.

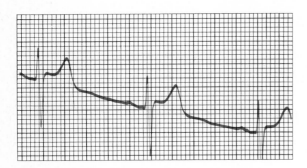

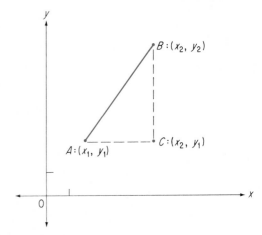

Example 5 Find the length of the line segment with the given endpoints.
(a) $A: (2, -3)$ and $B: (5, 1)$
(b) $D: (5, 3)$ and $E: (-1, 3)$
(c) $F: (4, 7)$ and $G: (4, -2)$

Solution (a) $AB = \sqrt{(5 - 2)^2 + [1 - (-3)]^2} = \sqrt{3^2 + 4^2} = \sqrt{25} = 5$
(b) $DE = \sqrt{[(-1) - 5]^2 + (3 - 3)^2} = \sqrt{(-6)^2 + 0^2} = \sqrt{36} = 6$
(c) $FG = \sqrt{(4 - 4)^2 + [(-2) - 7]^2} = \sqrt{0^2 - (-9)^2} = \sqrt{81} = 9$

The distance formula may be used to obtain an equation for any circle on a coordinate plane. The circle with center $C: (h, k)$ and radius r is the set of points $P: (x, y)$ at a distance r from the center C.

348

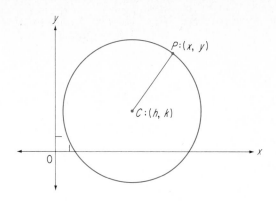

$$\sqrt{(x - h)^2 + (y - k)^2} = r$$
$$(x - h)^2 + (y - k)^2 = r^2$$

Example 6   Describe the set of points (x, y) such that   (a) $x^2 + y^2 = 4$ (b) $(x - 1)^2 + (y - 3)^2 = 1$ (c) $(x + 2)^2 + (y + 1)^2 = 9$.

Solution   (a) Since $x = x - 0$ and $y = y - 0$, the given equation can be expressed as

$$(x - 0)^2 + (y - 0)^2 = 2^2$$

Each point of the set is 2 units from the origin. Thus the set of points is a circle with center $(0, 0)$ and radius 2.

(b) The set of points is a circle with center $(1, 3)$ and radius 1.

(c) Since $x + 2 = x - (-2)$ and $y + 1 = y - (-1)$, the given equation can be expressed as

$$[x - (-2)]^2 + [y - (-1)]^2 = 3^2$$

and the set of points is a circle with center $(-2, -1)$ and with radius 3.

Example 7   Give an algebraic representation for the points that are on a coordinate plane and are at most 2 units from the point $(3, -5)$.

Solution   The points form a circular region (the points of a circle and the interior points of the circle) with center $(3, -5)$ and radius 2. Thus we have

$$(x - 3)^2 + (y + 5)^2 \leq 4$$

Find the length and the midpoint of the line segment with the given endpoints.

1. (0, 3) and (4, 3)
2. (2, 1) and (5, 1)
3. (2, 1) and (2, 5)
4. (5, −2) and (5, 7)
5. (−3, 2) and (−3, 7)
6. (−3, 2) and (−7, 2)

Find the midpoint of the line segment with the given endpoints.

7. (0, 0) and (6, −4)
8. (−4, −2) and (0, 0)
9. (−4, −2) and (−6, 8)
10. (−6, 4) and (4, 2)
11. (5, 3) and (−1, 7)
12. (−3, −7) and (5, 7)

Find the coordinates of the indicated point.

13. Endpoint B of line segment AB with A: (1, 3) and midpoint (2, 5).
14. Endpoint B of line segment AB with A: (3, 0) and midpoint (−1, 4).
15. Endpoint A of line segment AB with B: (−5, 6) and midpoint (−3, −1).
16. Vertex C of the square with A: (0, 0), B: (a, 0), and D: (0, a).
17. Vertex S of rectangle $QRST$ with Q: (0, 0), R: (a, 0), and T: (0, b).
18. Vertex C of quadrilateral $ABCD$ with A: (0, 0), B: (7p, 0), and D: (2p, q) with positive numbers p and q, for $ABCD$ (a) a parallelogram (b) an isosceles trapezoid that is not a parallelogram.
19. The vertex G of a quadrilateral $EFGH$ with E: (0, 0), F: (a, b), and H: (0, 5b) with positive numbers a and b, for $EFGH$ (a) a parallelogram (b) an isosceles trapezoid that is not a parallelogram.
20. The vertex S of an isosceles trapezoid $QRST$ that is not a parallelogram, with Q: (0, 0), R: (a, 0), and T: (b, c), where $0 < 2b < a$.

Find the length of the line segment with the given endpoints.

21. A: (1, 2) and B: (4, 6)
22. C: (−1, 6) and D: (4, −6)
23. E: (2, 16) and F: (9, −8)
24. G: (−2, 5) and H: (7, 11)
25. I: (−1, −3) and J: (2, 0)
26. L: (4, −5) and M: (−7, −11)

Describe the specified set of points (x, y).

27. $x^2 + y^2 = 16$
28. $(x + 3)^2 + (y - 2)^2 = 25$
29. $x^2 + y^2 < 36$
30. $(x - 5)^2 + (y + 1)^2 < 49$
31. $(x - 1)^2 + (y - 2)^2 \leq 4$
32. $(x + 1)^2 + (y + 2)^2 \leq 25$

Give an algebraic representation for the specified set of points on a coordinate plane.

33. The points of the circle with center (2, 5) and radius 3.
34. The points of the circle with center $(-3, 4)$ and radius 2.
35. The points of the circular region with center $(-3, -1)$ and radius 4.
36. The interior points of the circle with center $(6, -4)$ and radius 5.
37. The exterior points of the circle with center $(-4, 3)$ and radius 4.
38. The exterior points of the circle with center $(-3, -2)$ and radius 6.
39. The points at most 5 units from the point $(-1, 2)$.
40. The points at least 2 units and at most 5 units from the origin.
41. The points at least 3 units and at most 4 units from the point $(1, -2)$.
42. The points (a) 5 units from the y-axis (b) 2 units from the x-axis.
43. The points (a) 3 units from the line $x = 2$ (b) 4 units from the line $y = -1$.
44. The points at most 2 units from the y-axis and at most 3 units from the origin.

Use properties of a coordinate plane and prove the given statement.

45. If A: (0, 0), B: $(a, 0)$, and C: (b, c) are the vertices of a triangle, then the midpoints of the sides AC and BC are endpoints of a line segment that is parallel to, and half as long as, the side AB.
46. If P: (0, 0), Q: $(a, 0)$, R: (a, b), and S: (a, b) are the vertices of a trapezoid $PQRS$ where $0 < c < a$ and $0 < b$, then the midpoints of the sides PS and QR are the endpoints of a

line segment that is parallel to the bases and has length equal to one-half the sum of the lengths of the bases.

*47. The diagonals of a parallelogram bisect each other.

*48. The midpoints of the sides of any quadrilateral *ABCD* are vertices of a quadrilateral *PQRS* such that the diagonals of *PQRS* bisect each other.

*49. Write a BASIC program to print the coordinates of the endpoints and the midpoint of the line segment with the given endpoints on a number line.

(a) 7, 11   (b) 75, 123
(c) -567, 891   (d) 4357, -5437

BASIC was introduced in Section 2-4.

*50. Write a BASIC program to print the coordinates of the endpoints and the midpoint of the line segment with the given endpoints on a coordinate plane.

(a) (2, 5), (8, 17)   (b) (75, 17), $(-21, 93)$
(c) (537, 693), $(-23, 4537)$ (d) (756, 898), $(-75, -94)$

Explorations

Taxicab geometry may be obtained by defining the distance between any two locations (x_1, y_1) and (x_2, y_2) to be

$$|x_2 - x_1| + |y_2 - y_1|$$

1. Use an integral coordinate plane to represent a modern city with streets one unit apart and intersecting at right angles. Prepare for a middle school class an introduction of the taxi-distance between **(a)** any two given street corners **(b)** any two given points on streets.

2. Make a drawing of the city considered in Exploration 1. Select, and enclose in a small circle, a particular street corner as the location of a school. Label with *A* each of the street corners that are at taxi-distance 1 from the school. Label with *B*, *C*, and *D* each of the street corners that are at taxi-distances 2, 3, and 4, respectively, from the school. Draw the quadrilateral determined by the points labeled *B*. Draw the quadrilaterals for the points labeled *A*, *C*, and *D*. What is the shape of these "taxi-circles" of points at a taxi-distance *n* from the school? Express the number of street corners on each taxi-circle as a function of *n*. Express the area α of each taxi-circle as a function of *n*.

3. Select a point on a street but in the middle of a block as the location of the school. For *n* = 1, 2, 3, and 4, label points on streets as in Exploration 2, draw the taxi-circles;

8 ELEMENTS OF GEOMETRY

and for n greater than 1, express the number of labeled points on each taxi-circle as a function of n.

4. Suppose that Mary and Doris live at street corners with coordinates $(-3, 3)$ and $(3, -3)$ in the city considered in Exploration 1. Make a drawing of the city and mark at least ten street corners that are at equal taxi-distances from the homes of the two girls.

5. Prepare a report on taxicab geometry. For example, see "Taxicab geometry offers a free ride to a non-Euclidean locale" in Martin Gardner's "Mathematical Games" starting on page 18 of the November 1980 issue of *Scientific American*.

Describe and, if necessary, graph and label on a coordinate plane, the set of all possible endpoints of line segments that have length two units and have one endpoint on the given figure.

6. The origin.

7. The x-axis.

8. The graph of $xy = 0$.

9. The graph of $(x - y)(x + y) = 0$.

10. The lines $|x|$ equal to (a) 1 (b) 2 (c) 3.

11. The circle with center at the origin and radius (a) 1 (b) 2 (c) 3.

12. The square with vertices at $A:$ $(0, 0)$, $B:$ $(4, 0)$, $C:$ $(4, 4)$, and $D:$ $(0, 4)$.

13. The rectangle with vertices at $P:$ $(0, 0)$, $Q:$ $(10, 0)$, $R:$ $(10, 20)$, and $S:$ $(0, 20)$.

8-5
Figures in Space

The recognition of such common three-dimensional figures (solids) as *spheres*, cubes, triangular pyramids, and right circular cylinders preceded the art of writing. Five thousand years ago in the great river valleys of the area that became Iraq and

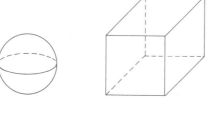

Sphere Cube Triangular pyramid Right circular cylinder

Iran, it was a common business practice to use a set of solid figures as an invoice with a shipment of goods. Clay was abundant, could be molded easily when wet, and hardened well in the sun. Sets of solid clay figures were made in easily recognized geometric shapes. A messenger conveying a shipment of sheep, loaves of bread, or other objects would also carry an invoice. The invoice was a hollow clay ball about the size of a golf ball. Inside the ball would be a set of tiny geometric solids packed in straw. The number of these solids would be the same as the number of objects shipped. The person who received the shipment could check that all items were present by breaking the ball and matching the geometric solids, one-to-one, with the objects received. Geometric solids were used so that they would not be confused with the fragments of the broken hollow ball. The shipment could be checked without a knowledge of counting, without words for numbers, and without a written language. At one time the shape of the solid and the marks on the solid indicated the kind of object shipped. Over many centuries pictures on the balls to indicate the kinds of objects shipped and the numbers of objects shipped gradually evolved into the earliest known writing of words.

The geometric solids were simply "recognized." There were no formal definitions. The shapes of the solids were identifiable "attributes" of the figures. Even today the initial recognition of a solid figure is "sensed"—often by both visual and tactile (touch, handling) senses. In order to identify a solid figure visually, the figure must be viewed from several directions (front, top, side, and so on). Suppose that you are shopping for a spherical lamp shade and that there are two types at the store. Each type has a small hole at the top for the lamp cord. One type is otherwise spherical; the other has a portion cut away so that the lamp provides extra light directly below it.

If the two lamp shades were viewed from the top looking directly at the small holes for the lamp cord, the lamp shades would look alike. But if the lamp shades were viewed from the side instead of the top, their differences would be apparent.

The complete identification of the shape of an unknown object requires that the object be viewed from all possible directions. Suppose that a bright moon were directly overhead

8 ELEMENTS OF GEOMETRY

and that shadows of an otherwise invisible Martian spaceship were visible on level ground. In order to identify the shape of the spaceship from its shadows, we would need to observe the changes in its shadow as the spaceship rolled and several silhouettes were viewed. Even then the identification would be difficult.

To illustrate the process of identifying space figures by their shadows, we shall consider a related game. An object is placed behind a screen, a bright light (such as the sun) is in the distant background, and on the screen only the shadow of the object is visible. To simplify the game we assume that the unknown object is one of the four solids mentioned at the beginning of this section, that is, a sphere, a cube, a triangular pyramid, or a right circular cylinder. The solid figures may be located in any position so that the shadow may be a top view, a side view, an end view, or a view from some other position. Opposite views, such as the top view and the bottom view, have the same shadow and therefore are considered to be the same, that is, *not different,* views. The light source is assumed to be very far away and the shadows are all assumed to be on planes that are perpendicular to the rays of light. The same effect can be achieved by placing an object on a overhead projector and observing the shadow on the screen. The shadows obtained using perpendicular rays resemble the shadows obtained when the sun is directly overhead. The long shadows of the early morning or late evening provide excellent geometric illustrations, but of a different geometry.

Suppose that the object has a circular region *(disk)* as its shadow. Can the object be

    the sphere?   the cube?   the pyramid?   the cylinder?

Suppose that the object has a square region as its shadow. Can the object be

    the sphere?   the cube?   the pyramid?   the cylinder?

Before the last question can be answered we need to know whether the height of the cylinder is equal to its diameter. For future discussion of these particular figures, we make four assumptions.

The sphere has diameter one meter.

The cube has edge one meter.

Each edge of the pyramid is one meter.

The cylinder has diameter one meter and height 125 centimeters.

Now the previous questions can be answered. Only the cube can have a square region as its shadow.

Which of the four solid figures can have a rectangular region that is not a square region as its shadow? Which of the solid figures can have a triangular region as its shadow? Both the cube and the cylinder can have shadows that are rectangular regions but not square regions. Only the pyramid can have a shadow that is a triangular region.

Example 1 A sphere with radius 10 centimeters is rolled around on a plane. What changes, if any, occur among the shadows of the sphere from light rays that intersect the plane of the shadows at right angles?

Solution All shadows are circular regions with radius 10 centimeters.

Example 2 A cube with edge 4 centimeters is rolled with an edge or face on the plane at all times. What changes, if any, occur among the shadows of the cube from light rays that intersect the plane at right angles?

Solution Each shadow is a square or a rectangular region with one side 4 centimeters long. The other side of the rectangular shadow may have any length from 4 centimeters (when the cube has a face on the plane) to $4\sqrt{2}$ centimeters (when the cube is balanced on an edge, that is, has a diagonal of a face parallel to the plane) as in the adjacent figure.

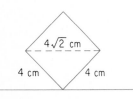

Identify a few objects that have approximately the shapes of common space figures such as spheres, rectangular boxes, cubes, pyramids, or cones.

After we can recognize solid figures such as triangular pyramids, cubes, and spheres a few definitions are useful in describing solid figures and objects. On a plane we may think of a polygon as a union of line segments. In space we may think of a **polyhedron** (plural **polyhedra**) as a union of polygonal regions. The polygonal regions are the **faces** of the polyhedron; the sides of the polygonal regions are **edges** of the polyhedron; the vertices of the polygonal regions are **vertices** of the polyhedron. A polyhedron with four triangular regions as faces is a triangular pyramid, also called a tetrahedron. A polyhedron with rectangular faces is a rectangular box, which is also called a rectangular parallelepiped since the opposite faces are on parallel planes, that is, planes that have no points in common. A cube is a rectangular box with square faces.

 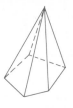

Triangular pyramid   Square pyramid   Pentagonal pyramid   Circular cone

A polyhedron with all but one of its vertices on a plane is a pyramid. The vertices on a plane are vertices of the base of the pyramid. Pyramids are often classified as triangular, rectangular, square, and so forth, by the shapes of their bases. Since a circle is not a polygon (a circle does not have a finite number of sides), a circular cone is not a pyramid or a polyhedron. However, a cone may be considered to be obtained from pyramids by increasing the number of sides of the base indefinitely.

Example 3 The pyramid *ABCDE* is a rectangular pyramid. Name its
(**a**) vertices   (**b**) edges   (**c**) faces.

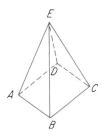

Solution   (**a**) Points *A*, *B*, *C*, *D*, and *E*.
(**b**) Line segments *AB*, *BC*, *CD*, *DA*, *AE*, *BE*, *CE*, and *DE*.
(**c**) Rectangular region *ABCD* and triangular regions *ABE*, *BCE*, *CDE*, and *DAE*.

We use the edges and faces of a rectangular box to explore some relationships among lines and planes in space before considering other polyhedra. Suppose that the edges of the

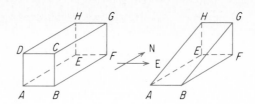

box shown in the figure are located so that AB is an east-west line, AE is a north-south line, and AD is a vertical line. (If you are in a rectangular room, its walls, floor, and ceiling may provide a better example than the box.) The lines AB, EF, HG, and DC are all east-west lines and therefore parallel lines. The fact that the lines AB and GH are parallel implies that they are on a plane $ABGH$, as shown in the second figure. In general, two lines in space are

>**intersecting lines** if they have exactly one point in common
>
>**parallel lines** if they are coplanar and do not intersect
>
>**skew lines** if they are not coplanar.

For example, the lines AB and CG are skew lines. The vertical line BC is perpendicular to the line BA and also to the line BF. If a line intersects a plane so that the line is perpendicular to at least two lines on the plane, then the line is perpendicular to the plane. Thus the line BC is perpendicular to the plane ABF. This implies that the line BC is perpendicular to every line that contains the point B and is on the plane ABF.

Example 4 The figure represents a rectangular box. Name the edges that (**a**) are parallel to the line SZ (**b**) intersect SZ (**c**) form skew lines with SZ (**d**) are perpendicular to the plane WXY.

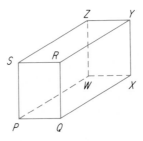

Solution (**a**) PW, QX, and RY (**b**) SP, SR, ZW, and ZY
(**c**) PQ, RQ, WX, and XY (**d**) SZ, RY, PW, and QX

8 ELEMENTS OF GEOMETRY

The visualization of space figures can be greatly enhanced by the use of *computer graphics*. For example, a picture of a space figure such as a proposed new building may be shown on the visual display of a computer. The display is similar to a television picture. The computer can be programmed so that new pictures are shown representing views of the building from several different positions or sides. Architects and others find computers very useful for producing different views of objects without first producing a physical model.

Any two planes in space either are **parallel planes** (no points in common) or intersect in a line. (It is not possible in three-dimensional space for two planes to have one and only one point in common.) If a polyhedron has all of its vertices in two parallel planes, then the edges in those planes are the edges of the two bases of the polyhedron. The edges that are not edges of a base are lateral edges. If the lateral edges are on parallel lines, then the polyhedron is a prism. Prisms are often classified as triangular, square, rectangular, and so forth according to the shapes of their bases.

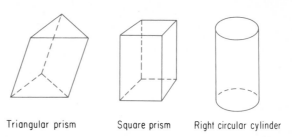

Triangular prism   Square prism   Right circular cylinder

A prism with a lateral edge perpendicular to the plane of a base is a **right prism.** The square prism in the figure is a right prism; the triangular prism is not a right prism. A **cylinder** is not a prism (or polyhedron) but may be considered to be obtained by increasing the number of sides of the base of a prism indefinitely. A circular cylinder with the line through the centers of its bases perpendicular to the plane of a base is a **right circular cylinder.**

Example 5   The adjacent figure represents a portion of a square pyramid viewed from above.

Identify the given statement as true or false.

(a) Edges *AD* and *BC* intersect.
(b) Lines *AD* and *BC* intersect.
(c) Line *AD* intersects plane *BCE*.
(d) Line segment *AD* intersects plane *BCE*.
(e) Face *ADGH* intersects plane *BCE*.
(f) Plane *ADG* intersects plane *BCE*.

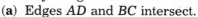

Solution   (a) False   (b) True   (c) True   (d) False   (e) False
(f) True

EXERCISES   Throughout these exercises all shadows are assumed to be formed by light rays that intersect the plane of the shadow at right angles.

Assume that the unknown figure is one of the four discussed in this section (sphere, cube, triangular pyramid, cylinder). Specify which of these figures is then identified by the listed shadows of two different views of the figure.

1. Two square regions.
2. A circular region and a rectangular region.
3. A square region and a rectangular region.
4. Two circular regions.
5. Two triangular regions.
6. A triangular region and the region of a quadrilateral that is not a rectangle.

Identify as true or false.

7. Every shadow of a sphere is a circular region.
8. Every shadow of a cube is a square region.
9. Every shadow of a triangular pyramid is a triangular region.
10. Every shadow of a cube is either a square region or a rectangular region.

11. The tetrahedron *MNOP* is a triangular pyramid. Name its (**a**) vertices (**b**) edges (**c**) faces.

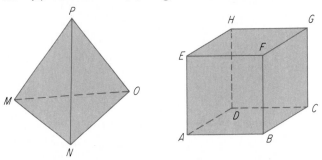

12. Repeat Exercise 11 for the cube shown in the figure.
13. Think of the lines along the edges of the cube in Exercise 12 and identify the lines that appear to be (**a**) parallel to *AB* (**b**) skew to *AB* (**c**) parallel to the plane *ABFE* (**d**) perpendicular to the plane *ABFE*.

Use pencils, tabletops, and so on to represent lines and planes and determine whether each statement appears to be true or false in our ordinary (Euclidean) geometry.

14. Given any line *m* and any point *P* that is not a point of *m*, there is exactly one line *t* that is parallel to *m* and contains the point *P*.

15. Given any line m and any point P is that not a point of m, there is exactly one line q such that q contains P and m and q are intersecting lines.

16. Given any line m and any point P that is not a point of m, there is exactly one line s such that s contains P and s and m are skew lines.

17. Given any line m and any point P that is not a point of m, there is exactly one plane that is parallel to m and contains the point P.

18. Given any plane ABC and any point P that is not a point of ABC, there is exactly one plane that contains P and intersects ABC.

19. Given any plane ABC and any point P that is not a point of ABC, there is exactly one plane that contains P and is parallel to ABC.

20. Given any plane ABC and any point P that is not a point of ABC, there is exactly one line that contains P and intersects ABC.

21. Given any plane ABC and any point P that is not a point of ABC, there is exactly one line that contains P and is parallel to ABC.

22. Given any plane ABC and any line m that is parallel to ABC, there is exactly one plane that contains m and intersects ABC.

23. Given any plane ABC and any line m that is parallel to ABC, there is exactly one plane that contains m and is parallel to ABC.

24. Given any plane ABC and any line m that is parallel to ABC, there is for any given point P that is not on m and not on ABC exactly one line that is parallel to m and also parallel to ABC.

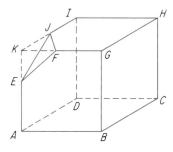

A triangular pyramid KEFJ has been removed from a cube as indicated in the adjacent figure. Identify the given statement as true or false.

25. Line segment EF intersects plane (a) GHI (b) ABC (c) DCH.

26. Line EF intersects plane (a) GHI (b) ABC (c) DCH.

27. Triangular region EFJ intersects plane (a) GHI (b) ABC (c) DCH.

28. Plane EFJ intersects plane (a) GHI (b) ABC (c) DCH.

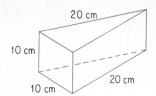

The adjacent figure represents a triangular wedge. The bases are on parallel planes and are isosceles triangles. One face is a square 10 centimeters on each edge. Each of the edges of the wedge is either 10 centimeters or 20 centimeters long.

29. If the wedge can have a square shadow, what is (are) the possible lengths of the edges of the shadow?

30. Repeat Exercise 29 for rectangular shadows.

31. Repeat Exercise 29 for triangular shadows.

32. If the wedge can have a shadow that is not one of those considered in Exercises 29 through 31, sketch another possible type of shadow.

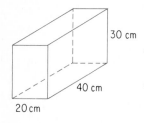

The adjacent figure represents a cylindrical log with diameter one meter and length three meters.

33. Describe the shadows that would be obtained if the log were (**a**) rolled on its cylindrical side, along the plane of the shadows (**b**) stood on end.

34. If for example, the log is in the process of being stood on end, can it have a shadow of a different shape than those considered in Exercise 33? If so, sketch one such shadow.

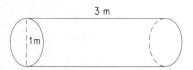

The adjacent figure represents a rectangular box with edges 20 centimeters, 30 centimeters, and 40 centimeters long.

*35. Can the box have a square shadow? If so, what is the length of the edge of the smallest possible square shadow?

*36. Think of the box as rolling end over end about its edges of length 20 centimeters in the plane of the shadow.
(**a**) What sorts of figures are all the possible shadows as the box turns?
(**b**) What are the dimensions of the smallest such shadow?
(**c**) What are the dimensions of the largest such shadow?

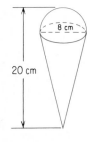

The adjacent figure represents an ice cream cone with a hemispherical cap of ice cream. The diameter of the cap is 8 centimeters; the total height is 20 centimeters.

37. Describe two possible shadows of the figure.

*38. (**a**) Describe the shadow with the smallest possible area.
(**b**) Describe the shadow with the largest possible area.
(**c**) Sketch the shape of a shadow that is neither the smallest nor the largest.

Explorations

The representation of space figures on a plane is both an art and a science. One of the common procedures for representing space figures, orthographic projection, makes use of views from three mutually perpendicular directions (top view, front view, and side view) and includes dashed lines for hidden edges. The pattern for presenting the views, *orthographic projections,* is always the same; the side view is conventionally the right side.

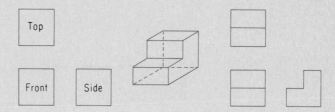

Suppose that one quarter of a cube were removed as in the figure; then the three views would be as shown. Each of the following figures has the same top and front views as the previous example but the side view would be different as shown with the figure.

Sketch the indicated figure and its three orthographic projections.

1. A cube.
2. A sphere.
3. A cylinder with $h = 2r$.
4. A square pyramid.

Sketch a space figure that has the three given orthographic projections.

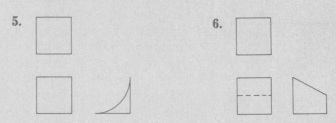

5.

6.

7.

8.

Sketch at least three different figures and their side views such that all figures have the given top and front views.

9.

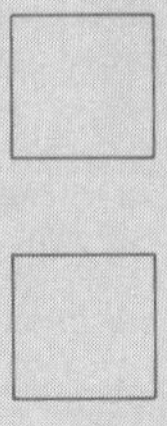

10.

11.

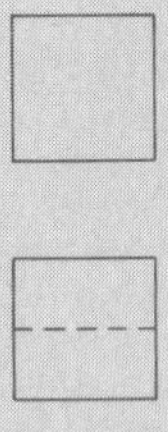

12.

13. Sketch at least five other space figures and their orthographic projections.

14. Read and prepare a summary of "The Earliest Precursor of Writing" by Denise Schmandt-Besserat on pages 50 through 59 of the June 1978 issue of *Scientific American*. See also the comment by Stephen J. Lieberman on pages 10 and 15 of the November 1978 issue of *Scientific American*.

8-6
Measures of Space Figures

Any polyhedron or other simple closed surface is the common boundary of a bounded solid region (its *interior*) and an unbounded solid region (its *exterior*). The **volume** V of the polyhedron is the number of solid unit cubes that are "equivalent" to the interior of the polyhedron. It is customary to speak of the "volume of the polyhedron" rather than specifying the "volume of the solid region bounded by the polyhedron." The surface area S of any polyhedron is the sum of the areas of its faces. In drawings of space figures, hidden edges are indicated by dashed line segments, as in the adjacent drawing of a unit cube.

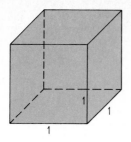

Volumes are considered in a manner very similar to areas. We define any rectangular box with edges of length a, b, and c on a common endpoint to have volume abc (see adjacent figure). A cube with edge of length e has volume e^3 and surface area $6e^2$. The rectangular box with edges a, b, and c has surface area $2ac + 2ab + 2bc$. As for lengths and areas the word *volume* is used to refer to both the *volume measure* (number) and the *volume measurement* (number of units). The units are specified when they are known.

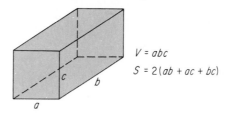

$V = abc$

$S = 2(ab + ac + bc)$

Example 1   For a rectangular box 20 centimeters by 10 centimeters by 4 centimeters, find   **(a)** the volume   **(b)** the surface area.

Solution   **(a)** Sketch the box. The number of cubic centimeters that could be placed in the box can be counted. The height of 4 centimeters indicates that there could be four layers of unit cubes. Each layer must cover a 10 centimeter by 20 centimeter rectangular region and thus contain 200 unit cubes. Then the four layers contain 4×200, that is, 800 cm³. As in the formula, $V = 20 \times 10 \times 4 = 800$ cm³.

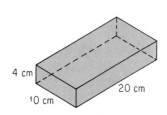

4 cm

20 cm

10 cm

(b) The box has six rectangular regions as faces. We may think of the top and bottom as each 10 centimeters by 20 centimeters, that is, with area 200 cm² each; the two ends as 10 centimeters by 4 centimeters, that is, with area 40 cm² each; and the two sides as 20 centimeters by 4 centimeters, that is, with area 80 cm² each. Thus the surface area of the box is, as in the formula,

$S = 2ac + 2ab + 2bc$

$$2(10 \times 20) + 2(10 \times 4) + 2(20 \times 4) \quad \text{that is} \quad 640 \text{ cm}^2.$$

Several solid figures were considered in Section 8-5. In advanced courses it can be proven that the bases of any prism or cylinder are congruent figures; we assume that such is the case. The height h of a prism or cylinder is the distance between its bases. The volume V of a prism or cylinder is

$$V = Bh$$

where B is the area of the base and h is the height. If the lateral edges of a prism are perpendicular to the plane of a base, then the prism is a right prism and the length of an edge is the height. The surface areas S of three common solids are shown in the figures.

Think of ways in which you use measures of space figures. Consider, for example, distances in space (such as the distance between two classes), areas of space figures (such as the surface of a lawn or the walls of a room that is to be painted), and volumes or capacities (such as the amount of gasoline now in the car).

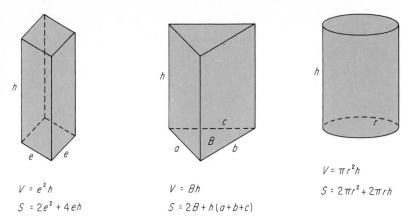

h

e e

h

a B b

c

h

r

$V = e^2h$

$S = 2e^2 + 4eh$

$V = Bh$

$S = 2B + h(a+b+c)$

$V = \pi r^2 h$

$S = 2\pi r^2 + 2\pi rh$

Example 2 For a right circular cylinder with radius 3 centimeters and height 10 centimeters, find **(a)** the volume **(b)** the surface area.

Solution **(a)** $V = \pi(3^2)(10) = 90\pi \text{ cm}^3$
  (b) $S = 2\pi(3^2) + 2\pi(3)(10) = 78\pi \text{ cm}^2$
    Note that the lateral (curved) surface is like the paper label around a cylindrical can. If it were cut along a line perpendicular to the base, it would unroll into a rectangular region with the circumference of the base as one side and the height of the cylinder as the other side.

Pyramids and cones were also considered in Section 8-5. Each has a base and a vertex that is not on the plane of the base. The height is the distance of that vertex from the plane of the base as measured along a line that is perpendicular to the plane of the base. The volume of a pyramid or cone is

$$V = \frac{1}{3}Bh$$

where B is the area of the base and h is the height. Formulas for volumes of pyramids, cones, and spheres are shown with the following figures and accepted here without formal proof.
  The surface area of a pyramid is the sum of the areas of its faces. The volumes and surface areas of many other space figures may be found by adding or subtracting those measures of common figures.

366

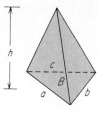

$V = \frac{1}{3} Bh$

$V = \frac{1}{3} \pi r^2 h$

$S = \pi r^2 + \pi r \sqrt{r^2 + h^2}$

$V = \frac{4}{3} \pi r^3$

$S = 4 \pi r^2$

Example 3 A certain building has the shape of a rectangular box with a hemispherical dome on top. The box is 20 meters by 30 meters by 10 meters. The dome has radius 6 meters and the height of the building is 16 meters. Find **(a)** the volume of the building **(b)** the exposed surface area of the building, that is, the area above the level ground on which the building is located.

Solution **(a)** The rectangular box has volume $20 \times 30 \times 10$, that is, 6000 m³. The dome has volume $(1/2)[(4/3)\pi \times 6^3]$, that is, 144π m³. Thus the building has volume $(6000 + 144\pi)$ m³.

(b) The sides of the building have area $200 + 300 + 200 + 300$, that is, 1000 m². The flat roof has area $(600 - 36\pi)$ m². The dome has area $(1/2)(4\pi \times 36)$, that is, 72π m². Thus the exposed surface area is $1000 + (600 - 36\pi) + 72\pi$, that is, $(1600 + 36\pi)$ m².

EXERCISES *For a cube with edges of the given length, find* **(a)** *the volume* **(b)** *the surface area.*

1. 5 cm	2. 8 dm	3. 7 cm
4. 2 m	5. 15 mm	6. 25 cm

For the given rectangular box, find **(a)** *the volume* **(b)** *the surface area.*

7.

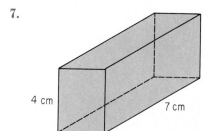

4 cm   7 cm   3 cm

8.

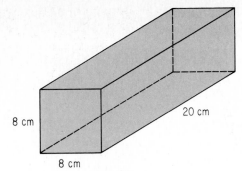

8 cm   20 cm

8 cm

9.

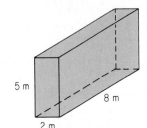

5 m

8 m

2 m

*10.

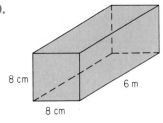

8 cm   6 m

8 cm

Find the volume of the indicated pyramid.

11.

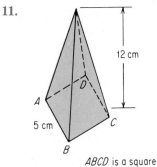

12 cm

A

5 cm

C

B

ABCD is a square

12.

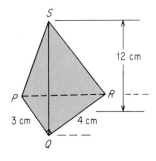

S

12 cm

P

3 cm   4 cm

R

Q

For the indicated space figure, find **(a)** *the volume*
(b) *the surface area.*

13.

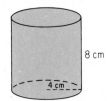

8 cm

4 cm

14.

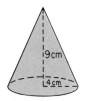

9 cm

4 cm

15. A spherical ball with radius 10 centimeters.

16. A right circular cylinder 20 centimeters long, with radius 3
centimeters, and with hemispherical caps on both ends.

17. A cone with radius 5 centimeters and with height 12 centimeters.

18. A cone with radius 4 centimeters, a hemispherical base of the same radius, and total height of cone and hemisphere 20 centimeters.

In Exercises 19 through 22, describe the change in volume and the change in the surface area of the given figure.

19. A cube when the length of each edge is multiplied by (a) 2   (b) 3   (c) 1/2   (d) a positive number k.

20. Repeat Exercise 19 for a rectangular box.

21. Repeat Exercise 19 for the radius and height of a right circular cylinder.

22. Repeat Exercise 19 for the radius and height of a cone.

*23. All of the formulas for volumes that we have considered are special cases of the **prismoidal formula**.

$$V = \frac{h}{6}(B_1 + 4M + B_2)$$

where B_1 and B_2 are the areas of the bases and M is the area of the intersection of the figure with a plane halfway between the bases. In this case the vertex of a pyramid or a cone is considered a base of area zero. Show that the prismoidal formula gives the usual formula for the volume of a cube.

*24. Repeat Exercise 23 for   (a) a prism with base of area B and height h (b) a cylinder with radius r and height h.

*25. Think of a sphere as in the figure and use the prismoidal formula to find an expression for the volume of a sphere of radius r.

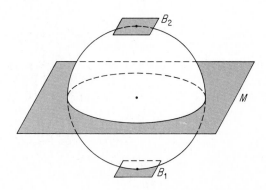

For a square pyramid with height h and a base with an edge of linear measure e the midsection is a square with an edge $\frac{1}{2}e$. Use the prismoidal formula to find an expression for the volume of the square pyramid.

Use the prismoidal formula to find an expression for the volume of a circular cone with height h if the base has radius r and the midsection has radius $\frac{1}{2}r$.

Explorations

Each of the Explorations 1 through 5 involves the determination of a shortest path a spider has to crawl along the outside of a box to reach a fly F that is assumed to stay in one place. As in the figure, we consider a closed rectangular box 12 centimeters wide, 18 centimeters long, and 8 centimeters tall.

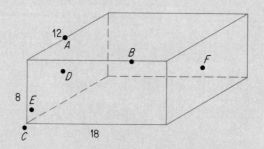

For each exploration assume that the fly is at the center F (intersection of the diagonals) of an end of the box. Copy the figure, draw the shortest path(s) for the spider, and find the length of the shortest path. If there are two or more paths of minimal length, draw them all.

1. The spider is at position A in the center of the top edge of the opposite end of the box.

2. The spider is at position B at the top of an adjoining side and 4 centimeters from the end containing the fly. (*Hint:* Think of the box as flattened out in some way.)

3. The spider is in the corner C at the bottom of the opposite end of the box.

4. The spider is at the middle D of the end of the box opposite the fly.

5. The spider is on the opposite end of the box at position E 1 centimeter from the vertical edge of the box over C and 2 centimeters from the bottom of the box.

6. Cut out a circular region (disk) and remove a sector by cutting along two radii. Use the remaining part of the disk as a model for the lateral surface of a cone. Explore ways of predicting the radius and the height of the resulting cone.

Volumes of space figures may be used to develop the concept of large numbers. Conjecture answers for the following questions relative to your mathematics classroom or another room that is accessible to you. Then develop procedures for verifying, or correcting, your conjectures.

7. Would 1 million pennies fit in the room? One thousand? One billion?
8. Repeat Exploration 1 for quarters.
9. Would 1 million ping-pong balls fit in the room? Tennis balls? Basketballs?
10. Would 1 million jelly beans fit in the room? Eggs? Quart cartons of milk?

Chapter 8 Review

Solutions to the following exercises may be found within the text of Chapter 8. Try to complete each exercise without referring to the text.

Section 8-1   Figures on a Plane

Use paper folding and demonstrate the given statment.

1. There is one and only one line through two given points.
2. On a plane there is one and only one line that is perpendicular to a given line at a given point of the line.
3. Any two coplanar lines that are perpendicular to the same line are parallel lines.

Section 8-2   Measures of Plane Figures

4. State the unit of precision for the given measurement.
   (a) 5.2 kilometers
   (b) $3\frac{1}{8}$ units

5. Identify the given measurement as exact or approximate.
   (a) June lives *2 kilometers* from school.
   (b) Susan bought *two new tires*.
   (c) Bob bought *1 kilogram* of grapes.

6. Find the area of the given region.

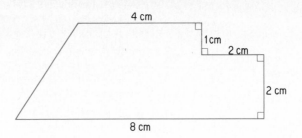

7. Find the length and the area of the given curve formed by line segments and semicircles. The line segments intersect at right angles.

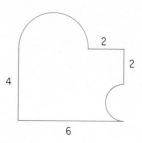

Section 8-3   The Golden Ratio

8. Explain why you would expect the decimal parts of G and G^2 to be the same, where G is the golden ratio.

9. List the first six terms of the Fibonacci sequence 1, 1, ...

Section 8-4   Figures on a Coordinate Plane

10. Find the length and the midpoint of the line segment AB for A: (1, 5) and B: (7, 5).

11. Find the midpoint M of the line segment AB with the given endpoints.
    (a) A: (6, 1) and B: (−2, 5).
    (b) A: (−3, 4) and B: (1, −2)

12. Find the length of the line segment with the given endpoints.
    (a) A: (2, −3) and B: (5, 1)
    (b) D: (5, 3) and E: (−1, 3)
    (c) F: (4, 7) and G: (4, −2)

13. Describe the set of points (x, y) such that
    (a) $x^2 + y^2 = 4$ (b) $(x − 1)^2 + (y − 3)^2 = 1$
    (c) $(x + 2)^2 + (y + 1)^2 = 9$.

14. Give an algebraic representation for the points that are on a coordinate plane and are at most 2 units from the point (3, −5).

Section 8-5   *Figures in Space*

15. A sphere with radius 10 centimeters is rolled around on a plane. What changes, if any, occur among the shadows of the sphere from light rays that intersect the plane of the shadows at right angles?

16. A cube with a 4-centimeter edge is rolled with an edge or face on the plane at all times. What changes, if any, occur among the shadows of the cube from light rays that intersect the plane at right angles?

17. The pyramid *ABCDE* is a rectangular pyramid. Name its **(a)** vertices   **(b)** edges   **(c)** faces.

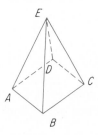

18. The figure represents a rectangular box. Name the edges that **(a)** are parallel to the *SZ*   **(b)** intersect *SZ* **(c)** form skew lines with *SZ*   **(d)** are perpendicular to the plane *WXY*.

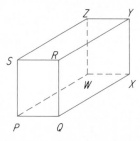

Section 8-6   *Measures of Space Figures*

19. For a rectangular box 20 centimeters by 10 centimeters by 4 centimeters find **(a)** the volume   **(b)** the surface area.

20. For a right circular cylinder with radius 3 centimeters and height 10 centimeters find **(a)** the volume   **(b)** the surface area.

21. A certain building has the shape of a rectangular box with a hemispherical dome on top. The box is 20 meters by 30 meters by 10 meters. The dome has radius 6 meters and the height of the building is 16 meters. Find **(a)** the volume of the building **(b)** the exposed surface area of the building, that is, the area above the level ground on which the building is located.

Chapter 8 Test

Use paper folding to demonstrate the given statement.

1. Any given line segment has one and only one perpendicular bisector.

2. For any given triangle the three lines that contain a vertex and the midpoint of the opposite side are concurrent.

3. State the unit of precision for the given measurement.

   (a) 2.5 cm **(b)** $2\frac{1}{2}$ cm

4. Identify the indicated measurement as exact or approximate.
   (a) Spot weighs *22 kilograms*
   (b) Jack is carrying *three books*.

5. Sketch a quadrilateral with diagonals *PQ* and *CD*.

Find the area of the given figure.

6.
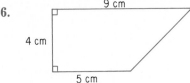

7.
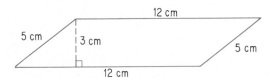

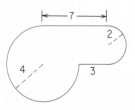

8. For the curve in the adjacent figure, find **(a)** the length **(b)** the area. The curve is formed by line segments and arcs of circles.

9. Identify at least two instances in which golden ratios have been, or are being, used or may be observed.

10. Find the coordinates of the endpoint *B* of the line segment *AB* with *A*: (2, 6) and midpoint *M*: (5, 1).

8 ELEMENTS OF GEOMETRY

11. Describe the specified set of points on a coordinate plane.
    (a) $x^2 + y^2 \le 81$ (b) $(x - 1)^2 + (y + 2)^2 > 4$

Give an algebraic representation for the specified set of points on a coordinate plane.

12. The points (a) 3 units from the y-axis (b) 2 units from the line $x = 2$.

13. The points at least 1 unit and at most 2 units from the point (2, 3).

14. The points equidistant from the point $(-3, 0)$ and $(5, 0)$.

Sketch the indicated figure and two of its possible shadows, with their dimensions labeled.

15. A pyramid 6 centimeters high with a square base having an edge of 3 centimeters.

16. A right circular cylinder with height 10 centimeters and radius 2 centimeters.

17. A right circular cone with height 10 centimeters and radius 2 centimeters.

Identify as true (T, always true) or false (F, not always true).

18. (a) There is one and only one line that is perpendicular to a given plane and contains a given point.
    (b) There is one and only one plane that is perpendicular to a given plane and contains a given line.

19. (a) Any two lines intersect or are parallel.
    (b) Any two planes intersect or are parallel.

20. (a) Every shadow of a cube is a square region.
    (b) Every shadow of a sphere is a circular region.

For each figure find (a) the volume (b) the surface area.

21. A cube with an edge of length 3 meters.

22. A rectangular box with edges of lengths 3 meters, 5 meters, and 7 meters.

23. A right circular cylinder of radius 2 centimeters and height 7 centimeters.

24. A spherical ball with radius 5 centimeters.

25. If each of the edges of a cube is multiplied by 5, describe the change in (a) the surface area (b) the volume.

An Introduction to Probability

We make frequent reference to counting problems in everyday language.

> The probability of rain today is 20%.
> The odds are in her favor.
> His chances are 50–50.

One of the authors of this book has two children. One of these children is a boy. What is the probability that they are both boys? Believe it or not the answer is not 1/2. This problem appears in the Explorations for Section 9-3.

Predictions of probabilities are usually based upon some form of counting of past experiences and a comparison of the number of favorable outcomes with the total number of outcomes. For example, since 20% = 1/5, the statement that the probability of rain is 20% is a statement that in past situations in which conditions were like the present, the ratio of the number of such days in which it has rained to the total number of these days is 1 to 5. In general, the *probability* of an event is a number, usually expressed as a ratio:

$$\frac{\text{number of favorable cases}}{\text{total number of possible cases}}$$

The simple task of counting is an essential part of the study of probability. To illustrate various problems in this chapter, we shall invent a fictitious club consisting of a set M of members, $M = \{\text{Betty, Doris, Ellen, John, Tom}\}$

Let us form a committee that is to consist of one boy and one girl, each selected from the set M of club members. How many such committees are possible? One way to find the answer to this question is by means of a *tree diagram*.

The use of tree diagrams is an extremely effective way to solve many problems that involve probabilities.

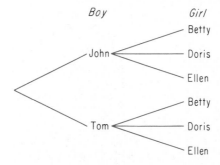

For each of the two possible choices of a boy, there are three possible choices of a girl. Thus the following six distinct possible committees can be formed and can be read from the tree diagram.

John-Betty	Tom-Betty
John-Doris	Tom-Doris
John-Ellen	Tom-Ellen

Suppose that we had selected a girl first. Then the tree diagram would be as shown in the next figure and there would still be six possibilities, the same six committees as before.

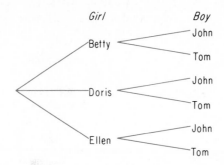

Example 1 How many different selections of two officers, a president and a vice-president, can be elected from the set M of club members?

Solution Let us select the officers in two stages. There are five possible choices for the office of president. Each of these five selections may be paired with any one of the remaining four members. Thus there are 20 possible choices. These can be read from the following diagram.

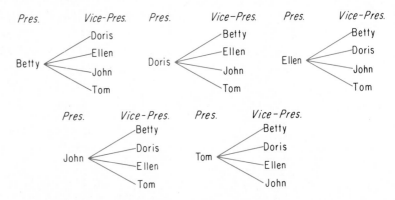

Here is a problem for you to explore. One hundred students enter a tennis tournament. How many games must be played to determine a single winner? (*Hint:* Try the strategy of solving a simpler problem. For example, how many games must be played for a tournament of eight students?

In general, if one task can be performed in m different ways and a second task can be performed in n different ways, then the first and second tasks together can be performed in $m \times n$ different ways. This **general principle of counting** can be extended if there are additional tasks.

$$m \times n \times r \times \cdots \times t$$

9 AN INTRODUCTION TO PROBABILITY

Example 2   The club M must send a delegate to a meeting tomorrow and also a delegate to a different meeting next week. How many different selections of these delegates may be made if any member of the club may serve as a delegate to either, or both, of these meetings?

Solution   There are five possible choices of a delegate to the first meeting. Since no restriction is made, we assume that the same member may attend each of the two meetings. Thus, there are five choices for the delegate to next week's meeting. In all, there are 5×5, that is, **25** choices.

Example 3   How many three-letter "words" may be formed from the set of vowels $V = \{a, e, i, o, u\}$ if no letter may be used more than once? (A word in this sense is any arrangement of three letters, such as *aeo*, *iou*, and so on.)

Solution   There are five choices for the first letter, four for the second, and three for the third. In all, there are $5 \times 4 \times 3$, that is, **60** possible words. Notice that 125 words are possible if repetitions of letters are permitted.

EXERCISES

1. If no letter may be used more than once in a given "word," how many three-letter "words" may be formed from the given set?
   (a) $\{m, a, t, h\}$ (b) $\{m, e, t, r, i, c\}$

2. Repeat Exercise 1 for four-letter "words."

3. Repeat Exercise 1 if repetitions of letters are allowed.

4. If repetitions of letters are allowed, how many four-letter "words" may be formed from the given set?
   (a) $\{h, o, p, e\}$ (b) $\{a, e, i, o, u\}$

5. How many different batteries consisting of a pitcher and a catcher may a baseball team form from the specified group?
   (a) Four pitchers and two catchers.
   (b) Five pitchers and three catchers.
   (c) Six pitchers and three catchers.

6. How many different outfits can Roberto assemble if he can wear any combination of the specified shirts and slacks?
   (a) Three sport shirts and four pairs of slacks.
   (b) Five sport shirts and four pairs of slacks.
   (c) Six sport shirts and four pairs of slacks.

7. How many different outfits can Maria assemble if she can wear any combination of the specified, on the next page, dresses, hats, and pairs of shoes?

(a) Four dresses, three hats, and two pairs of shoes.

(b) Five dresses, three hats, and two pairs of shoes.

(c) Six dresses, four hats, and three pairs of shoes.

8. Assume that no person is allowed to hold more than one office at a time and the number of members of the Portland Swim Club is (a) 15 (b) 20 (c) 50 (d) 100. How many different sets of officers consisting of a president, a vice-president, and a secretary are possible?

How many two-digit numbers may be formed from the given set of digits (a) *if repetitions are not allowed* (b) *if repetitions are allowed?*

9. {1, 2, 3, 4}

10. {1, 2, 3, 4, 5}

11. {1, 2, 3, 4, 5, 6}

12. {1, 2, 3, 4, 5, 6, 7}

13. {1, 2, 3, 4, 5, 6, 7, 8}

14. {1, 2, 3, 4, 5, 6, 7, 8, 9}

How many three-digit numbers may be formed from the given set of digits if zero is not to be used as the first digit and (a) *repetitions are not allowed* (b) *repetitions are allowed?*

15. {0, 1, 2, 3, 4}

16. {0, 1, 2, 3, 4, 5}

17. {0, 1, 2, 3, 4, 5, 6}

18. {0, 1, 2, 3, 4, 5, 6, 7}

19. {0, 1, 2, 3, 4, 5, 6, 7, 8}

20. {0, 1, 2, 3, 4, 5, 6, 7, 8, 9}

Hint for Exercise 21(a): Think of the words in this way:
$$\underset{i}{\underline{}}\ \underline{}\ \underline{}.$$

21. Find the number of "words" of three different letters that may be formed from the set of vowels $V = \{a, e, i, o, u\}$ if (a) the first letter must be i (b) the first letter must be e and the last letter must be i.

22. How many different license plates can be made using a letter from our alphabet followed by four decimal digits (a) if the first digit must not be zero (b) if the first digit must not be zero and no digit may be used more than once?

In Exercises 23 through 28, how many three-digit numbers may be formed from the set $\{0, 1, 2, 3, \ldots, 9\}$ *if zero is not an acceptable first digit and the given conditions must be satisfied?*

*23. Repetitions are allowed and the number must be divisible by (a) 5 (b) 25.

*24. Repetitions are allowed and the number must be (a) even (b) divisible by 10.

*25. Repetitions are not allowed and the number must be (a) divisible by 10 (b) even.

*26. Repetitions are not allowed and the number must be divisible by (a) 5 (b) 25.

*27. The number must be odd and less than 600 with repetitions (a) allowed (b) not allowed.

*28. Repeat Exercise 27 for numbers less than 800.

*29. Find the number of different possible license plates if each one is to consist of two letters of our alphabet followed by four decimal digits; the first digit may not be zero, and no repetitions of letters or numbers are permitted.

*30. Repeat Exercise 29 for license plates consisting of four consonants followed by three decimal digits.

*31. In a certain combination lock, there are 60 different positions. To open the lock you move to a certain number in one direction, then to a different number in the opposite direction, and finally to a third number in the original direction.
(a) What is the total number of such "combinations" if the first turn must be clockwise?
(b) What is the total number of such "combinations" if the first turn may be either clockwise or counterclockwise?

Explorations

Find (or make) two boxes to use as two cubes. Cover each cube with paper so that the cubes are of different colors, such as red and green. To introduce the vocabulary of dice games think of one cube as a red die and the other cube as a green die. Number the faces of each die 1 through 6. Have two students hold the "dice" so that the rest of the class can see only one number on each die.

There is evidence of dice having been used as early as 3000 B.C.

1. Suppose that the number on the green die is 5. If possible identify a number that can be shown on the red die so that the two numbers have the indicated sum.
(a) 5   (b) 6   (c) 7
(d) 11   (e) 3   (f) 12

2. If the number on the green die is 2, what is the set of numbers that can be obtained as sums by suitable selections of numbers on the red die?

3. What is the set of numbers that can be obtained as sums by suitable selections of a number on each of the two dice?

4. Which of the sums obtained in Exploration 3 can be obtained in only one way? List this way for each of the numbers.

5. Which of the sums obtained in Exploration 3 can be obtained in only two ways? List the ways for each of the numbers.

6. Which of the sums obtained in Exploration 3 can be obtained in the largest number of ways? List these ways.

Counting problems arise in many mathematical puzzles. See the adjacent figure and consider, for example, the following dart puzzles.

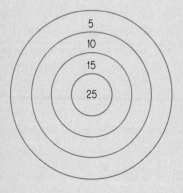

7. You are allowed to throw four darts, and we shall assume that there are no misses. In how many different ways can you obtain a score of 60?

8. Draw six concentric circles similar to the four circles in the figure for Exploration 7. Starting at the center and working outward, write the numbers 20, 19, 13, 12, 8, and 7 in order, one in each region. Use four darts, assume no misses, and show two ways of scoring 40.

9. Find a floor with boards of the same width (or draw equally spaced parallel lines on the floor) and cut a piece of wire so that the distance between the lines separating the boards is twice the length of the wire. This exploration is a special case of the **Buffon needle problem** with the wire serving as the needle. Each trial consists of dropping the needle and observing whether the needle touches a line. Let T be the number of trials and N the number of these trials in which the needle touches a line. It is known from advanced mathematics that for very large values of T the probability of the needle touching a line is $1/\pi$ and is approximately N/T. Make at least 100 trials and compute T/N as an approximation for π.

10. Ask at least 100 people to select an integer between 1 and 10. Tabulate the number of times that each integer is selected. Discuss the results relative to a recent conjecture that 7 is the most frequently selected number under such circumstances.

11. Repeat Exploration 10 for a two-digit number that is less than 50 and has odd numbers for both of its digits. Compare your results with the conjecture that the number most frequently selected is 37.

9-2
Definition of Probability

When a normal coin is tossed, we know that there are two distinct and **equally likely** ways in which it may land, heads or tails. We say that the probability of getting a head is one out of two, or simply 1/2.

In throwing one of a pair of normal dice, there are six equally likely ways in which the die may land. We say that the probability of throwing a 5 on one throw of a die is one out of six, or 1/6.

In each of these two examples, the **events** that may occur are said to be **mutually exclusive.** That is, one and only one of the events can occur at any given time. When a coin is tossed, there are two possible events (heads and tails); one and only one of these may occur. When a single die is thrown, there are six events {1, 2, 3, 4, 5, 6}; one and only one of these may occur.

Informally, we define the probability of success as the ratio of the number of successes of an event to the number of possible outcomes of that event.

> If an event can occur in any one of n mutually exclusive and equally likely ways, and if m of these ways are considered favorable, then the **probability** $P(A)$ that a favorable event A will occur is given by the formula
>
> $$P(A) = \frac{m}{n}$$

The probability m/n satisfies the relation $0 \le m/n \le 1$, since m and n are integers and $m \le n$.

> When success is inevitable, $m = n$ and the probability is 1.
>
> When an event cannot possibly succeed, $m = 0$ and the probability is 0.

For example, the probability of getting either a head or a tail on a single toss of a coin is 1, assuming that the coin does not land on an edge. The probability of throwing a sum of 13 with a single throw of a pair of normal dice is 0. (Always assume, unless otherwise instructed, that normal dice are used.)

The sum of the probability that an event occurs and the probability that the same event does not occur is 1.

> If $P(A) = \frac{m}{n}$, then $P(\text{not } A) = 1 - \frac{m}{n}$.

Example 1   A single card is selected from a deck of 52 bridge cards. What is the probability that it is a spade? What is the probability that it is not a spade? What is the probability that it is an ace or a spade?

Of the 52 cards, 13 are spades. Therefore, the probability of selecting a spade is 13/52, that is, 1/4. The probability that the card selected is not a spade is $1 - 1/4$, that is, 3/4. There are four aces and 12 spades besides the ace of spades. Therefore, the probability that the card selected is an ace or a spade is $(4 + 12)/52$, that is, 16/52, which we express as 4/13.

It is very important that only equally likely events be considered when the probability formula is applied; otherwise, faulty reasoning can occur.

Consider again the first question asked in Example 1. One might reason that any single card drawn from a deck of cards is either a spade or is not a spade; thus there are two possible outcomes, and the probability of drawing a spade must therefore be 1/2. It is correct to say that there are these two possible outcomes, but of course they are *not* equally likely since there are 13 spades in a deck of cards and 39 cards that are not spades. If all possible events are equally likely, the events are said to occur **at random.**

Example 2 A committee of two is to be selected at random from the set

$$M = \{\text{Betty, Doris, Ellen, John, Tom}\}$$

by drawing names out of a hat. What is the probability that both members of the committee will be girls?

Solution We solve this problem by first listing all of the possible committees of two that can be formed from the set M.

Betty – Doris	Doris – John
Betty – Ellen	Doris – Tom
Betty – John	Ellen – John
Betty – Tom	Ellen – Tom
Doris – Ellen	John – Tom

Making a list continues to be a very important problem-solving strategy.

Of the ten possible committees, there are three (those boxed) that consist of two girls. Thus the probability that both members selected are girls is 3/10. What is the probability of selecting a committee to consist of two boys?

EXERCISES *What is the probability that the specified event will occur in a single throw of one die?*

1. An even number.
2. An odd number.
3. A number greater than 2.
4. A number less than 4.
5. A number different from 4.
6. A number different from 0.
7. The number 0.
8. A number less than 7.

What is the probability that the specified event will occur in a single draw from an ordinary deck of 52 bridge cards?

9. An ace.   10. A king.
11. A spade.   12. A red card.

We will consider naming a committee of two from the set
$N = \{Alice, Bob, Carolyn, Doug, Ellen, Frank\}$.

13. How many different committees of two can be formed?
14. What is the probability that a committee of two will consist of two boys?
15. What is the probability that a committee of two will consist of two girls?
16. What is the probability that a committee of two will consist of one boy and one girl?

17. What is the probability that the first-named author of this text was born in the month of December?
18. What is the probability that your instructor's telephone number has a 7 as its final digit?
19. The probability of obtaining all heads in a single toss of three coins is 1/8. What is the probability that not all three coins are heads in such a toss?
20. What is the probability that the next person you meet was not born on a Sunday?
*21. Two distinct integers from the set 1 through 7 are selected at random.
    (a) What is the probability that the first integer selected is even?
    (b) What is the probability that both integers are even?
*22. Repeat Exercise 21 for the set of integers 1 through 13.

Explorations

1. Identify at least one item that appears regularly in the daily newspaper that is based upon probability.
2. Observe and list in order the last two digits of 20 license plates on automobiles in any large parking lot. Repeat this procedure for at least five sets of 20 two-digit numbers. For each set of 20 numbers note how often you find a repetition of pairs of digits (the same two-digit number appearing at least twice).
3. Repeat the preceding exploration by opening a telephone book and selecting 20 telephone numbers at random. Re-

cord the last two digits only, and note the frequency with which you find a repetition of pairs of digits within a set of 20 two-digit numbers.

If you are enrolled in a class of 30 students, what would you guess is the probability that there are two members of the group that celebrate their birthdays on the same date? Normally you would expect the probability to be low, inasmuch as there are at least 365 different days of the year on which an individual's birth date could occur, and 30 is small relative to 365.

Although the mathematical explanation is beyond our scope at this time, it can be shown that the actual probability of at least two birth dates occurring on the same day of the year in a random group of 30 students is 0.71. That is, you can safely predict that 71% of the time such an occurrence will take place. Or, alternatively, if you bet that at least two members of the group have the same birth date, then you can expect to win such a bet approximately 71 times out of 100.

The graph shows the probability of two common birth dates for people in groups of different sizes. Note that the probability in a group of 23 is approximately 0.50. That is, for a group of 23 individuals, there is a 50% chance of having such an occurrence take place. For a group of 60 the probability is 0.99, or almost certainty!

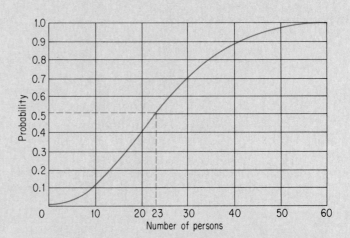

4. Use the given graph to determine the probability of two or more common birth dates for the members of a class of (a) 20 students (b) 40 students.

5. Use the members of your class to test the results shown in the graph.

6. Test the results shown in the graph by going to an encyclopedia and using the dates of birth of 30 individuals selected at random.

9-3
Sample Spaces

It is often convenient to solve problems of probability by making a list of all possible outcomes. Such a listing is called a **sample space.** Consider first the problem of tossing two coins. The sample space for this problem is given by the following set of all possible outcomes.

| H, H | H, T | T, H | T, T |

These four possible outcomes may be obtained by using a tree diagram to list all possible cases or by listing the possibilities as in the following array.

From the chart we may observe the following.

Two heads occur in one event.

One head and one tail occur in two events.

No heads, that is, two tails, occur in one event.

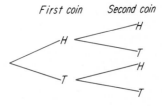

First coin	Second coin
H	H
H	T
T	H
T	T

Event	Probability
2 heads	$\frac{1}{4}$
1 head	$\frac{2}{4}$
0 heads	$\frac{1}{4}$

Then we may list the various probabilities regarding the tossing of two unbiased coins. Since all possibilities have been considered, the sum of the probabilities is 1. This provides a check on our computation. The list of probabilities is sometimes called a **probability distribution.**

For the case of three coins, the following tree diagram and array may be made.

First coin	Second coin	Third coin	First coin	Second coin	Third coin
			H	H	H
			H	H	T
			H	T	H
			H	T	T
			T	H	H
			T	H	T
			T	T	H
			T	T	T

From the tree diagram we have the following sample space for the tossing of three coins.

{*HHH, HHT, HTH, HTT, THH, THT, TTH, TTT*}

Event	Probability
0 heads	$\frac{1}{8}$
1 head	$\frac{3}{8}$
2 heads	$\frac{3}{8}$
3 heads	$\frac{1}{8}$

We may also list the probabilities of specific numbers of heads as in the accompaning array. Both for two coins and for three coins the sum of the probabilities is 1. That is, all possible events have been listed and these events are mutually exclusive. Note also that for two coins there were four possible outcomes and for three coins there were eight possible outcomes. For n coins there would be 2^n possible outcomes.

Example   A box contains two red and three white balls. Two balls are drawn in succession without replacement. List a sample space for this experiment.

Solution   To identify individual balls, we denote the red balls as R_1 and R_2, and the white balls as W_1, W_2, W_3. Then the sample space is

Note in the example that a sample space may be presented as the elements of an array as well as in set notation.

R_1R_2	R_2R_1	W_1R_1	W_2R_1	W_3R_1
R_1W_1	R_2W_1	W_1R_2	W_2R_2	W_3R_2
R_1W_2	R_2W_2	W_1W_2	W_2W_1	W_3W_1
R_1W_3	R_2W_3	W_1W_3	W_2W_3	W_3W_2

EXERCISES   *Use the sample space of the preceding example to find the probability of each event.*

1. Both balls are red.
2. Both balls are white.
3. The first ball is red.
4. The first ball is red and the second ball is white.
5. One ball is red and the other is white.

Use the sample space for the outcomes when three coins are tossed to find the probability of each event.

6. All three coins are heads.
7. At least two coins are heads.
8. At most one coin is tails.

9. Make a tree diagram and give the sample space for the tossing of four coins.

In Exercises 10 through 12 use the sample space for the outcomes when four coins are tossed (Exercise 9) to find the probability of each event.

10. **(a)** No coins are heads.
    (b) All four coins are tails.

11. **(a)** At least three coins are heads.
    (b) At most one coin is tails.

12. **(a)** At most two coins are tails.
    (b) At least two coins are heads.

13. Make a tree diagram and give a sample space for the throwing of a pair of dice. Represent each outcome by an ordered pair of numbers. For example, let (1, 3) represent a 1 on the first die and a 3 on the second die.

In Exercises 14 through 20 use the sample space for the outcomes when a pair of dice are thrown (Exercise 13) to find the probability of each event.

14. **(a)** The number on the first die is 2.
    (b) The number on the first die is not 2.

15. **(a)** The number 1 is on both dice.
    (b) It is not true that the number 1 is on both dice.

16. **(a)** The same number is on both dice.
    (b) There are different numbers on the two dice.

17. **(a)** The sum of the numbers obtained is 11.
    (b) The sum of the numbers obtained is not 11.

18. **(a)** The sum of the numbers obtained is 7.
    (b) The sum of the numbers obtained is not 7.

19. **(a)** The number on the second die is twice the number on the first die.
    (b) The number on one die is twice the number on the other die.

20. **(a)** The number on one die is three more than the number on the other die.
    (b) The number on one die is two less than the number on the other die.

21. A box contains two red balls R_1 and R_2 and two white balls W_1 and W_2. List a sample space for the outcomes when two balls are drawn in succession without replacement. Find the probability that both balls are red.

22. Repeat Exercise 21 for the case in which the first ball is replaced before the second ball is drawn.

23. Repeat Exercise 21 for a box that contains four red balls and two white balls.

24. Repeat Exercise 23 for the case in which the first ball is replaced before the second ball is drawn.

25. Make a tree diagram and a give a sample space for a toss of a coin followed by the throw of a die.

In Exercises 26 through 28 use the sample space for the outcomes when a coin is tossed and a die is thrown (Exercise 25) to find the probability of each event.

26. (a) The coin is heads and the number on the die is even.
    (b) The coin is tails and the number on the die is odd.

27. (a) The coin is heads and the number on the die is 5.
    (b) The coin is tails and the number on the die is not 5.

28. (a) The coin is tails and the number on the die is greater than 4.
    (b) The coin is heads and the number on the die is prime.

Explorations

Assume that a red die and a green die are thrown and the sum S of the numbers shown on the dice is noted.

1. Give the sample space of the possible values of S.

2. List the number r on the red die and the number g on the green die as (r, g) and make a table of all possible such ordered pairs of numbers for each possible value of S.

3. List each possible value of S with its probability.

4. Throw a pair of dice at least forty times and see if the probabilities of the values of S are approximately illustrated by the results obtained.

There are many interesting probability questions whose answers are not intuitively obvious. Here are two that are best solved by means of a sample space.

5. Three cards are in a box. One is red on both sides, one is white on both sides, and one is red on one side and white on the other. A card is drawn at random and placed on a table. The card has a red side showing. What is the probability that the side not showing is also red? (Contrary to popular belief, the answer is not 1/2.)

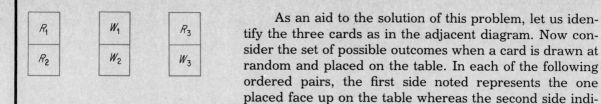

As an aid to the solution of this problem, let us identify the three cards as in the adjacent diagram. Now consider the set of possible outcomes when a card is drawn at random and placed on the table. In each of the following ordered pairs, the first side noted represents the one placed face up on the table whereas the second side indicates the face hidden from view.

$$\{R_1R_2, \quad R_2R_1, \quad R_3W_3, \quad W_1W_2, \quad W_2W_1, \quad W_3R_3\}$$

Inasmuch as we are told that a red side is showing, we may narrow down the sample space to the first three pairs only. Of these three possibilities, if the first side is red can you now tell the probability that the second side is also red?

Hint for Exploration 6: List the sample space using *B* for boy and *G* for girl. Then note that *GG* is *not* a possibility.

6. One of the authors of this book has two children. One of these children is a boy. What is the probability that they are both boys? Believe it or not, the answer is *not* 1/2!

7. Find several newspaper articles that involve sample spaces and note the importance of the selection of the sample space. For example, suppose that the first paragraph of an article contains a report that a small company employs 10 men and 10 women with average salaries at $18,000 and $14,000, respectively (discrimination?). Suppose also that a detailed reading of the article reveals that when length of service is considered, the salaries may be described as in the array.

Sex / Years of service	Male Number	Male Average salary	Female Number	Female Average salary
Less than 5	2	$10,000	8	$12,000
At least 5	8	20,000	2	25,000
Totals	10	18,000	10	14,600

Although the women's salaries appear better under this more detailed comparison, the purpose of this example is not to enter the arguments about discrimination. Rather, our goal is to emphasize the need for considering background details in selecting a sample space for use in a comparison. In this regard discuss the apparent reliability of the newspaper articles that you find.

9-4

Computation of Probabilities

If A and B represent two mutually exclusive events, then

$$\boxed{P(A \text{ or } B) = P(A) + P(B)}$$

That is, the probability that one event or the other will occur is the sum of the individual probabilities. Consider, for example, the probability of drawing an ace or a picture card (a jack, queen, or king) from an ordinary deck of 52 bridge cards.

The probability of drawing an ace, $P(A)$, is 4/52.

The probability of drawing a picture card, $P(B)$, is 12/52.

Then $P(A \text{ or } B) = \dfrac{4}{52} + \dfrac{12}{52} = \dfrac{16}{52} = \dfrac{4}{13}.$

Example 1   A bag contains three red, two black, and five yellow balls. Find the probability that a ball drawn at random (each is equally likely) will be red or black.

Solution   The probabilty of drawing a red ball, $P(R)$, is 3/10. The probability of drawing a black ball, $P(B)$, is 2/10. Then

$$P(R \text{ or } B) = P(R) + P(B) = 5/10 = 1/2$$

This process can be extended to find the probability of any finite number of mutually exclusive events.

$$\boxed{P(A_1 \text{ or } A_2 \text{ or } A_3 \text{ or } \cdots A_n) = P(A_1) + P(A_2) + P(A_3) + \cdots + P(A_n)}$$

Example 2   A single die is thrown. What is the probabiliy that either an odd number or a number greater than 3 appears?

Solution   There are three odd numbers possible, (1, 3, 5), so the probability of throwing an odd number is 3/6. The probability of getting 4, 5, or 6, that is, a number greater than 3, is also 3/6. Adding these probabilities gives 1. Something is obviously wrong, since a probability of 1 implies certainty and we can see that an outcome of 2 is neither odd nor greater than 3. The difficulty lies in the fact that the events are *not* mutually exclusive; a number may be both odd and also greater than 3. In particular, 5 is both odd and greater than 3. Thus $P(5)$ has been included twice. Since $P(5) = 1/6$, the answer is

$$\frac{3}{6} + \frac{3}{6} - \frac{1}{6}, \quad \text{that is,} \quad \frac{5}{6}.$$

Example 2 can be illustrated by means of the adjacent Venn diagram that lists all possible outcomes when a single die

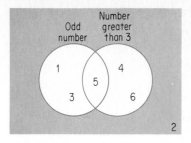

Odd number / Number greater than 3

is thrown. Note that the outcome 5 is listed in the intersection of the two sets since it is both odd and greater than 3. On the other hand, the outcome 2 fits into neither of these descriptions and is thus placed outside the circles.

In situations like that of Example 2 we need to subtract the probability that both events occur. Thus, if A and B are *not* mutually exclusive events, we have

$$P(A \text{ or } B) = P(A) + P(B) - P(A \text{ and } B)$$

Note that in Example 2, we had the following probabilities.

$P(A)$, the probability of an odd number, is 3/6.

$P(B)$, the probability of a number greater than 3, is 3/6.

$P(A \text{ and } B)$, the probability of an odd number greater than 3, is 1/6.

$$P(A \text{ or } B) = \frac{3}{6} + \frac{3}{6} - \frac{1}{6} = \frac{5}{6}.$$

By an actual listing we can see that five of the six possible outcomes in Example 2 are either odd or greater than 3, namely 1, 3, 4, 5, and 6. The only "losing" number is 2. Thus the probability $P(A \text{ or } B)$ must be 5/6.

In general, the two situations $P(A \text{ or } B)$ and $P(A \text{ and } B)$ that we have considered can be described by means of diagrams, where the points of the circular regions represent probabilites of events.

The first of the two adjacent diagrams shows mutually exclusive events.

$$P(A \text{ or } B) = P(A) + P(B)$$

In the second figure, the region that is shaded both horizontally and vertically represents $P(A \text{ and } B)$, $P(A \text{ and } B) \neq 0$, the events are not mutually exclusive, and

$$P(A \text{ or } B) = P(A) + P(B) - P(A \text{ and } B)$$

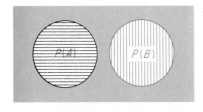

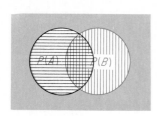

We must subtract $P(A \text{ and } B)$, since we have counted it twice in the sum $P(A) + P(B)$.

Next we turn our attention to the probability that several events will occur, one after the other. Consider the probability of tossing a coin twice and obtaining heads on the first toss and tails on the second toss. From a sample space we see that the probability is 1/4.

$$\{HH, \boxed{HT}, TH, TT\}$$

Furthermore, we see that the probability $P(A)$ that the first coin is heads is 1/2. The probability $P(B)$ that the second coin is tails is 1/2. Then

$$P(A \text{ and } B) = \frac{1}{2} \times \frac{1}{2} = \frac{1}{4}$$

Note that the outcome of the first toss does not affect the second toss.

In general, let the probability that an event A occurs be $P(A)$. Let the probability that a second event occurs after A has occurred be $P(B$ given $A)$. Then

$$\boxed{P(A \text{ and } B) = P(A) \times P(B \text{ given } A)}$$

Example 3 Urn A contains three white and five red balls. Urn B contains four white and three red balls. One ball is drawn from each urn. What is the probability that they are both red?

Solution Let $P(A)$ be the probability of drawing a red ball from urn A and $P(B)$ be the probability of drawing a red ball from urn B. Then

$$P(A) = \frac{5}{8} \qquad P(B) = \frac{3}{7} \qquad P(A \text{ and } B) = \frac{5}{8} \times \frac{3}{7} = \frac{15}{56}$$

Example 4 Two cards are selected in succession, without replacement, from an ordinary bridge deck of 52 cards. What is the probability that they are both aces?

Solution The probability that the first card is an ace is 4/52. If it is an ace, then the probability that the second card is an ace is 3/51. The probability that both cards are aces is (4/52) × (3/51), that is, 1/221.

What is the probability that the specified event will occur in a single throw of one die?

1. An even number or a number greater than 4.
2. An odd number or a number less than 4.
3. An odd number or a number greater than 4.
4. An even number or a number greater than 6.
5. An odd number or a number greater than 7.
6. An odd number or a number less than 6.
7. An even number or a number less than 7.

A single card is drawn from an ordinary deck of 52 bridge cards. Find the probability that the card selected is as described.

8. An ace or a queen.
9. A spade or a diamond.
10. A spade or a queen.
11. A spade and an ace.
12. A spade and a queen.
13. A heart or a king or a queen.
14. A club or an ace or a king.
15. A club and a spade.

Two cards are drawn in succession from an ordinary deck of 52 bridge cards without the first card's being replaced. Find the probability of the specified event.

16. (a) Both cards are spades.
    (b) Both cards are the ace of spades.
17. (a) The first card is a spade and the second card is a heart.
    (b) The first card is an ace and the second card is the king of hearts.
18. (a) The two cards are of the same suit.
    (b) The two cards are of different suits.

Two cards are drawn in succession from an ordinary deck of 52 bridge cards with the first card replaced before the second card is drawn. In Exercises 19 and 20 find the probability of the specified event.

19. (a) Both cards are diamonds.
    (b) Both cards are the ace of diamonds.
20. (a) The first card is a heart and the second card is a club.
    (b) The first card is an ace and the second card is the king of hearts.

21. A coin is tossed six times. What is the probability that all six tosses are heads?

22. A coin is tossed six times. What is the probability that at least one head is obtained? (*Hint*: First find the probability of getting no heads.)

23. A coin is tossed and then a die is thrown. Find the probability of obtaining (**a**) a head and a 3 (**b**) a head and an even number (**c**) a head or a 3 (**d**) a head or an even number.

24. A bag contains four red balls and seven white balls.
    (**a**) If one ball is drawn at random, what is the probability that it is white?
    (**b**) If two balls are drawn at random, without replacement, what is the probability that they are both white?

25. Repeat Exercise 24 for a bag with five red balls and seven white balls.

26. Repeat Exercise 24 for a bag with five red balls and eight white balls.

27. Five cards are drawn at random, without replacement, from an ordinary bridge deck of 52 cards. Find the probability that all five cards drawn are spades.

28. A box contains three red, four white, and six green balls. Three balls are drawn in succession, without replacment. Find the probability that (**a**) all three are red (**b**) the first is red, the second is white, and the third is green (**c**) none are green *(**d**) all three are of the same color.

29. Repeat Exercise 28 if each ball is replaced after it is drawn.

30. Repeat Exercise 28 for a box that contains four red, three white, and five green balls.

31. Repeat Exercise 28 for a box that contains four red, three white, and six green balls.

32. A die is thrown three times. Find the probability that (**a**) a 6 is obtained on the first throw (**b**) a 6 is obtained on each of the first two throws (**c**) a 6 is obtained on each of three throws (**d**) a 6 is obtained on the first throw and not obtained on the second or third throws.

33. A die is thrown three times. Find the probability that (**a**) an even number is obtained on all three throws (**b**) an even number is obtained on the first two throws and an odd number on the third throw (**c**) an even number is obtained on the first throw and an odd

number on the second and third throws   **(d)** exactly one even number is obtained.

*34. A die is thrown four times. Find the probability that an even number is obtained on **(a)** all throws   **(b)** exactly one throw   **(c)** exactly two throws.

*35. Repeat Exercise 34 for a die that is thrown five times.

*36. A die is thrown three times. Find the probability that **(a)** at least one 6 is obtained   **(b)** exactly one 6 is obtained.

Explorations

These explorations include several experiments that can be performed to illustrate basic concepts of probability.

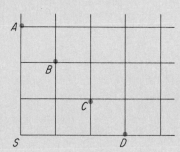

1. Consider the network of streets shown in the adjacent figure. You are to start at the point *S* and move three "blocks." Each move is determined by tossing a coin. If the coin lands *tails,* move one block to the right; for *heads,* move one block up. Your terminal point will be at *A, B, C,* or *D.* Try to predict the number of times you will land at each point if the experiment is to be repeated 16 times. Then complete 16 trials, keeping a tally of the number of times you land at each point. Compare your actual results with your predictions as well as with the results obtained by your classmates. Finally, if necessary, revise your predictions on the basis of your experimentation.

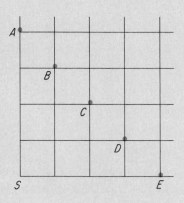

2. Repeat Exploration 1 for the adjacent network. This time you are to make four moves and will land at point *A, B, C, D,* or *E.*

3. If a pair of normal dice is thrown repeatedly, a sum of 7 can be expected theoretically to occur in one out of six throws. Thus if a pair of dice is thrown 36 times, on the average six of the throws will give a sum of 7. Throw a pair of dice 36 times and count the frequency with which a sum of 7 appears. Compare your actual results with the theoretical probability of 1/6.

4. As the number of throws becomes very large, that is, **in the long run,** the experimental results approach the theoretical results. Collect the data for Exploration 3 from each member of your class. Tabulate the results of your own experiment as for one student, your results and those of one of your classmates as for two students, your results and those of two classmates as for three students, and so

forth as suggested in the array. Note the tendency of S/N to approach 1/6 as N increases.

Number of students		1	2	3	4	5	6	. . .
Number N of throws		36	72	108	144	180	216	. . .
Number S of sums of 7		——	——	——	——	——	——	. . .
S/N		——	——	——	——	——	——	. . .

5. Make 72 throws of a pair of dice. On graph paper record each throw by placing an X in the column that represents the sum. The final result should be a bar graph that is fairly "normal" in shape. In particular the height of each bar should be approximately equal to the theoretical expectation of the corresponding sum for 72 throws. The following figure shows the results obtained after one experiment of 18 throws.

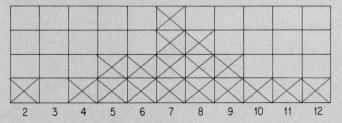

6. Make bar graphs as in Exploration 5 for the results obtained by **(a)** you and one classmate **(b)** you and two classmates **(c)** you and three classmates **(d)** you and four classmates. Compare your bar graph for Exploration 5 and these bar graphs. Describe the shape of the curve and the effect of increasing the number of throws.

7. Shuffle an ordinary deck of cards. Conjecture the number of cards that one must turn over, on the average, to reach the first ace. Then test your conjecture by performing the experiment at least 20 times, thoroughly shuffling the deck each time. Report on your results.

9-5
Odds and Mathematical Expectation

Consider the problem of finding the odds against obtaining a 3 in one throw of a die. Since the probability of obtaining a 3 is known to be 1/6, most people would say that the odds are therefore 6 to 1 against obtaining a 3. This is not correct—because out of every six throws of the die, in the long run, we expect to obtain one 3. The other five throws are not expected to produce 3's. Hence, the correct odds against obtaining a 3 in one throw of a die are 5 to 1. The odds in favor of obtaining a 3 are 1 to 5. Formally, we define odds as follows.

The sports section of a newspaper often includes statements regarding the "odds" in favor of, or against, a particular team or individual's winning or losing some encounter. For example, we may read that the odds in favor of the Cardinals' winning the pennant are "4 to 1." In this section we shall attempt to discover just what such statements really mean.

The **odds in favor** of an event are the ratio of the probability that an event will occur to the probability that the event will not occur.

$$\frac{P(A)}{1 - P(A)}$$

The **odds against** the occurrence of the event are the reciprocal of this ratio.

Thus the odds in favor of an event that may occur in several equally likely ways are the ratio of the number of favorable ways to the number of unfavorable ways.

Notice that odds and probabilities are very closely related. Indeed, if either the odds for or the probability of an event is known, then the other can be found. For example, if the odds for an event are 1 to 2, then we have the ratio

$$\frac{\text{number of favorable ways}}{\text{number of unfavorable ways}} = \frac{1}{2}$$

and the probability is the ratio

$$\frac{\text{number of favorable ways}}{\text{total number of all ways (favorable and unfavorable)}} = \frac{1}{3}$$

Similarly, if the probability is 2/5, then the odds are 2 to 3.

Example 1 Find the *odds in favor* of drawing a spade from an ordinary deck of 52 bridge cards and the *odds against* drawing a spade from an ordinary deck of 52 bridge cards.

Solution Since there are 13 spades in a deck of cards, the probability of drawing a spade is 13/52, that is, 1/4. The probability of failing to draw a spade is 3/4. The odds in favor of obtaining a spade are (1/4) ÷ (3/4), that is, 1/3.

The *odds in favor* of drawing a spade are stated as 1/3. They may also be stated as 1 to 3 or as 1:3. Similarly, the *odds against* drawing a spade are 3/1, which may be written as 3 to 1 or 3:1.

Mathematical expectation is closely related to odds and is defined as the product of the probability that an event will occur and the amount to be received upon such occurrence. Suppose that you are to receive $2.00 each time you obtain two heads on a single toss of two coins. You do not receive anything for any other outcome. Then your mathematical expectation will be one-fourth of $2.00 and three-fourths of $0. In other words, your mathematical expectation from the game is

$$\frac{1}{4}(\$2.00) + \frac{3}{4}(\$0) \quad \text{that is} \quad \$0.50.$$

This means that you should be willing to pay $0.50 each time you toss the coins if the game is to be a fair one. *In the long run* both you and the person who is running the game would break even. Note that for each game that you pay $0.50 and win $2.00, your net gain is $1.50; for each game that you pay $0.50 and lose, your net gain is a loss of $0.50. The game has two possible outcomes.

Pay $0.50 and win $2.00 with probability 1/4.
Pay $0.50 and receive nothing with probability 3/4.

If you pay $0.50 to play, your mathematical expectation of net gain in playing the game is

$$\frac{1}{4}(\$2.00 - \$0.50) + \frac{3}{4}(\$0 - \$0.50) \quad \text{that is} \quad \$0.$$

The zero expectation indicates that the game is a fair game to both you and the person running the game.

If an event has several possible outcomes that occur with probabilities p_1, p_2, p_3, and so forth, and for each of these outcomes one may expect the amounts m_1, m_2, m_3, and so on, then the mathematical expectation E may be defined as

$$\boxed{E = m_1 p_1 + m_2 p_2 + m_3 p_3 + \ldots}$$

Whenever you use the formula for mathematical expectation, it is worthwhile to check that all possible outcomes have been considered. To do so, show that the sum of the probabilities is equal to 1.

Example 2   Suppose that you pay 5¢ to play a game in which a coin is tossed twice. You receive 10¢ if two heads are obtained, 5¢ if exactly one head is obtained, and nothing if there are no heads. What is the expected value to you of this game?

Solution   Remember that you pay 5¢ to play the game. Thus for two heads you get your nickel back and gain another 5¢; for one head you break even; for no heads you lose the 5¢ that you have paid to play the game. In other words, the possible events, their probabilities, and their values to you are

Two heads	1/4	10 − 5,	that is,	5¢
One head	1/2	5 − 5,	that is,	0¢
No heads	1/4	0 − 5,	that is,	−5¢

Your expected value E, in cents, is

$$E = (5)\left(\frac{1}{4}\right) + (0)\left(\frac{1}{2}\right) + (-5)\left(\frac{1}{4}\right) = 0$$

The zero value implies that *if you played a large number of games,* your gains should equal your losses. If you paid less than 5¢ per game, you could expect to make a profit. If you paid more than 5¢ per game, you should expect to have a loss.

In Example 2, if you were not told the price for playing this game, then you could determine the fair price by computing your expectation in this way:

$$E = (10)\left(\frac{1}{4}\right) + (5)\left(\frac{1}{2}\right) + (0)\left(\frac{1}{4}\right) = 5$$

Thus, as previously observed, 5¢ is the fair price to pay to play this game.

EXERCISES *What are the odds in favor of the event?*

1. Two heads in a single toss of two coins.
2. At least two heads in a single toss of three coins.
3. Two heads when a single coin is tossed twice.
4. At least two heads when a single coin is tossed three times.

In Exercises 5 through 8 what are the odds against the event?

5. Two heads in a single toss of two coins.
6. An ace in a single draw from a deck of 52 bridge cards.
7. An ace or a king in a single draw from a deck of 52 bridge cards.
8. A 7 or an 11 in a single throw of a pair of dice.

9. For the event of obtaining a 7 or an 11 in a single throw of a pair of dice, what are the odds in favor of the event?
10. One hundred tickets are sold for a lottery. The grand prize is $2000. What is your mathematical expectation if you are given a ticket?
11. Repeat Exercise 10 for the case where 250 tickets are sold.
12. What is your mathematical expectation when you are given one of 400 tickets for a single prize worth $1200?
13. In the "long run" what is your expected profit, or loss, per game if you pay $2 to play each game and receive $10 if you throw a "double" (the same number on both dice) on a single toss of a pair of dice?

14. A box contains three dimes and two quarters. You are to reach in and select one coin, which you may then keep. Assuming that you are not able to determine which coin is which by its size, what would be a fair price for the privilege of playing this game?

15. There are three identical boxes on a table. One contains a five-dollar bill, one contains a one-dollar bill, and the third is empty. A man is permitted to select one of these boxes and to keep its contents. What is his expectation?

16. Three coins are tossed. What is the expected number of heads?

*17. Two bills are to be drawn without replacement from a purse that contains three five-dollar bills and two ten-dollar bills. What is the mathematical expectation for this drawing?

18. If there are two pennies, a nickel, a dime, a quarter, and a half-dollar in a hat, what is the mathematical expectation of the value of a random selection of a coin from the hat?

*19. Repeat Exercise 18 for the selection of two coins from the hat without replacing the first coin selected.

20. Suppose that the probability of your obtaining an A in this course is 0.3. What are the odds (a) in favor of your obtaining an A (b) against your obtaining an A?

*21. If the probability of an event is p, what are the odds (a) in favor of the event (b) against the event?

*22. Suppose that n tickets are sold at $1 each for a single prize of $1000. Express the theoretically expected value of each ticket in terms of n. (*Hint:* This value is the algebraic sum of the expected loss and the expected gain.)

*23. Repeat Exercise 22 if there is a first prize of $5000 and a second prize of $1000.

Explorations

Many games and experiments may be simulated using a table of random numbers instead of actually playing the game or performing the experiment.

Tables of *random numbers* may be generated by selecting numbers from a set with replacement and at random. Here is a table of 100 two-digit random numbers.

80	68	30	67	70	21	62	01	79	75
18	53	29	65	19	85	68	11	62	56
63	64	39	34	88	25	76	42	66	21
82	25	11	76	63	67	55	03	57	77
27	14	60	76	72	25	64	62	12	64
11	73	60	93	75	07	05	77	42	57
78	61	69	29	36	65	82	92	16	28
45	92	63	01	62	95	91	92	13	18
11	84	97	48	73	95	49	84	34	65
23	76	77	14	15	10	12	58	79	93

Three steps are involved in the use of a table of random numbers to simulate a game or an experiment.

(i) Decide on a method for selecting numbers from the table. For example, select the first 40 numbers left to right by rows top to bottom.

(ii) Assign sets of numbers to each element of the sample space of the experiment to be performed. This is done so that the probability of a number from the set from which the table was formed being in the assigned set is the same as the probability of the element to which the set is assigned. For example, assign even numbers to heads and odd numbers to tails for the tossing of a coin.

(iii) Consider the numbers selected in (i). If the number is not one of those assigned in (ii), proceed to the next number.

1. Use the given table of random numbers to simulate the tossing of a coin 100 times. Assign even numbers to heads and odd numbers to tails. How many heads are obtained?

2. Repeat Exploration 1 using the assignment of numbers 00 through 49 to heads and 50 through 99 to tails.

Use the given table of random numbers and simulate each experiment.

3. Toss two coins 100 times.
4. Throw a die 72 times.
5. Throw two dice 72 times.
6. Throw three dice 72 times.
7. Throw four dice 72 times.
8. Toss a coin and throw a die 72 times.

9. Try to obtain information from one of the state or provincial lotteries and estimate the mathematical expectation of a ten-dollar purchase of tickets.

9-6
Permutations

Suppose that three people, Ruth, Joan, and Debbie, are waiting to play singles at a tennis court. Two of the three can be selected in six different ways if the order in which they are named is significant, for example, if the first person named is to serve first. We may identify these six ways from a tree diagram.

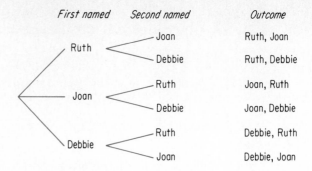

First named	Second named	Outcome
Ruth	Joan	Ruth, Joan
	Debbie	Ruth, Debbie
Joan	Ruth	Joan, Ruth
	Debbie	Joan, Debbie
Debbie	Ruth	Debbie, Ruth
	Joan	Debbie, Joan

We say that there are 3×2, that is, 6, *permutations* of the set of three people selected two at a time. In each case the *order* in which the two people are named is significant. A **permutation** of a set of elements is an *arrangement* of certain of these elements *in a specified order*. In the problem just discussed, the number of permutations of three things taken two at a time is 6. In symbols we write

Note that *order* is important in this discussion.

$$_3P_2 = 6, \quad \text{read as ``the number of permutations of three things taken two at a time is six.''}$$

To find a general formula for $_nP_r$, the number of permutations (arrangements) of n things taken r at a time, we use the *general principle of counting* (Section 9-1). Note that we can fill the first of the r positions in any one of n different ways. Then the second position can be filled in $n - 1$ different ways, and so on.

Position:	1	2	3	4	$\cdots$	r
	$\downarrow$	$\downarrow$	$\downarrow$	$\downarrow$		$\downarrow$
Number of choices:	n	$n-1$	$n-2$	$n-3$	$\cdots$	$n-(r-1)$ that is, $n-r+1$

The product of these r factors gives the number of different ways of arranging r elements selected from a set of n elements, that is, the permutations of n things taken r at a time.

No repetitions are allowed when n things are taken r at a time.

$$_nP_r = (n)(n - 1)(n - 2) \cdots (n - r + 1)$$

where n and r are integers and $n \geq r$.

Example 1   Find $_8P_4$.

Solution   Here $n = 8$, $r = 4$, and $n - r + 1 = 5$. Thus

$$_8P_4 = 8 \times 7 \times 6 \times 5 = 1680$$

Note that there are r, in this case 4, factors in the product.

The solution to Example 2 is based on the *general principle of counting.*

Guess how large 10! is. Then use a calculator to check your estimate.

Example 2 How many different three-letter "words" can be formed from the 26 letters of the alphabet if each letter may be used at most once?

Solution We wish to find the number of permutations of 26 things taken three at a time.

$$_{26}P_3 = 26 \times 25 \times 24 = 15,600$$

A special case of the permutation formula occurs when we consider the permutations of n things taken n at a time. For example, let us see in how many different ways we may arrange in a row the five members of a given set. Here we have the permutations of five things taken five at a time.

$$_5P_5 = 5 \times 4 \times 3 \times 2 \times 1$$

In general, for n things n at a time, $n = n$, $r = n$, and $n - r + 1 = 1$.

$$_nP_n = (n)(n - 1)(n - 2) \cdots (3)(2)(1)$$

We use a special symbol, $n!$, read "n factorial," for this product of integers 1 through n. The following examples should illustrate the use of this new symbol.

$1! = 1$	$5! = 5 \times 4 \times 3 \times 2 \times 1$
$2! = 2 \times 1$	$6! = 6 \times 5 \times 4 \times 3 \times 2 \times 1$
$3! = 3 \times 2 \times 1$	$7! = 7 \times 6 \times 5 \times 4 \times 3 \times 2 \times 1$
$4! = 4 \times 3 \times 2 \times 1$	$8! = 8 \times 7 \times 6 \times 5 \times 4 \times 3 \times 2 \times 1$

Also, we *define* $0! = 1$ so that $(n - r)!$ may be used when $r = n$.

Using this **factorial notation**, we are now able to provide a different, but equivalent, formula for $_nP_r$.

$$_nP_r = n(n - 1)(n - 2) \ldots (n - r + 1)$$
$$\times \frac{(n - r)(n - r - 1)(n - r - 2) \ldots (3)(2)(1)}{(n - r)(n - r - 1)(n - r - 2) \ldots (3)(2)(1)} = \frac{n!}{(n - r)!}$$

Then for $r = n$ we have $_nP_n = n!$

Example 3 Evaluate $_7P_3$ by two different methods.

Solution (a) $_7P_3 = 7 \times 6 \times 5 = 210$

(b) $_7P_3 = \dfrac{7!}{4!} = \dfrac{7 \times 6 \times 5 \times 4 \times 3 \times 2 \times 1}{4 \times 3 \times 2 \times 1}$

$$= 7 \times 6 \times 5 = 210$$

Example 4   A certain class consists of 10 girls and 12 boys. They wish to elect officers so that the president and treasurer are girls and the vice-president and secretary are boys. How many such sets of officers may be obtained?

Solution   The number of selections of the president and the treasurer is $_{10}P_2$. The number of selections of the vice-president and secretary is $_{12}P_2$. By the general principle of counting, the total number of possible selections of a set of officers is

$$(_{10}P_2) \times (_{12}P_2) = (10 \times 9) \times (12 \times 11) = 11{,}880$$

Example 5   (a) Find the number of three-digit numbers that can be formed using the digits 7, 8, 9 if no digit may be used more than once in a number.   (b) How many of these numbers will be even?   (c) What is the probability that a number selected at random will be odd?

Solution   (a) $_3P_3 = 3! = 6$

(b) For the number to be even, the units digit must be even. We have only one choice for this digit, that is, 8. Thus two of the numbers will be even, namely, 798 and 978.

(c) Since two of the numbers are even, the other four numbers will be odd. Thus the probability that a number selected at random will be odd is 4/6, that is, 2/3.

At times there may be items in a set that are not distinguishable from one another. Consider, for example, the number of different arrangements possible using the letters of the word ERROR. If the three R's were distinguishable letters, then the answer would be $_5P_5 = 5! = 120$. In that case, each of the following six different words listed below would appear in the listing.

$ER_1R_2OR_3$ In the absence of subscripts,
$ER_1R_3OR_2$ these six words would appear
$ER_2R_1OR_3$ as the single word ERROR. That is,
$ER_2R_3OR_1$ they could not be distinguished
$ER_3R_1OR_2$ from one another.
$ER_3R_2OR_1$

Now note that the three R's, if distinguishable by the subscripts, can be arranged in $_3P_3 = 3!$ ways. Therefore, if the subscripts are erased, the number of possible distinguishable arrangements of the letters of the word ERROR is $5!/3! = 5 \times 4 = 20$. In general:

Try to find a systematic manner for listing all 20 possible arrangements.

For a set of n objects of which r are alike, the number of distinguishable permutations of the n objects is $\dfrac{n!}{r!}$.

Example 6 Find the number of distinguishable permutations of the letters of the word OHIO. List all such possible arrangements.

Solution We have $n = 4$ and $r = 2$. Thus $\dfrac{n!}{r!} = \dfrac{4!}{2!} = 12$. These may be listed as follows:

OHIO	HOIO	OOHI	OOIH
IHOO	HIOO	OIHO	OIOH
OHOI	HOOI	IOHO	IOOH

The preceding discussion can be extended to form this general result.

> Given a set of n objects of which n_1 are alike, n_2 are another kind that are alike, . . . , and n_k are yet another kind that are alike, then the number of distinguishable permutations of the n objects is
>
> $$\frac{n!}{n_1!\, n_2! \cdots n_k!}$$

EXERCISES

1. Consider the set $S = \{\text{bat, ball}\}$ and list the permutations of the elements of S taken (a) one at a time (b) two at a time.

2. Consider the set $R = \{\text{reading, writing, arithmetic}\}$ and list the permutations of the elements of R taken (a) one at a time (b) two at a time (c) three at a time.

3. Consider the set $T = \{A, B, C, D\}$ and list the permutations of the elements of T taken (a) one at a time (b) two at a time (c) three at a time (d) four at a time.

Evaluate.

4. $5!$

5. $6!$

6. $\dfrac{8!}{6!}$

7. $\dfrac{11!}{7!}$

8. $_7P_2$

9. $_7P_3$

10. $_{10}P_1$

11. $_{10}P_{10}$

12. $_{12}P_{12}$

13. $_{12}P_3$

14. $_{10}P_3$

15. $_{10}P_7$

Express in terms of n.

16.	$_nP_1$	17.	$_nP_{n-1}$	18.	$_nP_2$
19.	$_nP_{n-2}$	20.	$_nP_n$	21.	$_nP_0$

Solve for n.

22.	$_nP_1 = 7$	23.	$_nP_1 = 21$	24.	$_nP_{n-1} = 6$
25.	$_nP_{n-1} = 120$	26.	$_nP_2 = 20$	27.	$_nP_2 = 30$

28. Find the number of different arrangements of the set of five letters $V = \{a, e, i, o, u\}$ taken (a) two at a time (b) five at a time.

29. (a) Find the number of four-digit numbers that can be formed using the digits 1, 2, 3, 4, 5 if no digit may be used more than once in a number.
    (b) How many of these numbers will be even?
    (c) What is the probability that such a four-digit number will be even?
    (d) What is the probability that such a four-digit number will be odd?

30. Find the number of different signals that can be formed by running up three flags on a flagpole, one above the other, if seven different flags are available.

31. How many different signals can be formed by running up seven flags on a flagpole, one above another, if four of the flags are of the same color?

32. Repeat Exercise 31 if three of the flags are red, two are blue, and two are green.

33. Find and list the number of distinguishable permutations of the letters of the word PAPA.

34. Repeat Exercise 33 for the word EERIE.

35. How many eleven-letter "words" can be formed from the letters of the word MISSISSIPPI?

36. Repeat Exercise 35 for the word MATHEMATICS.

37. An interior designer wishes to use books of different colors side by side on a shelf as a decoration. Considering color only, how many different arrangements can be formed with 10 books if one is black, two are red, three are green, and four are blue?

38. Repeat Exercise 37 for 12 books if two are red, two are green, four are blue, and four are black.

Explorations

Permutations are assumed to be linear (along a line) unless otherwise specified. Suppose that we consider the permutation **(circular permutation)** of seating people at a circular table. The various places at the table are assumed to be indistinguishable and only the relative positions of the people are considered. Then there is only one arrangement for seating one person at a circular table and only one arrangement for seating two people at a circular table with two equally spaced chairs.

Use a circular region to represent a circular table, identify people by letters A, B, C, . . . , and sketch each of the possible arrangements for seating the indicated number of equally spaced chairs.

1. Three.

2. Four.

In Explorations 3 and 4, how many arrangements are possible for seating the indicated number of people at a circular table with the indicated number of equally spaced chairs?

3. Five.

4. Six.

5. Explain the appropriateness of the expression $_nP_n/n$, that is, $(n - 1)!$, for the number of permutations of n people in n seats around a circular table.

Unlike the seating of people around a table, a key ring can be turned over. Consider the effect of this phenomenon on the number of distinguishable permutations of keys on a circular key ring.

6. How many arrangements are possible for five keys on a circular key ring?

7. How many arrangements are possible for seven keys on a circular key ring?

8. Give an expression for the number of possible arrangements of n keys on a circular key ring.

9-7
Combinations

A fictitious club with five members was considered in Section 9-1. The set M of members is {Betty, Doris, Ellen, John, Tom}. The number of ways in which a president and a vice-president can be selected is essentially the number $_5P_2$ of permutations of five things taken two at a time. Here order is important, since

Betty as president and Doris as vice-president is a different set of officers than Doris as president and Betty as vice-president.

Now, suppose we wish to select a committee of two members from the set M without attaching any meaning to the order in which the members are selected. Then the committee consisting of Betty and Doris is certainly the same as the one consisting of Doris and Betty. In this case, we see that *order is not important,* and we call such a set a **combination.** One way to determine the number of possible committees of two to be formed from the set M is by a listing of all possible *subsets* of two. There are ten possible subsets (committees) of two.

Betty-Doris	Doris-John
Betty-Ellen	Doris-Tom
Betty-John	Ellen-John
Betty-Tom	Ellen-Tom
Doris-Ellen	John-Tom

We summarize this discussion by saying that the number of combinations of five things taken two at a time is ten. In symbols we write

$$_5C_2 = 10, \quad \text{read as "the number of combinations of five}$$
$$\text{things taken two at a time is ten."}$$

Note that $_5P_2 = 20$ and $_5C_2 = 10 = {}_5P_2 \div 2$ since each combination of two elements such as {Betty, Doris} could have come from either of two permutations of those elements, in this case Betty-Doris or Doris-Betty. Here is the basic distinction between permutations and combinations.

A **permutation** is an *ordered set* of elements and *order is important.*

A **combination** is a *set* of elements and *order is not important.*

Consider a set of three elements; then

$$_3P_3 = 3 \times 2 \times 1 = 6 \quad \text{and} \quad _3C_3 = 1$$

In general $_nP_n = n!$ but $_nC_n = 1$ for any counting number n. Also, for n greater than or equal to 2

$$_nC_2 = \frac{_nP_2}{_2P_2} = \frac{n(n-1)}{2 \times 1}$$

where $_2P_2$ is the number of permutations (arrangements) associated with each combination of two elements.

#18

To find $_5C_3$ consider again the specific problem of selecting committees of three from the set M. There are ten such possibilities, and we list them for $M = \{B, D, E, J, T\}$ using only the first initial of each name.

B, D, E	B, D, J	B, D, T	B, E, J	B, E, T
B, J, T	D, E, J	D, E, T	D, J, T	E, J, T

Note that selecting committees (subsets) of three is equivalent to selecting subsets of two to be omitted. That is, omitting J and T is the same as selecting B, D, and E. Therefore, we find that $_5C_3 = {_5C_2} = 10$.

Inasmuch as we wanted only committees, and assigned no particular jobs to the members of each committee, we see that order is not important. However, suppose that each committee is now to elect a chairperson, secretary, and historian. In how many ways can this be done within each committee? This is clearly a problem in which order is important; we must therefore use permutations. The number of such possible arrangements within each committee is $_3P_3$, that is, 3!. For example, the committee consisting of B, D, and E can rearrange themselves as chairperson, secretary, and historian, respectively, as follows.

B, D, E	B, E, D	D, E, B
D, B, E	E, B, D	E, D, B

All six of these permutations are associated with just one combination. Similarly, the elements of each of the 10 combinations can be arranged in 6 ways. Thus

$$_5C_3 \times {_3P_3} = {_5P_3} \quad \text{and} \quad {_5C_3} = \frac{_5P_3}{_3P_3}$$

In general, each combination of three elements is associated with $_3P_3$ permutations of these elements.

$_3P_3 = 3! = 3 \times 2 \times 1$

$$_nC_3 = \frac{_nP_3}{_3P_3} = \frac{n(n-1)(n-2)}{3 \times 2 \times 1}$$

As in the cases of $_nC_2$ and $_nC_3$, the general form of $_nC_r$ may be expressed as a quotient and is frequently written in symbols in any one of these forms.

$$_nC_r = \binom{n}{r} = \frac{_nP_r}{_rP_r} = \frac{_nP_r}{r!} = \frac{n(n-1)(n-2)(n-3)\ldots(n-r+1)}{r(r-1)(r-2)(r-3)\ldots 1} = \frac{n!}{r!(n-r)!}$$

#19,20

Note that when $_nC_r$ was first expressed as a quotient of numbers, the denominator was $r!$, the product of r successive integers starting with r and decreasing; also the numerator was the product of r successive integers but in this case starting with n and decreasing.

Example 1   Evaluate $_7C_2$ by two different methods.

Solution   Use $_nC_r = \dfrac{_nP_r}{r!}$ and $_nC_r = \dfrac{n!}{r(n-r)!}$.

(a) $_7C_2 = \dfrac{_7P_2}{2!} = \dfrac{7 \times 6}{2 \times 1} = 21$

(b) $_7C_2 = \dfrac{7!}{2!5!} = \dfrac{7 \times 6 \times 5 \times 4 \times 3 \times 2 \times 1}{2 \times 1 \times 5 \times 4 \times 3 \times 2 \times 1} = 21$

Example 2   A box contains six red beads and four green beads. How many sets of four beads can be selected from the box so that two of the beads are red and two are green?

Solution   We are dealing with a problem that involves combinations since the order in which the beads are selected does not affect the set that is obtained. The number of subsets of two red beads is $_6C_2$; the number of subsets of two green beads is $_4C_2$. By the general principle of counting, the total number of subsets of four beads with two red beads and two green beads is the product $_6C_2 \times _4C_2$.

$$_6C_2 = \dfrac{6!}{2!4!} = \dfrac{6 \times 5 \times 4 \times 3 \times 2 \times 1}{2 \times 1 \times 4 \times 3 \times 2 \times 1} = 15$$

$$_4C_2 = \dfrac{4!}{2!2!} = \dfrac{4 \times 3 \times 2 \times 1}{2 \times 1 \times 2 \times 1} = 6$$

$$_6C_2 \times _4C_2 = 15 \times 6 = 90$$

Ninety different sets of four beads with two red and two green can be selected.

Example 3   How many different hands of five cards each can be dealt from a deck of 52 cards? What is the probability that a particular hand contains four aces and the king of hearts?

Solution   The order of the five cards is unimportant, so this is a problem involving combinations.

$$_{52}C_5 = \dfrac{52!}{5!47!} = \dfrac{52 \times 51 \times 50 \times 49 \times 48}{5!} \times \dfrac{(47!)}{(47!)} = 2{,}598{,}960$$

412

The probability of obtaining any one particular hand, such as that containing the four aces and the king of hearts, is $1/(2,598,960)$.

Many problems in probability are most conveniently solved through the use of the concepts of combinations presented here. For example, let us consider again Example 4 of Section 9-4. There we were asked to find the probability that both cards would be aces if two cards are drawn from a deck of 52 cards. This problem can be solved by noting that $_{52}C_2$ is the total number of ways of selecting two cards from a deck of 52 cards. Also, $_4C_2$ is the total number of ways of selecting two aces from the four aces in a deck. The required probability is then given as

$$\frac{_4C_2}{_{52}C_2} = \frac{\dfrac{4!}{2!2!}}{\dfrac{52!}{2!50!}} = \frac{4 \times 3}{52 \times 51} = \frac{1}{221}$$

EXERCISES *Evaluate.*

1. $\dfrac{7!}{4!3!}$ 2. $\dfrac{8!}{4!4!}$ 3. $\dfrac{10!}{3!7!}$ 4. $\dfrac{12!}{8!4!}$

5. $\dfrac{20!}{18!2!}$ 6. $\dfrac{52!}{50!2!}$ 7. $_8C_2$ 8. $_8C_4$

9. $_8C_5$ 10. $_9C_5$ 11. $_{11}C_3$ 12. $_{15}C_4$

13. List the $_3P_2$ permutations of the set $\{r, s, t\}$. Then identify the permutations associated with each combination and find $_3C_2$.

14. List the $_4P_3$ permutations of the set $\{w, x, y, z\}$. Then identify the permutations associated with each combination and find $_4C_3$.

15. List the elements of each of the $_4C_3$ combinations of the set $\{a, b, c, d\}$. Then match each of these combinations with a $_4C_1$ combination to illustrate the fact that $_4C_3 = {_4C_1}$.

16. Find a formula for $_nC_n$ for any positive integer n.

17. Find the value and give an interpretation of $_nC_0$ for any positive integer n.

18. Evaluate $_3C_0$, $_3C_1$, $_3C_2$, $_3C_3$ and check that the sum of these combinations is 2^3, the number of possible subsets that can be formed from a set of three elements.

19. Evaluate $_5C_0$, $_5C_1$, $_5C_2$, $_5C_3$, $_5C_4$, $_5C_5$ and check the sum as in Exercise 18.

20. Use the results obtained in Exercises 18 and 19 and conjecture a formula for any positive integer n for
$$_nC_0 + \,_nC_1 + \,_nC_2 + \,_nC_3 + \cdots + \,_nC_{n-1} + \,_nC_n$$

21. How many sums of money (include the case of no money) can be selected from a penny, a nickel, a dime, a quarter, a half-dollar, and a one-dollar bill?

22. A man has a penny, a nickel, a dime, a quarter, and a half-dollar in his pocket. How many different amounts can he leave as a tip if he wishes to use exactly two coins?

23. A class consists of 8 boys and 12 girls. How many different committees of four can be selected from the class if each committee is to consist of two boys and two girls?

24. How many different hands of 13 cards each can be selected from a bridge deck of 52 cards?

25. How many choices of three books to read can be made from a set of nine books?

26. Explain why a combination lock should really be called a permutation lock.

27. Urn A contains five balls and urn B contains eight balls. How many different selections of ten balls each can be made if three balls are to be selected from urn A and seven from urn B?

28. An urn contains three black balls and three white balls.
   (a) How many different selections of four balls each can be made from this urn?
   (b) How many of these selections will include exactly three black balls?

29. Repeat Exercise 28 for an urn that contains ten black balls and six white balls.

30. Three arbiters are to be chosen by lot from a panel of 12. What is the probability that a certain individual on the panel will be one of those chosen?

31. Six cards are drawn at random from an ordinary bridge deck of 52 cards. Find the probability that all four aces are among the cards drawn.

State whether each question involves a permutation, a combination, or a permutation and a combination. Then answer each question.

32. The 30 members of the Rochester Tennis Club are to play on a certain Saturday evening. How many different pairs can be selected for playing singles?

33. How many different selections of a set of four records can be made by a disc jockey who has 12 records?

34. In how many orders can eight people line up at a single theater ticket window?

35. How many lines are determined by 12 points if no three points are collinear?

36. How many selections of nine students to play baseball can be made from a class of (a) nine students (b) 12 students (c) 20 students?

37. How many different hands can be dealt from a deck of 52 bridge cards if each hand contains (a) four cards (b) seven cards?

38. A committee of five is to be selected from a group that includes six Democrats and five Republicans. In how many ways can this be done if the committee is to consist of three Democrats and two Republicans?

39. How many sets of five cards can be selected from a deck of 52 cards if two of the cards are to be spades and three of the cards are to be hearts?

40. Five cards are dealt from a deck of 52 cards. What is the probability that the hands contain (a) four aces (b) three aces and two kings (c) five hearts?

41. All 25 members of a class shake hands with one another on the opening day of class.
    (a) How many handshakes are there in all?
    (b) How many will there be if, in addition, each shakes hands with the instructor as well?

*42. A class is to be divided into two committees of at least one student each. How many different pairs of committees are possible from a class of eight students?

*43. How many different pairs of committees of four students each can be formed from a class of eight students?

Explorations

Blaise Pascal (1623–1662) was a young French mathematical genius, having invented the first "computing machine" before he was 20. Poor health convinced him to devote most of his life to religion. Although Pascal's work on the arithmetic triangle was printed in 1665, the triangle was actually referred to in 1303 by the Chinese algebraist Chu Shih-chieh.

Recall that the notation $_nC_r$ may also be written in the form $\binom{n}{r}$. Consider the following array for sets of n elements where $n = 1, 2, 3, \ldots$.

$n = 1$: $\quad\binom{1}{0} \quad \binom{1}{1}$

$n = 2$: $\quad\binom{2}{0} \quad \binom{2}{1} \quad \binom{2}{2}$

$n = 3$: $\quad\binom{3}{0} \quad \binom{3}{1} \quad \binom{3}{2} \quad \binom{3}{3}$

$n = 4$: $\quad\binom{4}{0} \quad \binom{4}{1} \quad \binom{4}{2} \quad \binom{4}{3} \quad \binom{4}{4}$

$n = 5$: $\quad\binom{5}{0} \quad \binom{5}{1} \quad \binom{5}{2} \quad \binom{5}{3} \quad \binom{5}{4} \quad \binom{5}{5}$

$\dots\dots\dots\dots\dots\dots\dots\dots\dots\dots\dots\dots\dots\dots$

If we replace each symbol by its equivalent number, we may write the following array, known as *Pascal's triangle*.

```
n = 1:              1    1
n = 2:           1    2    1
n = 3:        1    3    3    1
n = 4:     1    4    6    4    1
n = 5:  1    5   10   10    5    1
```

$\dots\dots\dots\dots\dots\dots\dots\dots\dots\dots\dots\dots\dots\dots$

We read each row in this array by noting that the first entry in the nth row is $\binom{n}{0}$, the second is $\binom{n}{1}$, the third is $\binom{n}{2}$, and so

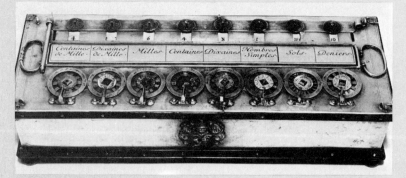

Pascal's adding machine

on until the last entry, which is $\binom{n}{n}$. Since $\binom{n}{0} = \binom{n}{n} = 1$, each row begins and ends with 1.

There is a simple way to continue the array with very little computation. In each row the first number is 1 and the last number is 1. Each of the other numbers may be obtained as the sum of the two numbers appearing in the preceding row to the right and left of the position to be filled. Thus, to obtain the sixth row, begin with 1. Then fill the next position by adding 1 and 5 from the fifth row. Then add 5 and 10 to obtain 15, add 10 and 10 to obtain 20, and so forth as in this diagram.

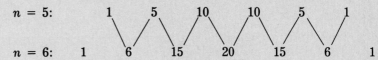

Pascal's triangle may be used in a routine manner to compute probabilities as follows. The elements of the second row are the numerators for the probabilities when two coins are tossed; the elements of the third row are the numerators when three coins are tossed; and so on. The denominator in each case is found as the sum of the elements in the row used. For example, when three coins are tossed, we examine the third row (1, 3, 3, 1). The sum is 8. The probabilities of 0, 1, 2, and 3 heads are then given, respectively, as

$$\frac{1}{8} \qquad \frac{3}{8} \qquad \frac{3}{8} \qquad \frac{1}{8}$$

Note that the sum of the entries in the second row is 4, the sum in the third row is 2^3 or 8, the sum in the fourth row is 2^4 or 16, and, in general, the sum in the nth row will be 2^n.

1. Use the fourth row of Pascal's triangle to find the probability of 0, 1, 2, 3, 4 heads in a single toss of four coins.
2. Construct Pascal's triangle for $n = 1, 2, 3, \ldots, 10$.
3. List the entries in the eleventh row of Pascal's triangle.
4. Repeat Exploration 5 for the twelfth row.
5. See how many different patterns of numbers you can find in Pascal's triangle. For example, do you see the sequence 1, 2, 3, 4, 5, . . . ?
6. Compare the numbers in the nth row of Pascal's triangle and the coefficients in the expression for $(a + b)^n$. Demonstrate this comparison for $n = 6$. Note that

$$(a + b)^2 = 1a^2 + 2ab + 1b^2$$
$$(a + b)^3 = 1a^3 + 3a^2b + 3ab^2 + 1b^3$$

7. Pascal's triangle is used as the basis for a very interesting and mystifying card trick. Learn how to perform this trick by reading Chapter 15, "Pascal's Triangle," in Martin Gardner's book *Mathematical Carnival*, published by Alfred A. Knopf, Inc., 1975. Then perform the trick in class and explain how it works.

Chapter 9 Review

Solutions to the following exercises may be found within the text of Chapter 9. Try to complete each exercise without referring to the text.

Section 9-1   Counting Problems

1. How many different selections of two officers, a president and a vice-president, can be elected from the set {Betty, Doris, Ellen, John, Tom} of club members?
2. The club with members listed in Exercise 1 must send a delegate to a meeting tomorrow and also a delegate to a different meeting next week. How many different selections of these delegates may be made if any member of the club may serve as a delegate to either, or both, of these meetings?
3. How many three-letter "words" may be formed from the set of vowels $V = \{a, e, i, o, u\}$ if no letter may be used more than once?

Section 9-2   Definition of Probability

4. A single card is selected from a deck of 52 bridge cards. What is the probability that it is a spade? What is the probability that it is not a spade? What is the probability that it is an ace or a spade?
5. A committee of two is to be selected at random from the list of members given in Exercise 1. What is the probability that both members of the committee will be girls?

Section 9-3   Sample Spaces

6. List a sample space for the problem of tossing two coins.
7. A box contains two red and three white balls. Two balls are drawn in succession without replacement. List a sample space for this experiment.

Section 9-4   Computation of Probabilities

8. A single die is thrown. What is the probability that either an odd number or a number greater than 3 appears?

9. Urn A contains three white and five red balls. Urn B contains four white and three red balls. One ball is drawn from each urn. What is the probability that they are both red?

10. Two cards are selected in succession, without replacement, from an ordinary bridge deck of 52 cards. What is the probability that they are both aces?

Section 9-5 Odds and Mathematical Expectation

11. Find the odds in favor of drawing a spade from an ordinary deck of 52 bridge cards and the odds against drawing a spade from an ordinary deck of 52 bridge cards.

12. Suppose that you pay 5¢ to play a game in which a coin is tossed twice. You receive 10¢ if two heads are obtained, 5¢ if exactly one head is obtained, and nothing if no heads are obtained. What is the expected value to you of this game?

Section 9-6 Permutations

13. How many different three-letter "words" can be formed from the 26 letters of the alphabet if each letter may be used at most once?

14. Evaluate $_7P_3$ by two different methods.

15. A certain class consists of 10 girls and 12 boys. They wish to elect officers so that the president and treasurer are girls and the vice-president and secretary are boys. How many such sets of officers may be obtained?

16. (a) Find the number of three-digit numbers that may be formed using the digits 7, 8, and 9 if no digit may be used more than once in a number.
    (b) How many of these numbers will be even?
    (c) What is the probability that a number selected at random will be odd?

17. Find the number of distinguishable permutations of the letters of the word OHIO. List all such possible arrangements.

Section 9-7 Combinations

18. Evaluate $_7C_2$ by two different methods.

19. How many different hands of five cards each can be dealt from a deck of 52 cards? What is the probability that a particular hand contains four aces and the king of hearts?

Chapter 9 Test

1. Evaluate (a) $\dfrac{17!}{13!}$ (b) $\dfrac{101!}{99!}$.

2. Evaluate (a) $_{20}P_2$ (b) $_{20}C_2$.

3. Consider the set $S = \{c, o, w\}$ and make a tree diagram to show the permutations of the letters of S taken 2 at a time.

4. How many different four-letter "words" can be formed from the set $B = \{p, e, n, c, i, l\}$ if no letter may be used more than once?

5. Repeat Exercise 4 if repetitions of letters are permitted.

6. One card is selected from a deck of 52 bridge cards. Find the probability that the selected card is (a) not a spade (b) a spade or a heart.

7. Find the number of even three-digit numbers that may be formed from the set of digits $\{1, 2, 3, 4, 5, 6, 7\}$ if repetitions are not allowed.

8. A man has a nickel, a dime, a quarter, and a half-dollar. If he leaves a tip of exactly two coins, how many values are possible for such a tip?

9. Use a sample space to represent the possible outcomes described in Exercise 8 and find the probability that the tip is at least 30¢ when the selection of the two coins is at random.

10. What is the probability of obtaining on a single throw of two dice (a) a 7 (b) a number greater than 7?

11. Two integers between 0 and 10 are selected at random without replacement. What is the probability that at least one of the numbers is odd?

A single card is drawn from an ordinary deck of 52 cards. Find the probability that the card selected is as described.

12. (a) The ace of spades.

    (b) An ace or a spade.

13. (a) An ace or a red card.

    (b) An ace and a red card.

Two cards are selected in succession without replacement from an ordinary deck of 52 bridge cards. Find the probability of the specified event.

14. (a) Both cards are black.

    (b) Either both cards are black or both are red.

15. (a) Both cards are kings.

    (b) Either both cards are kings or both are queens.

16. What are the odds in favor of obtaining an ace when one card is drawn from an ordinary deck of 52 bridge cards?

17. What is your mathematical expectation when you are given 3 tickets out of a total of 6000 tickets for a single price worth $1000?

18. Four coins are tossed.
    (a) What is the expected number of heads?
    (b) What are the odds in favor of at least one head?

19. How many eight-letter "words" can be formed from the letters of the word ASSASSIN?

20. A bag contains five red balls and three green balls. Two balls are drawn in succession without replacement. Find the probability that (a) both balls are red (b) the first ball is red and the second one is green.

21. Assume that n is greater than or equal to 5 and given an expression for (a) $_nP_5$ (b) $_nC_5$.

22. Find and simplify a formula for the given expression.

    (a) $_nP_0$ (b) $\dfrac{_nC_r}{_nP_r}$

23. Three judges are to be selected from panel of ten. What is the probability that a particular member of the panel will not be chosen?

24. Nine people draw lots for their positions (1 through 9) in a line for concert tickets. How many outcomes are possible?

25. A class of ten students is divided into pairs of committees of at least one student each. How many such pairs of committees are possible?

An Introduction to Statistics

10

10-1
Uses and Abuses of Statistics

Because of the widespread use of statistics for the consumer, we find that many adults are ready to quote "facts" that they read or hear, often without real understanding. Therefore, in this chapter we make an effort to acquaint you with a sufficient number of basic concepts so that you may better understand and interpret statistical data.

Reminder:

LATIN		ENGLISH
datum	~	fact
data	~	facts

The average consumer is besieged daily with statistical data that are presented over radio and television, in newspapers, and in various other media. We are urged to buy a certain commodity because of statistical evidence presented to show its superiority over other brands. We are cautioned *not* to consume a particular item because of some other statistical study carried out to show its danger. We are told to watch certain programs, read certain magazines, see certain movies, and eat certain foods because of evidence produced to indicate the desirability of these acts as based on data gathered concerning the habits of others.

Almost every issue of the daily newspaper presents data in graphical form to help persuade the consumer to follow certain courses of action. Thus we are told what the "average" citizen eats, reads, earns, and even how the "average" person's leisure time is spent. Unfortunately, too many of us are impressed by statistical data regardless of their source.

Statistical data and concepts affect each of us and often have a profound influence upon the opportunities available to us. For example, among the items considered when you applied for admission to college were

Your rank (first, second, third, . . .) in your school class.
Your verbal Scholastic Aptitude Test score.
Your mathematics Scholastic Aptitude Test score.

Statistical computations determine the cost of your life insurance, health insurance, and automobile insurance—to mention only a few such items.

It has become fashionable, as well as informative and impressive, to support all sorts of predications and assertions with statistical computations. Our main purposes in this chapter are to help you become aware of the extensive use of statistics and to recognize some of its abuses. Many of the most blatant abuses of statistics occur in advertisements. For example, we frequently read or hear statements such as

Brush your teeth with GLUB and you will have fewer cavities.

The basic question that this statement fails to answer is: "Fewer cavities than what?" Regardless of the merits of GLUB, it is probably safe to say the following of every brand of toothpaste: "Brush your teeth with **XXXX** and you will have fewer cavities than *if you never brush your teeth at all.*" In other words, the example given is misleading in that it implies the superiority of GLUB without presenting sufficient data to warrant comparisons.

Many types of statements must be examined very carefully to separate their statistical implications from the impressions they are intended to make. Consider these statements.

More people are killed in automobile accidents than in wars. Therefore, it is safer to be on the battlefield.

Although the first statement may be true, the second one does not necessarily follow. Since there are generally many more people at home driving cars than there are in combat, the *percent* of deaths on the battlefield is actually much higher than in automobiles.

Frequently, graphs are presented in a style that misleads the reader. There are two common techniques for misleading casual readers of graphs. A scale may be started at a number different from zero. For example, if the scale on a graph to show trends in the circulation of a magazine does not start at zero, then the relative changes may appear much more significant than they really are. The first graph has been cut off so that the changes or trends appear to have greater significance than is actually the case. The same facts are shown in the second graph where the increase in circulation is recognizable as a very modest one of 400 readers, that is, 0.4%.

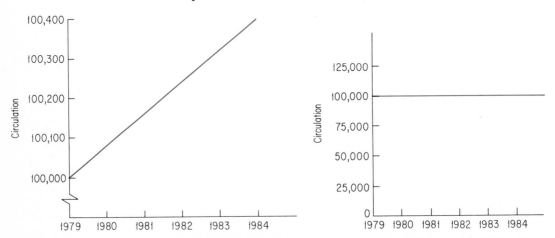

Another common abuse of graphical presentations is based upon the fact that similar figures have linear ratios that are very different from their area and volume ratios. For example, if the linear ratio is 3 to 2, then the area ratio is 9 to 4 and the volume ratio is 27 to 8. Thus area and volume measures are frequently misused to provide a visual misinterpretation of the actual data.

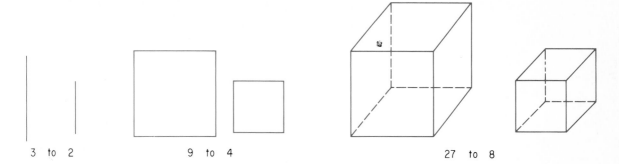

3 to 2   9 to 4   27 to 8

Collecting actual examples of uses of statistics to mislead readers and analyzing advertising claims can be enlightening. The intelligent consumer needs to ask basic questions about statistical statements that are presented to the public. Unfortunately, too many people tend to accept as true any statement that contains numerical data, probably because such statements sound impressive and authentic. Consider, for example, this statement:

Two out of three adults no longer smoke.

Before accepting and repeating such a statement, you should ask such questions as the following.

Does the statement sound reasonable?
What is the source of these data?
Can the facts be verified?
Do the facts appear to conform to your own observations?

You will undoubtedly think of other questions to raise. The important thing is that you raise these questions and do not accept all statistical facts as true in general. This is not to imply that all such facts are biased or untrue. Rather, as we have seen, and shall see later in this chapter, even true statements can be presented in many ways so as to provide various impressions and interpretations.

You should examine the logical structure of each statement; for example, is the statement or its converse suggested? You should also examine each statement to search for ambiguous words or meaningless comparisons. For example, consider the following assertion.

If you take vitamin C daily, your health will be better.

In addition to the questions you might raise about the scientific basis for this statement, note that the word *better* is ambiguous. We need to ask the question: "Better than what?" That is, we need to know the basis for comparison. As an extreme case, we certainly can say that your health will be better than the health of someone who is desperately ill! The exercises that follow will provide you with an opportunity to test your powers of reasoning and questioning.

EXERCISES

Consult available newspapers and try to locate similar statements that present numerical data. Analyze these statements for possible misuses of statistics.

Discuss each of the following examples and tell what possible misuse or misinterpretation of statistics each one displays.

1. Most automobile accidents occur near home. Therefore, one is safer taking long trips than short ones.

2. Over 95% of the doctors interviewed endorse SMOOTHIES as a safe cigarette to smoke. Therefore, it is safe to smoke SMOOTHIES.

3. More college students are now studying mathematics than ever before. Therefore, mathematics must be a very popular subject.

4. Professor X gave out more A's last semester than Professor Y. Therefore, one should try to enroll in X's class next semester rather than in Y's class.

5. At a certain college 100% of the students bought *Introduction to Mathematics* last year. Therefore, this must be a very popular book at that school.

6. Most accidents occur in the home. Therefore, it is safer to be out of the house as much as possible.

7. Over 75% of the people surveyed favor the Democratic candidate. Therefore, that candidate is almost certain to win the election.

8. A psychologist reported that the American girl kisses an average of 79 men before getting married.

9. A magazine reported the presence of 200,000 stray cats in the city of New York.

10. A recent report cited the statistic of the presence of 9,000,000 rats in the city of New York.

11. Arizona has the highest death rate from asthma in the nation. Therefore, if you have asthma, you should not go to Arizona.

12. During World War II more people died at home than on the battlefield. Therefore, it was safer to be on the battlefield than to be at home.

13. In a pre-election poll, 60% of the people interviewed were registered Democrats. Thus the Democratic candidate will surely win the election.

14. Over 90% of the passengers who fly to a certain city do so with airline X. It follows that most people prefer airline X to other airlines.

15. A newspaper exposé reported that 50% of the children in the local school system were below average in arithmetic skills.

For the given statement, list two or three questions that you would wish to raise before accepting the statement as true. Identify words, if any, that are ambiguous or misleading.

16. Most people can swim.

17. Short men are more aggressive than tall men.

18. If you walk 3 miles a day, you will live longer.

19. Brand A aspirin is twice as effective as brand B.

20. Teenagers with long hair are happier individuals.

21. People who swim have fewer heart attacks.

22. If you sleep 8 hours the night before an exam, you will do better.

23. Most students have success with this textbook.

24. Teachers tend to be less conscious about social ills.

25. About three out of four college students marry within one year of graduation.

Use the following graph for Exercises 26 and 27.

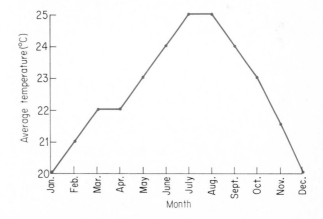

26. Disregarding the scale, what visual impression do you get from this graph about the fluctuation of temperatures during the year?

27. In what ways, if any, is the graph misleading?

Use the following bar graph to answer Exercises 28 through 30.

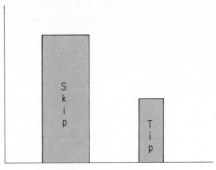

Circulation of two leading magazines

28. Can one deduce from the graph that the circulation of SKIP is greater than the circulation of TIP?
29. Can one deduce from the graph that the circulation of SKIP is approximately twice as great as the circulation of TIP?
30. What, if anything, could possibly be misleading or deceiving about the graph?

Explorations

Begin a collection from newspapers, magazines, and other media of examples of misuses of statistics. In particular find examples that use each of the following.

1. Scales that have been cut off to mislead the reader.
2. Areas or volumes used to exaggerate comparisons.
3. Parts of the figure enlarged to provide undue emphasis.

For Explorations 4 and 5, discuss the given classic quotations.

4. Facts are facts.
5. Figures don't lie but liars figure.

6. Read and report to your class on at least one chapter of a book such as *How To Use (and Misuse) Statistics* by Gregory A. Kimble and published by Prentice-Hall, Inc., 1978.

10-2
Descriptive Statistics

Descriptive statistics is concerned with characterizing or summarizing a given set of data. Such condensations of data need to be carefully examined, as we have just observed. The consideration of methods of presentation of data helps us interpret the numerous statistical presentations that we encounter in newspapers and elsewhere. Accordingly, the present introduction to statistics is primarily concerned with descriptive statistics. The other main branch of statistics, **inferential statistics,** is concerned with the prediction of future events (election results, weather, stock market activity, and so on), and is a relatively advanced aspect of the study of statistics.

Norway issued these stamps on the 100th anniversary of its Central Bureau of Statistics.

As consumers our basic statistical needs are for the interpretation of statistical presentations made by others. Several of the methods used for statistical presentations are illustrated for the tossing of four coins. Assume that the coins have been tossed 32 times and that each time the number of heads obtained has been recorded. These **raw data** are listed in the order of the occurrence of the events.

NUMBER OF HEADS

1	2	3	2	4	2	2	1
0	2	4	2	2	1	3	3
2	3	1	1	2	2	3	0
4	1	3	2	4	2	1	2

The information is somewhat more meaningful when the number of heads is tallied and summarized in tabular form as a **frequency distribution.**

Use the given raw data to verify the entries in this table.

Number of heads	Tally	Frequency
0				2
1									7
2															13
3								6
4						4

This information can now be treated graphically in a number of different ways. One common form of presenting data is by means of a **histogram,** a bar graph without spaces between the bars.

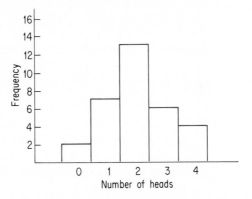

Frequently, a histogram is used to construct a **frequency polygon** or **line graph**. Thus the histogram can be approximated by a line graph obtained by connecting the midpoints of the tops of successive bars. Then the graph is extended to the base line as in the next figure.

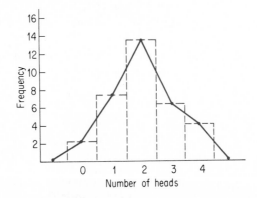

Line graphs appear to associate values with all points on a continuous interval rather than just with points that have in-

10 AN INTRODUCTION TO STATISTICS

tegers as coordinates. Thus such line graphs are more appropriate for **continuous data** (all intermediate values have meaning) rather than with **discrete data** (only isolated values have meaning). For example, consider a graph of the temperatures in a particular city. Since it was 10 °C at 5 P.M. and 5 °C at 6 P.M., the temperature must have passed through *all* possible values between 10 and 5 in one hour. This, then, is an example of continuous data. Temperatures may be plotted and connected to produce a line graph as in the figure. Since gradual changes are expected, we "smooth out" the graph as a curve without "sharp turns."

Hour	6 A.M.	7	8	9	10	11	12	1	2	3	4	5	6 P.M.
Temperature	2	5	5	8	10	12	15	16	15	12	11	10	5

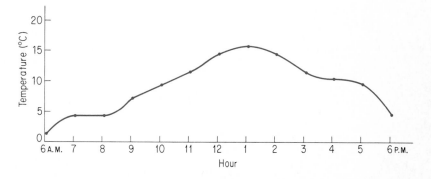

There are many different types of graphs used to present statistical data. One very popular type is the **circle graph**, which is especially effective when one wishes to show how an entire quantity is divided into parts. Local and federal government documents frequently use this type of graph to show the distribution of tax money and various budget distributions. As another example, consider the following family budget.

Food	40%
Household	25%
Recreation	5%
Savings	10%
Miscellaneous	20%
Total	100%

To draw a circle graph for these data, we first recognize that there are 360° in a circle. Then we find each of the given percents of 360, and with the aid of a protractor construct central angles of appropriate sizes as in the following figure.

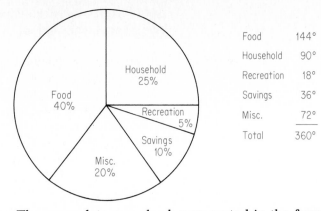

Food	144°
Household	90°
Recreation	18°
Savings	36°
Misc.	72°
Total	360°

Verify each entry in the table. For example, show that 40% of 360° is 144°.

The same data can also be presented in the form of a **divided bar graph.** Here we select an arbitrary unit of length to represent 100%, and divide this in accordance with the given percents. In practice, even though the rectangular region (bar) has an arbitrary length, it is highly desirable to select that length so that it can be subdivided easily to obtain the desired parts. In the case of the data for our example all parts are integral multiples of 5%, that is, of one-twentieth of the 100% total. Thus it is very convenient to select the length of the bar so that twentieths of that length are easily identified. For example, if the total length were 10 centimeters, then the lengths of the parts would be 4 centimeters, 2.5 centimeters, 0.5 centimeter, 1 centimeter, and 2 centimeters.

There is an almost endless supply of graphs that could be used to illustrate this method of presenting data. Indeed, the reader is constantly subjected to graphical presentations in the daily newspapers and in other periodicals. Other examples of graphs are considered in the exercises.

Food 40%	Household 25%	Rec. 5%	Savings 10%	Misc. 20%

EXERCISES

1. Toss four coins simultaneously and record the number of heads obtained. Repeat this for a total of 32 tosses of four coins. Present your data in the form of a frequency distribution.

2. Present the data for Exercise 1 in the form of a bar graph.

3. Present the data for Exercise 1 in the form of a line graph.

4. Repeat Exercise 1 for 32 tosses of a set of five coins.

5. Present the data for Exercise 4 in the form of a bar graph.

6. Present the data for Exercise 4 in the form of a line graph.

7. Here is the theoretical distribution of heads when four coins are tossed for a total of 64 times. Present these data in the form of a bar graph.

Number of heads:	0	1	2	3	4
Frequency:	4	16	24	16	4

8. Here is the theoretical distribution of heads when five coins are tossed for a total of 64 times. Present these data in the form of a bar graph.

Number of heads:	0	1	2	3	4	5
Frequency:	2	10	20	20	10	2

9. Construct a circle graph to show this distribution of time spent by one student.

Sleep	8 hours
School	6 hours
Homework	4 hours
Eating	2 hours
Recreation	4 hours
Total	24 hours

10. Construct a divided bar graph for the data of Exercise 9.

Use the following graphs to answer Exercises 11 through 17.

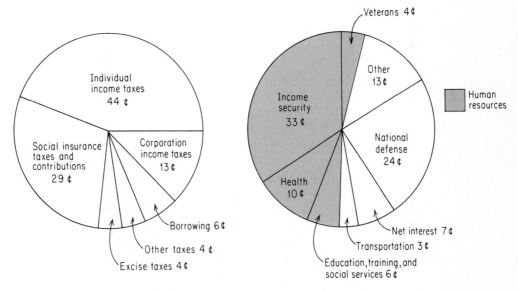

11. What percent of the government dollar comes from individual income taxes?

12. What percent of the government dollar goes to income security?

13. Can you tell from the graphs **(a)** what percent of the government expenditures was for health matters? **(b)** How many dollars were spent on health matters? Explain your answers.

14. What is the size of the central angle in the graph that shows the amount spent for transportation?

15. To the nearest degree, what is the size of the central angle that shows the amount obtained from individual income taxes?

16. Of every million dollars collected, how many dollars were collected from excise taxes?

17. Of every million dollars spent, how many dollars were spent for health?

Explorations

1. Report on contemporary statistical studies of a controversial subject, such as smoking, heart disease, or the safety of nuclear generators of electricity.

2. Read and report to your class on at least one of the following selections from *Statistics: A Guide to the Unknown.* This book was edited by Judith M. Tanur et al. for a joint committee of the American Statistical Association and the National Council of Teachers of Mathematics. The book is published by Holden-Day, Inc., 1972.

   "Setting Dosage Levels" by W. J. Dixon on pages 34 through 39.

   "Statistics, Scientific Method, and Smoking" by B. W. Brown, Jr. on pages 40 through 51.

   "The Importance of Being Human" by W. W. Howells on pages 92 through 100.

   "Parking Tickets and Missing Women: Statistics and the Law" by Hans Zeisel and Harry Kalven, Jr. on pages 102 through 111.

   "Deciding Authorship" by Frederick Mosteller and David L. Wallace on pages 164 through 175.

3. Begin a collection from newspapers, magazines, and other media of uses of statistics. In particular find examples of as many different kinds of graphs as possible.

10-3
Measures of Central Tendency

The word *average* is often used loosely and may have a number of different meanings.

Most of us have neither the ability nor the desire to digest large quantities of statistical data. Rather, we prefer to see a graphical presentation of such data, or some number cited as representing the entire collection of data. Thus we are often faced with statements such as these.

On the average, 9 out of 10 doctors recommend H_2O.
The average family earns $15,000 per year and has 1.8 children.
In an average college class 21% of the grades are A's.

In each case the word *average* is used in an effort to provide a capsule summary of a collection of data by means of a single number.

Assume that the following set of data represents the number of written reports required of ten students last year.

{6, 7, 8, 9, 10, 12, 15, 15, 20, 28}

Then each of the following statements is correct for this particular set of ten students.

1. *On the average each student was required to write 13 reports last year.*

In this case the average is computed by finding the sum of the given numbers and dividing the sum by 10, the number of students. This is the most commonly used type of average. It is referred to as the **arithmetic mean,** or simply the **mean,** of a set of data, and is generally denoted by the Greek letter μ (read as "mu").

6
7
8
9
10
12
15
15
20
<u>28</u>
130

$$\mu = \frac{130}{10} = 13$$

Arithmetic mean: 13

2. *On the average each student was required to write 11 reports last year.*

In this case the word *average* is used to denote the number that is in the middle of the ordered set of data or, if there is an even number of elements in the set, the number that is halfway between the two middle numbers. The number that divides a set of scores in this way is called the **median** of the set of data. In this case it is determined as the number midway between 10 and 12. Note that the median is not necessarily one of the scores and that the data must be considered in order of size before the median can be found.

$$
\left.
\begin{array}{c}
6 \\
7 \\
8 \\
9 \\
10
\end{array}
\right\}
\quad \text{Five scores below 11}
$$

— Median: 11

$$
\left.
\begin{array}{c}
12 \\
15 \\
15 \\
20 \\
28
\end{array}
\right\}
\quad \text{Five scores above 11}
$$

3. *On the average each student was required to write 15 reports last year.*

In this case the average is the number that appears most frequently in the set of data. That is, it is correct to say that more of these students were required to write 15 reports than any other number. Such an average is referred to as the **mode** of a set of data. The mode is the only type of average that is always an observable element of the set.

4. *On the average each student was required to write 17 reports last year.*

In this case the average is computed by first adding the smallest and largest numbers and then dividing by 2, as in finding the midpoint of an interval on a number line.

$$
\frac{6 + 28}{2} = \frac{34}{2} = 17
$$

In the example regarding written reports, the midrange would be the best type of average to use to persuade someone that too many reports are required. Actually, however, the median is more representative of the number of reports required of a typical student.

This is the **midrange**, the midpoint of the interval (range) from the smallest number to the largest number. In cold climates, the midrange of the daily temperature is found each day. If the midrange (M) is less than 65 °F on a particular day, then the difference $65 - M$ is the number of **degree-days** for that particular date. Fuel dealers use the number of degree days to estimate the amount of fuel that their customers are using and thus the time at which additional fuel should be delivered.

Each of the four types of averages that we have discussed is known as a **measure of central tendency**. That is, each average is an attempt to describe a set of data by means of a single representative number. Note, however, that not every distribution has a mode, whereas some may have more than one mode.

The set of scores {8, 12, 15, 17, 20} has no mode.

The set {8, 10, 10, 12, 15, 15, 17} is **bimodal** and has both 10 and 15 as modes.

Example 1 For the given set of test scores, find **(a)** the mean **(b)** the median **(c)** the mode **(d)** the midrange.

{72, 80, 80, 82, 88, 90, 96}

Solution **(a)** The sum of the seven scores is 588.

$$\mu = \frac{588}{7} = 84$$

(b) The median is 82. This is the middle score; there are three scores below 82 and three above it.
(c) The mode is 80, since this score appears more frequently than any other.
(d) The midrange is 84 since $(72 + 96)/2 = 168/2 = 84$.

Example 2 Repeat Example 1 for these scores.

{30, 80, 80, 82, 88, 90, 96}

Solution **(a)** $\mu = 546/7 = 78$.
(b) The median is still 82.
(c) The mode is still 80.
(d) The midrange is 63 since $(30 + 96)/2 = 126/2 = 63$.

A comparison of Examples 1 and 2 indicates that the test scores are the same except for the first one. The median and the mode are the same for the two sets of scores. The arithmetic mean and the midrange are not the same.

The mean and the midrange are affected by extreme scores and one of these two measures of central tendency should be used as a representative of a set of data when extreme (high or low) scores should be reflected in the average. Otherwise the median or mode should be used to describe a set of data that contains extreme scores unless deception is a major objective.

Consider this example of the earnings of the employees of a small business run by a supervisor and three other employees. The supervisor earns $50,000 per year. The others earn $9000, $10,000, and $11,000, respectively. To impress the union, the owner claims that the employees are paid an average salary of $20,000. The arithmetic mean is selected as the representative salary.

$$50,000 + 9000 + 10,000 + 11,000 = 80,000$$
$$80,000 \div 4 = 20,000$$

Actually, the median, $10,500, would present a fairer picture of the average, or typical, salary.

As a more extreme example, consider a group of 49 people with a mean income of $10,000. Let us see what happens to the average income of the group when an additional person with an income of $300,000 joins the group.

$$49 \times 10,000 = 490,000 \quad \text{(total income of 49 people)}$$
$$1 \times 300,000 = 300,000$$
$$\text{Sum} = 790,000 \quad \text{(total income of 50 people)}$$
$$\text{Average} = \frac{790,000}{50} = 15,800$$

In reading or hearing advertisements, the consumer must always raise the question of the type of average being used. Furthermore, the source and the plausibility of the data should be questioned.

The average (arithmetic mean) salary is now $15,800. The addition of one extreme salary raised the mean by $5800, whereas the median and the mode of the incomes are unaffected.

EXERCISES

For the given set of data, find (a) *the mean* (b) *the median* (c) *the mode* (d) *the midrange.*

1. {60, 61, 65, 65, 70, 73, 73, 79, 84}
2. {73, 79, 80, 82, 84, 84, 92}
3. {10, 15, 18, 19, 21, 24, 26, 27}
4. {10, 11, 14, 14, 14, 17, 18}
5. {85, 61, 68, 73, 91, 68, 93}
6. {6, 7, 7, 8, 8, 8, 9, 9, 10, 10}
7. {9, 8, 9, 8, 6, 7, 8, 9, 10, 8}
8. {61, 69, 73, 78, 81, 86, 88, 88, 92, 97}

In Exercises 9 through 11 state which one, if any, of the four measures of central tendency seems most appropriate to represent the data described.

9. The average salary in a shop staffed by the owner and five employees.
10. The average salary of the workers in a factory that employs 100 people.
11. The average number of cups of coffee ordered by individual diners in a restaurant.

12. What relationship does the mode have to the use of this word in everyday language?
13. The mean score on a set of 15 tests is 75. What is the sum of the 15 test scores?
14. The mean score on a set of 30 tests is 84. What is the sum of the test scores?

15. The mean score on a set of 40 tests is 86. What is the sum of the test scores?

16. The mean score on nine of a set of ten tests is 70. The tenth score is 50. What is the sum of the test scores?

17. The mean score on 25 of a set of 27 tests is 80. The other two scores are 30 and 35. What is the sum of the scores?

18. Two sections of a course took the same test. In one section the mean score on the 25 tests was 80. In the other section the mean score on the 20 tests was 75.
    (a) What is the sum of the 45 test scores?
    (b) To the nearest integer what is the mean of the 45 test scores?

19. The mean score on five tests is 85. The mean on three other tests is 78. What is the mean score on all eight tests?

20. A student has a mean score of 85 on seven tests taken to date. What score must the student achieve on the eighth test in order to have a mean score of 90 on all eight tests? Comment on your answer.

*21. An interesting property of the arithmetic mean is that the sum of the deviations (considered as signed numbers $x_i - \mu$) of each score x_i from the mean μ is 0. Show that this is true for the set of scores: {8, 10, 13, 17, 22}.

*22. Show that the sum of the deviations from the mean is 0 for the sets of data given in Exercises 2, 4, 6, and 8.

*23. Repeat Exercise 22 for the sets of data given in Exercises 1, 3, 5, and 7.

24. During a semester four students obtained the following grades on their hour tests.

    M: 60, 70, 70, 75, 80
    P: 50, 50, 80, 90, 100
    S: 40, 75, 80, 85, 95
    T: 40, 40, 85, 90, 100

    (a) Copy and complete the following summary table showing the students' averages for the semester.

	Mean	Median	Mode	Midrange
M				
P				
S				
T				

(**b**) Which student had the highest grade for each of these averages?

*(**c**) Which of these students would you rank as the best one of the four?

*(**d**) Provide an argument for each of the students to claim that each should be rated as the best of the four.

Explorations

Make any necessary inquiries and describe the way in which the given average is usually determined.

1. For your college class, the average grade on a particular test.
2. The batting average for a professional baseball player.
3. The average number of points per game during a particular season for a basketball player.
4. The average number of points per game during a particular season for a football team.
5. The average number of miles per gallon on the highway for a new car.
6. The average seasonal snowfall at a particular ski resort.
7. The average daily temperature at a weather station.
8. The average monthly rainfall at a weather station.

Consider newspapers, magazines, books that are not mathematics books, and radio and television programs to identify as many instances as you can in which the word *average* appears to be used in the specified sense.

9. For the arithmetic mean.
10. For the median.
11. For the mode.
12. For the midrange.
13. To encourage different interpretations by different readers or listeners.

10-4
Measures of Dispersion

A measure of central tendency describes a set of data through the use of a single number. However, as in the examples considered in Section 10-3, a single number without other information can be misleading. Some information can be obtained by comparing the arithmetic mean and the median, since the mean is affected by extreme scores and the median is relatively stable.

440

In this section we consider ways of obtaining information about sets of data without considering all of the elements of the data individually.

A **measure of dispersion** is a number that provides some information on the variability of a set of data. The simplest such measure to use for a set of numbers is the **range,** the difference between the largest number and the smallest number in the set.

Consider the test scores of these two students.

Betty: {68, 69, 70, 71, 72} $\mu = 70$
Jane: {40, 42, 70, 98, 100} $\mu = 70$

The range for Betty's test scores is $72 - 68$, or 4; the range for Jane's test scores is $100 - 40$, or 60. The averages and ranges provide a more meaningful picture of the two students' test results than the averages alone.

Betty: $\mu = 70$ range, 4
Jane $\mu = 70$ range, 60

Although the range is an easy measure of dispersion to use, it has the disadvantage of relying on only two extreme scores—the lowest and the highest scores in a distribution.

Consider Bob's test scores:

{40, 90, 95, 95, 100}

Bob is quite consistent in his work, and there may well be some good explanation for his one low grade. However, in summarizing his test scores one would report a range of 60, since $100 - 40 = 60$. The midrange 70 and the arithmetic mean 84 of Bob's test scores are also seriously affected by the extreme score. However, the mode 95 is not affected by the low score; also, the median 95 would be the same if the score of 40 were replaced by any other score.

To disperse is to spread out or to distribute. A measure of dispersion is a number that describes the amount of spreading out, or the density of the distribution, of the data.

There is another measure of dispersion that is widely used in describing statistical data. This measure is known as the **standard deviation,** and is denoted by the Greek letter σ (sigma). The average consumer of statistics needs to be able to understand and interpret standard deviations. This need is particularly acute for teachers since they need to interpret the performance of their students on standardized tests.

To understand the significance of standard deviations as measures of dispersion, we first turn our attention to a discussion of **normal distributions.** The line graphs of these distributions are the familiar bell-shaped curves that are used to describe distributions for so many physical phenomena. For example, the distribution of intelligence quotient (IQ) scores in the entire population of the United States can be pictured by a **normal curve.**

The area under a normal curve represents the entire population. According to psychologists, an IQ score over 130 is con-

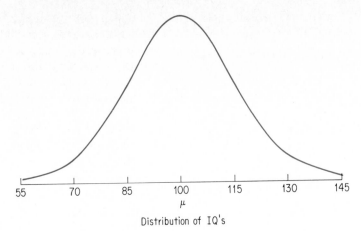

sidered to be superior. At the other end of the scale, a score below 70 generally indicates some degree of academic retardation.

A normal curve should be expected only for data that involve a large number of elements and are based upon a general population. For example, the IQ scores of the honor society members in a school should not be expected to fit the distribution for the school as a whole.

If three standard deviations are added to and subtracted from the mean of a normal distribution, practically all (99.7%) of the data will fall on the interval from $\mu - 3\sigma$ to $\mu + 3\sigma$. If an interval of two standard deviations from the mean is considered, approximately 95% of all data is included. An interval of one standard deviation about the mean includes approximately 68% of all data in a normal distribution. We may summarize these statements as follows.

In a normal distribution the mean, the median, and the mode all have the same value. This common value is associated with the axis of symmetry of the normal curve.

$$\mu \pm 1\sigma \approx 68\%$$
$$\mu \pm 2\sigma \approx 95\%$$
$$\mu \pm 3\sigma \approx 100\%$$

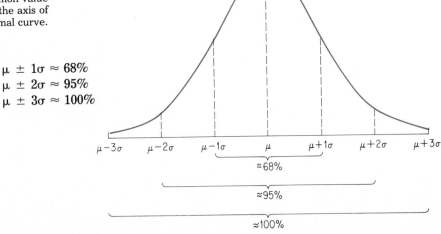

10 AN INTRODUCTION TO STATISTICS

Let us now return to the graph of IQ scores. Suppose we are told that for this distribution $\sigma = 15$. We may then show these standard deviations on the base line of the graph. According to our prior discussion, we may now say that approximately 68% of the population have IQ scores between 85 and 115, that is, on the interval $100 \pm 1\sigma$. Approximately 95% of the population have IQ scores between 70 and 130, that is, on the interval $100 \pm 2\sigma$. Finally, almost everyone has an IQ score between 55 and 145, that is, on the interval $100 \pm 3\sigma$.

Here "almost everyone" really means 99.7%; that is, we might expect 0.3% (3 in 1000) of the population to have scores below 55 or above 145.

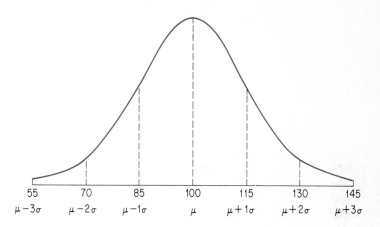

55   70   85   100   115   130   145

$\mu - 3\sigma$ $\mu - 2\sigma$ $\mu - 1\sigma$ μ $\mu + 1\sigma$ $\mu + 2\sigma$ $\mu + 3\sigma$

Example 1   What percent of the population have IQ scores below 115?

Solution   We know that 50% of the population have IQ scores below 100, the mean. Also, 68% have scores on the interval $\mu \pm 1\sigma$. Then by the symmetry of the normal curve, 34% have scores between 100 and 115, and, as in the figure, 84% have scores below 115.

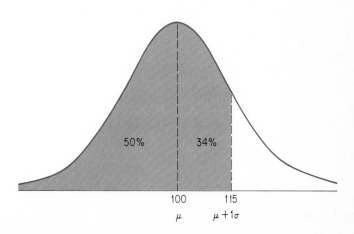

50%   34%

100   115

μ $\mu + 1\sigma$

Scores such as an IQ score of 115 or a College Entrance Examination Board score of 600 are often called *raw scores*. Each of these two raw scores is one standard deviation above the mean and is said to have *z*-score 1 in the scale −3, −2, −1, 0, 1, 2, 3 of standard deviations from the mean. In general a score x in a set of scores with mean μ and standard deviation σ has *z-score z* where

$$z = \frac{x - \mu}{\sigma}$$

Note that when considering questions such as those in Example 1 we do not concern ourselves with the number of people who make any particular score such as 100 or 115. The scale −3, −2, −1, 0, 1, 2, 3 in standard deviations enables us to use the properties of a normal distribution, such as the 68% within one standard deviation of the mean, for a wide variety of situations. For example, we found in Example 1 that 84% of the population have IQ scores below 115, which is one standard deviation above the mean. This same result can be used in many other situations. Scholastic Aptitude Test (SAT) scores have a mean of 500 and a standard deviation of 100. Thus 84% of the SAT scores on a particular test are below 600. College Entrance Examination Board (CEEB) scores also have a mean of 500, a standard deviation of 100, and 84% of the scores below 600.

Example 2   What percent of the population have IQ scores above 130?

Solution   Find the difference between the given score (130) and the mean (100). Then divide this difference by the standard deviation (15) to obtain the *z-score*.

$$z = \frac{130 - 100}{15} = 2$$

The scores above 130 are the scores above $\mu + 2\sigma$. The interval $\mu \pm 2\sigma$ includes the scores of 95% of the population, so 47.5% of the population have scores on the interval from 100 to 130. Thus 97.5% of the population have scores below 130, and 2.5% of the population have IQ scores above 130.

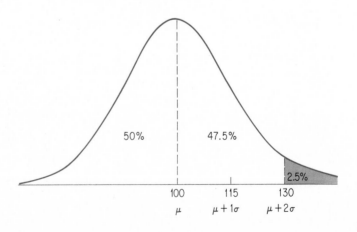

Note that we have discussed only normal distributions. However, the standard deviation can be computed for other distributions as well. In general, many sets of data tend to be approximately normal, and the standard deviation is a very useful measure of dispersion.

The result obtained in Example 2 as applied to SAT and CEEB scores shows that 97.5% of the scores are below 700 and 2.5% of the scores are over 700.

The standard deviation is relatively cumbersome to compute, especially for data grouped in a table. However, it may be instructive to note the computation of σ for a small set of scores. For such data, we generally use the formula

$$\sigma = \sqrt{\frac{\Sigma d^2}{n}}$$

Think of the Greek letter Σ (capital sigma) as an abbreviation for *sum of*. Then to find Σd^2 you determine the deviation of each score from the mean, square each deviation, and then *add* these squares.

where Σd^2 represents the sum of the squares of the deviations of each score from the mean, and n is the number of scores.

Consider the set of scores

{8, 10, 12, 16, 19}

We first compute the mean as 13. In the column headed d we find the differences when the mean μ is subtracted from each score. The sum of these deviations is 0, as noted in Section 10-3, Exercise 21. Thus to obtain a meaningful average of these deviations we may use either absolute values or the squares of the deviations. The standard deviation is based on the squares of the deviations, as in the column headed d^2.

SCORES	d	d^2	
8	-5	25	$\sigma = \sqrt{80/5} = \sqrt{16}$
10	-3	9	$\sigma = 4$
12	-1	1	
16	3	9	
19	6	36	

5) 65   80 = Σd^2
$\mu = 13$

Example 3 Compute σ for the set of scores {10, 11, 13, 14, 17}.

Solution

Recall that σ is a measure of dispersion. In Example 3 we have a set of scores with the same mean as the set above, but with a much smaller range. Note that the standard deviation is correspondingly smaller. The computation of $\sqrt{6}$ can be completed by using a calculator or a table of square roots.

SCORES	d	d^2	
10	-3	9	$\sigma = \sqrt{30/5} = \sqrt{6}$
11	-2	4	$\sigma \approx 2.5$
13	0	0	
14	1	1	
17	4	16	

5) 65   30 = Σd^2
$\mu = 13$

These two sets of five scores for which standard deviations have been computed are much too small to expect a close match with a normal distribution. In each of these two cases all scores are within one and one-half standard deviations of the mean.

EXERCISES

1. Find the range for the given set of scores.
   {55, 67, 80, 85, 90, 92, 98}

2. Give a set of the same number of scores with the same mean as in Exercise 1 but with a smaller range.

3. Repeat Exercise 2 for a larger range.

4. Give a set of the same number of scores with the same range as in Exercise 1 but with a smaller mean.

5. Repeat Exercise 4 for a larger mean.

6. Give a set of ten scores with the same mean and the range as in Exercise 1.

Repeat the indicated exercise for this set of scores.

   {45, 53, 69, 77, 85, 87, 90, 92, 95}

7. Exercise 1.   8. Exercise 2.   9. Exercise 3.
10. Exercise 4.   11. Exercise 5.   12. Exercise 6.

In Exercises 13 through 15, state which the distributions can be expected to be approximately normal.

13. The scores of all graduating high school seniors on a particular college board examination.

14. The weights of all college freshmen males.

15. The number of heads obtained if 100 coins are tossed by each college graduate in the country.

16. For a normal distribution of 10,000 test scores the mean is found to be 500 and the standard deviation is 100.
    (a) What percent of the scores will be above 700?
    (b) What percent of the scores will be below 400?
    (c) About how many scores will be above 600?
    (d) About how many scores will be below 300?
    (e) About how many scores will be between 400 and 700?

17. If 100 coins are tossed repeatedly, the distribution of the number of heads is a normal one with a mean of 50 and a standard deviation of 5. What percent of the number of heads will be (a) greater than 60 (b) less than 45 (c) between 40 and 60?

Compute the standard deviation for the given set of scores.

18. {7, 9, 10, 11, 13}

19. {11, 12, 13, 15, 20, 20, 21}

20. {81, 83, 85, 88, 93}

21. {78, 82, 83, 85, 92}

*22. {68, 74, 80, 82, 83, 85, 86, 88, 91, 93}

*23. {58, 62, 65, 67, 70, 75, 77, 80, 85, 91}

Explorations

In reading educational literature one frequently encounters the term **coefficient of correlation**, usually denoted by the letter r. The coefficient of correlation is given as a decimal on the interval -1.00 to $+1.00$ and provides an indication of how two variables are related.

A perfect positive correlation of 1.00 indicates that two sets of data are related so that the changes in one set are a positive multiple of the corresponding changes in the other set. For example, here is a set of ages and weights for a group of six individuals. Note that for each increase of 5 years of age there is a corresponding increase of 10 pounds of weight.

AGE	WEIGHT	
20	130	
25	140	
30	150	The coefficient of
35	160	correlation is 1.00.
40	170	
45	180	

Now consider the following table for six other individuals. For each increase of 5 years of age, there is a corresponding decrease of 10 pounds of weight.

AGE	WEIGHT	
20	180	
25	170	
30	160	The coefficient of
35	150	correlation is -1.00.
40	140	
45	130	

It is important to note that correlation does not imply causation. For example, there might be a positive correlation between the size of shoe that a student wears and the student's handwriting ability. This would not mean that big feet improve one's handwriting.

A coefficient of correlation of 0 indicates no uniform change of either variable with respect to the other. In general practice such extreme cases seldom occur. Furthermore, one has to read the literature accompanying any particular test or research study to determine whether any particular coefficient of correlation can be considered as significant.

1. There has been a high positive correlation between expenditures for alcohol and for higher education in recent years. Does this mean that drinking alcohol provides one with the thirst for knowledge? Does it mean that education leads one to drink? How can you explain this high correlation?

In each exploration explain why you would expect to find a high or a low correlation between the two sets.

2. IQ scores and scores on college entrance examinations of all graduating high school seniors.
3. Scores made by elementary school students in reading and in arithmetic.
4. Age and physical abilities of mentally retarded individuals.
5. Manual dexterity and age of normal elementary school children.
6. Grades in arithmetic and number of hours spent by elementary school children watching television.
7. Weights relative to normal for their heights, of elementary school children and their mothers.
8. Academic grades and extent of participation in extracurricular activities of college students.
9. Effectiveness in teaching and years of college training of teachers.

10-5
Measures of Position

Rank in class, or class rank, is determined by counting down from the top of the class.

Anne, Bill, and Carol were all classmates in a high school class of 400 students. Anne was 25th from the top of her class, that is, her **class rank** was 25. Bill's class rank was 100 but he preferred to think that three-quarters of his classmates were ranked below him. Thus Bill described his position in the class as at the *third quartile* or as at the 75th *percentile*. Imagine that all 400 students were lined up according to their rank in class.

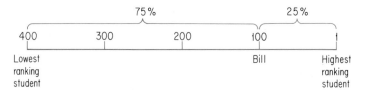

Then counting from the top of the class Ann would be 25th in line and Bill would be 100th. Bill is in the upper 25% of the class because 75% of the students are ranked below him.

The position of any element in an ordered set of data is often described in terms of class rank, quartiles, percentiles, or deciles. The class rank is the number of the element counting from the top of the list or class. The other three measures of position are determined by the number of quarters, hundredths, or tenths of the data that precede the specified element in the ordered set.

The ninety-nine **percentiles** are denoted by P_1, P_2, P_3, . . . , P_{98}, P_{99}. As in the case of Bill at the 75th percentile, the kth percentile P_k has k percent of the data preceding it. A score or element at the kth percentile is said to have **percentile rank** k. To find the percentile rank of a given score or element in an ordered set of data first express the number of elements at or above the given score as a percent of the total number of elements in the data and then subtract that percent from 100%. For example, Carol's rank in class was 40. The 40 students at or above Carol's rank are 10% of the class of 400. Thus Carol may describe her position as at the 90th percentile.

The three **quartiles** are denoted by Q_1, Q_2, Q_3 and may be defined in terms of percentiles.

$$Q_1 = P_{25} \qquad Q_2 = P_{50} \qquad Q_3 = P_{75}$$

The nine **deciles** are denoted by D_1, D_2, D_3, . . . , D_8, D_9 and may be defined in terms of percentiles.

$$D_1 = P_{10} \qquad D_2 = P_{20} \qquad D_3 = P_{30} \qquad . . . \qquad D_8 = P_{80} \qquad D_9 = P_{90}$$

The second quartile, Q_2, is the median (see Section 10-3). If the data have an even number of scores the median is taken halfway between the two middle scores. Although we usually avoid such situations, other quartiles, deciles, and percentiles are taken between scores when necessary.

Example 1   In a class of 500 students, how many students are ranked below the 80th percentile?

Solution   The 80th percentile is preceded by 80% of the students. Since 80% of 500 is 400, there are 400 students who rank below the 80th percentile.

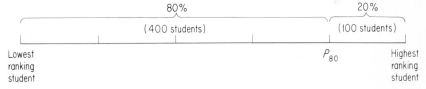

Example 2   Don is ranked 21st in a class of 300. What is Don's percentile rank in his class?

Solution   There are 279 members of the class ranked below Don. Since 279 is 93% of 300, there are 93% of the class ranked below Don. Thus Don's percentile rank is 93.

Verify that 279 is 93% of 300.

Percentiles for a normal distribution may be expressed in

terms of numbers of standard deviations from the mean. The most commonly used values are given in the following table.

Percentile	Deviations from the mean
1	−2.33
5	−1.65
10	−1.28
20	−0.84
30	−0.52
40	−0.25
50	0
60	0.25
70	0.52
80	0.84
90	1.28
95	1.65
99	2.33

Measures of position are important since individual scores mean very little. For example, a test score of 49 is excellent if the total possible score is 50 but disappointing if the total possible score is 100.

Example 3 Find the 95th percentile for a normal distribution of scores with mean 800 and standard deviation 200.

Solution From the table the 95th percentile is 1.65 standard deviations above the mean.

Reading a table is an important problem-solving skill. Be sure that you see how the preceding table is used in the solution for Example 3.

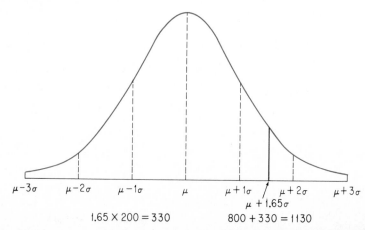

$\mu - 3\sigma$ $\mu - 2\sigma$ $\mu - 1\sigma$ μ $\mu + 1\sigma$ $\mu + 2\sigma$ $\mu + 3\sigma$
$\mu + 1.65\sigma$

$1.65 \times 200 = 330$ $800 + 330 = 1130$

The 95th percentile is 1130. That is, for this distribution 95% of the scores are below 1130.

In a high school class of 550 students how many students are ranked below the indicated position?

1. P_{90} 2. Q_3 3. D_8
4. D_9 5. Q_1 6. P_{80}

Find the percentile rank of the indicated student in a class of 400. Each student's rank in class is given.

7. Ann, 4th.   8. Barbara, 8th.   9. Charles, 12th.
10. David, 16th.   11. Eleanor, 24th.   12. Frank, 28th.

In each of the Exercises 13 through 15, which student ranks highest in a class of 300 students? A measure of position is given for each student.

13. Albert, P_{88}; Ben, D_9; Charles, 25th.
14. Debbie, D_9; Elizabeth, P_9; Florence, 35th.
15. George, Q_3; Harold, P_{70}; Irvine, 100th.

Repeat the indicated exercise for a class of 500 students.

16. Exercise 13.   17. Exercise 14.   18. Exercise 15.

Consider a normal distribution of SAT scores with arithmetic mean 500 and standard deviation 100. Use the table for percentiles and deviations from the mean.

19. Jack scored at the 95th percentile. What was his score?
20. Katherine scored 584. What was her percentile rank?
21. Find the percentile rank of each score.
    (a) 500   (b) 665   (c) 475   (d) 416   (e) 335   (f) 733
22. Find the score at each percentile rank.
    (a) 20   (b) 40   (c) 60   (d) 70   (e) 5   (f) 95

Explorations

Find as many examples as you can of each type of measure of position in newspapers, magazines, or books that are not mathematics books.

1. Quartiles.   2. Deciles.
3. Percentiles.   4. Rank in class.

5. Recent trends in SAT scores (see Section 10-4) and other test scores have caused numerous comments, suggestions, charges, and countercharges. Prepare a 10-minute report on some of the trends, causes, and implications during the last two decades.

10-6
Binomial Distributions

Many states require that all students take certain standardized tests. The test scores help the teachers identify the aptitudes and interests of the students. Then teachers have some objective bases for the informal day-to-day guidance and individual counseling that is an important part of teaching. For example, which students should be highly complimented and which should be challenged to seek a deeper understanding of a subject when a certain behavioral objective has been met? Which students are working close to their peak abilities and which are "coasting" unchallenged and unmotivated while performing well above the class average? (See Explorations 7 and 8.)

Many statistical results are based upon simple yes-no or true-false responses. A manufacturer may sample preferences of potential customers for a product versus a competitor's product. A teacher may give a true-false quiz or test. The result, the distribution of the data from any such experiment in which there are only two possible outcomes for each event, is a **binomial distribution**.

The classic binomial experiment is tossing a coin. For a "fair" coin, the ratio of the number of heads to the total number of tosses is expected to be about 1/2 since the probability of heads is 1/2 on each toss of the coin (Section 9-2). If the experiment of tossing a coin n times is repeated over and over, the expected mean of the distribution of heads on these trials of the experiment is $(1/2)n$. The result of tossing one coin n times is the same as that of tossing n coins once. For example, if one coin is tossed 100 times or 100 coins are tossed once, we have probability $p = 1/2$ and $n = 100$. Therefore, the mean or average number of heads expected for each experiment is $(1/2)(100)$, that is, 50.

The **standard deviation of a binomial distribution** can be found by the following formula, which we shall not prove,

Using this formula for heads and 100 coins, we have $p = 1/2$, $1 - p = 1/2$, $n = 100$, and

$$\sigma = \sqrt{(1/2)(1/2)(100)} = \sqrt{25} = 5$$

$$\sigma = \sqrt{p \times (1 - p) \times n}$$

where p is the probability of success and n is the number of trials.

For large values of n, a binomial distribution is approximately normal. It has been shown for a normal distribution that if three standard deviations are added to and subtracted from the mean, we have the limits within which almost all of the data will fall. For the example of 100 coins, the mean is 50, and three times the standard deviation is 15, thus giving the limits 35 to 65. This is frequently stated in terms of **confidence limits**: we may say with *"almost 100% confidence"* that the number of heads will be between 35 and 65. Furthermore, we may say with approximately 95% confidence that the number of heads will be between 40 and 60, that is, within two standard deviations of the mean. We may say with approximately 68% confidence that the number of heads will be between 45 and 55, that is, within one standard deviation of the mean.

A **control chart** is a graph on which lines are drawn to represent the limits within which all data are expected to fall a given percent of the time. In industry, for example, a chart like the following may be drawn on which data from samples are plotted.

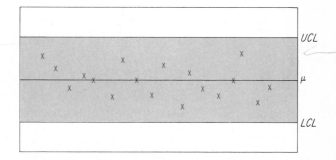

The middle horizontal line represents the **mean** (μ) or average around which these data are expected to lie. The other horizontal lines, called the **upper control limit** (UCL) and **lower control limit** (LCL), represent the limits within which all the data are expected to fall most of the time. Occasionally data may fall outside of these limits; as long as most of the samples tested produce data which lie between these lines, the process is said to be "in control."

For the case of 100 coins, the theoretical average is 50. The lower control limit would be $\mu - 3\sigma$, or 35. The upper control limit would be $\mu + 3\sigma$, or 65. We would suspect that something was wrong if a toss of 100 coins were to produce fewer than 35 or more than 65 heads. Such an event *can* take place just by chance, but it would be very unusual, occurring less than 3 times in 1000.

Example 1 An experiment consists of 180 throws of a single die.
(a) What is the mean number of fives to be expected?
(b) What is the standard deviation for the distribution?

Solution   (a) These are the possible results when a die is thrown.

For a "fair" die these six possibilities are equally likely and we say that the probability of a 5 is 1 out of 6, that is, 1/6. Thus we have $p = 1/6$ and $n = 180$. Therefore, the average number of fives expected is $(1/6)(180) = 30$. Theoretically, we can expect 1 out of every 6 throws to produce a 5 and expect 30 fives in 180 throws.

(b) The probability of obtaining a five is 1/6. Thus $p = 1/6$, $(1 - p) = 5/6$, $n = 180$, $\sigma = \sqrt{(1/6)(5/6)(180)} = \sqrt{25} = 5$.

Example 2   Suppose that a certain lecturer had a feeling that the 200 students in the lecture hall were not paying attention and asked two true-false questions about the lecture. Each student responded, possibly after tossing a coin, to each question with probability 1/2 of having a correct answer. Approximately how many students would give correct answers for (a) the first question   (b) both questions   (c) at least one question?

Solution   (a) $(1/2)(200)$, that is, 100.
(b) $(1/2)(1/2)(200)$, that is, 50.
(c) About 50 would have both answers wrong; the remaining 150 would have at least one answer correct.

If the instructor in Example 2 had asked only one question with a show of hands for one of the possible responses, the resulting show of hands might have appeared impressive without the students having the least idea what the question was. Note that the standard deviation is

$$\sqrt{(1/2)(1/2)(200)} = \sqrt{50} \approx 7$$

Therefore, we can assert with about 95% confidence that at least 86 and at most 114 hands would have been raised for the correct response, just by chance.

Most students have some idea of the correctness of answers to questions even if they have not been paying attention. Suppose that the students in Example 2 could answer correctly with probability 3/4. Then the solution would be

(a) (3/4)(200), that is, 150.

(b) (3/4)(3/4)(200), that is, 112 or 113.

(c) 200 − (1/4)(1/4)(200), that is, 187 or 188.

The distribution for a single student responding to 200 questions is the same as the distribution for 200 students responding to one question.

Statistical results must be interpreted very carefully. In the case of the fourth toss of a coin that has shown heads three times in a row, either the probability of tails is still 1/2 on the fourth toss or the coin has a bias. Similarly, in the case of test scores neither a sequence of successes nor a sequence of failures provides more than a mild indication of the capabilities of the individual. Physical limitations (exhaustion, sickness, allergies), emotional limitations (excitement or distress over past or forthcoming events), and attitudes (expectations, desires, compatibility with other students or the teacher) can drastically affect student performance.

Several studies have shown that students tend to perform at the level that they feel their teacher expects of them!

Statistical data have many constructive pedagogical uses. However, one fact deserves special recognition: *Standardized test scores are not exact rankings* of the students. Each score indicates that the student's actual position is probably on an interval about that score. The size of the interval will vary from one test to another and for any standardized test should be given in the instructions to the person who is to interpret the test. For example, a difference of test scores such as 598 and 601 in different subject areas is insufficient evidence for any major decisions as to the student's preferences or potential success. (Such a difference was once used by a student to decide upon a major when entering college. This was an outright abuse of the statistics.) A score of 598 might mean that the chances are 2 to 1 that the student's true score lies within the interval from 586 to 610; similarly, for 601 the chances might be 2 to 1 that the true score lies within the interval from 588 to 614. Note that these intervals, even for only 2 to 1 probability of correctness, have ranges of 24 and 26 points, respectively. Also the intervals overlap for about two-thirds of their extent. Even IQ scores may vary by 10, 15, or more points for the same person on tests taken as little as two weeks apart. Drastic differences are unusual but do occur.

The role of intervals may be illustrated by throws of a die. Consider the following graph of the theoretical results for the experiment in Example 1.

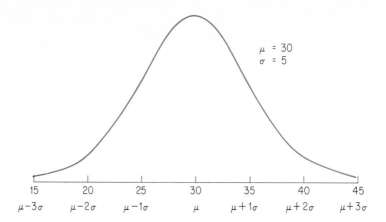

Thus, in a set of 180 throws of a die, we would expect with almost 100% confidence to have between 15 and 45 fives. With 95% confidence we would expect to have between 20 and 40 fives; and with 68% confidence we would expect between 25 and 35 fives.

These examples should serve to show that fluctuation is normal. That is, although the mean of the distribution is 30, we expect some variability. Not every set of 180 throws will produce exactly 30 fives—in fact, very few will. However, on the other hand, we expect this fluctuation to be within limits that can be described by our knowledge of the behavior of normal distributions. Someone could conceivably throw a die 180 times and produce 100 fives. However, an understanding of the principles of this chapter would dictate that you challenge the thrower of the die and examine the die quite carefully for evidence of foul play. In other words, although it is possible to obtain 100 fives, it is far from probable that this would happen.

EXERCISES   *In Exercises 1 through 5 consider a single die that is thrown 720 times.*

1. What is the mean number of sixes to be expected?
2. What is the standard deviation for the distribution of sixes?
3. What is the maximum number of sixes we can expect to obtain with almost 100% confidence?
4. What is the minimum number of sixes we can expect to obtain with almost 100% confidence?

5. Within what limits can we expect the number of sixes to fall approximately 95% of the time?

6. Suppose that each member of your class tosses an unbiased coin 100 times and counts the number of heads. What is the expected mean of the numbers of heads obtained?

7. As in Exercise 6 on what interval would you expect about two-thirds of the numbers of heads to occur?

8. Repeat Exercise 7 for 95% of the heads.

9. Repeat Exercise 7 for practically all of the heads.

10. Suppose that you assigned each member of your class the experiment of tossing an unbiased coin 80 times. Then suppose that they reported the numbers of heads obtained as follows:

Alice 45   Bob 40   Charles 30
Doris 35   Eve 25   Fred 50
Gwen 42   Harry 37   Ike 69

Explain why you would suspect that one or more of the reports reflected either a failure to do the experiment or the use of a biased coin.

Consider a single die that is thrown 180 *times.*

11. What is the mean number of threes to be expected?

12. What is the maximum number of threes we can expect, with almost 100% confidence, to obtain?

13. What is the minimum number of threes we can expect, with almost 100% confidence, to obtain?

14. Within what limits can we expect the number of threes to fall approximately 95% of the time?

Explain your answer for the given question.

15. Jane has an IQ score of 130 and SAT scores of 610 and 590. Does she appear to be working up to her potential?

16. Jack has an IQ score of 95 and SAT scores of 450 and 500. Does he appear to be working up to his potential?

17. A student once remarked to the teasing of a classmate: "I am one of the people that make it possible for you to be in the upper half of the class." How would you explain the situation to an irate parent who considered the fact that half of your students were below average in arithmetic skills to be sufficient evidence for firing you and finding a "good teacher?"

18. In a certain year the median SAT verbal score for students at Salem High School was 495. Was this cause for serious concern?

Find the approximate interval on which the indicated student can expect, with 95% confidence, a grade on a test of 25 questions.

19. José, who has an 80% chance of answering any given question correctly.

20. Dolores, who has a 90% chance of answering any given question correctly.

Explorations

1. Use the concepts of this section to establish the limits within which the total number of heads can be expected to fall when a coin is tossed 64 times, using the 95% confidence limits. Then test these results by tossing a coin 64 times, and asking as many other people as possible to repeat this experiment. (Note that instead of tossing a single coin for 64 tosses, you may also toss four coins for 16 tosses, 16 coins for 4 tosses, and so on.)

2. Repeat Exploration 1 for 180 throws of a single die, counting the number of sixes that appear.

Suppose that someone announces that he or she has just tossed a coin 19 times in a row and has obtained 19 heads. This person is about to toss the coin for the twentieth time. You are given an opportunity to place a bet on the outcome of this toss.

3. Would you bet on heads? On tails?

4. What is the probability of tossing 19 heads in a row with an unbiased coin?

5. What are the odds against tossing 19 heads in a row with an unbiased coin?

6. Discuss the probability that the coin and the method of tossing are both unbiased.

7. A student scores a grade of 127 on an aptitude test. What can you say about this student's aptitude as a result of this test score? Suppose that you are then given the additional information that the score of 127 is at the 95th percentile in a distribution of scores for all students who have taken this partiular test. What can you conclude then about this student's aptitude?

8. Why does percentile rank appear to be more significant than rank (first, second, third, . . .) in a class?

Chapter 10 Review

Solutions to the following exercises may be found within the text of Chapter 10. Try to complete each exercise without referring to the text.

Section 10-1 Uses and Abuses of Statistics

1. For the given statement, "Brush your teeth with GLUB and you will have fewer cavities," identify at least one word that is ambiguous or misleading and explain your answer.

2. Sketch a graph that is misleading and explain the basis for misinterpretation.

Section 10-2 Descriptive Statistics

3. Represent the following family budget by a circle graph: food 40%, household 25%, recreation 5%, savings 10%, miscellaneous 20%.

4. Represent the budget presented in Exercise 3 by a divided bar graph.

Section 10-3 Measures of Central Tendency

5. For the given set of test scores, find
    (**a**) the mean   (**b**) the median
    (**c**) the mode   (**d**) the midrange.

    {72, 80, 80, 82, 88, 90, 96}

6. Identify the two measures of central tendency that are usually most affected by extreme scores.

Section 10-4 Measures of Dispersion

7. What percent of the population have IQ scores below 115?

8. What percent of the population have IQ scores over 130?

9. Compute σ for the set of scores {10, 11, 13, 14, 17}.

Section 10-5 Measures of Position

10. In a class of 500 students, how many students are ranked below the 80th percentile?

11. Don is ranked 21st in a class of 300. What is Don's percentile rank in his class?

12. Find the 95th percentile for a normal distribution with mean 800 and standard deviation 200.

Section 10-6 Binomial Distributions

13. An experiment consists of 180 throws of a single die.
    (**a**) What is the mean number of fives to be expected?
    (**b**) What is the standard deviation for the distribution?

14. Suppose that a certain lecturer had a feeling that the 200 students in the lecture hall were not paying attention and asked two true-false questions about the lecture. Each student responded, possibly after tossing a coin, to each question with probability 1/2 of having the correct answer. Approximately how many students would give correct answers for
    (a) the first question   (b) both questions
    (c) at least one question?

Chapter 10 Test

1. Describe two common devices for making graphs misleading to the reader.

2. Make a frequency distribution for the grades of a class with grades: C, A, B, A, B, A, C, B, A, B, B, F, B, B, B, B, B, B, B, B, B, A, C, B, B, C, B, B, A, A, B, D, B, B, A, B, B.

Represent, in the form specified in Exercises 3, 4, and 5, the following distribution of grades given last semester by Professor X.

Grade	A	B	C	D	F
Per cent	20	25	40	10	5

3. Histogram.   4. Circle graph.   5. Divided bar graph.

The adjacent circle graph is based upon estimates of the population of mainland China and the population of the world. Use these data for Exercises 6 through 8.

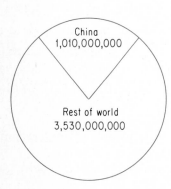

6. The population of China is approximately what percent (nearest whole number) of the population of the entire world? (Note that the population of the entire world includes the population of China as well.)

7. What is the size of the central angle (nearest degree) that represents the population of China?

8. Assume that you wish to draw a bar graph to show the same facts as does the circle graph. If the population of China is represented by a bar that is one centimeter high, approximately how long a bar (nearest 5 millimeters) would you need to show the population of the rest of the world?

Consider the set of scores {55, 62, 70, 74, 74, 79}. Then find the specified measure.

9. Arithmetic mean. 10. Median. 11. Mode.

For each of the following sets of scores, find (a) *the range* (b) *the midrange.*

12. {75, 82, 64, 98, 79}   13. {100, 45, 79, 82, 96, 68}

A collection of 1000 scores forms a normal distribution with a mean of 50 and a standard deviation of 10. What percent of the scores are expected to be on the specified interval?

14. Above 70.   15. Below 60.
16. Below 40.   17. Between 40 and 70.
18. Above 80.

19. Albert scored at the third quartile on a test in a class of 32 students. What was his class rank?

20. Beth ranked 30th in her high school class of 600. What is her percentile rank?

An unbiased coin is tossed 64 times.

21. What is the expected number of heads?
22. What is the standard deviation for results of this experiment when it is repeated many times?
23. Within what limits can we say, with almost 100% confidence, that we shall find the total number of heads?
24. If the experiment is repeated many times, what percent of the times should be expected to produce fewer than 36 heads?
25. As in Exercise 24 what percent of the times should be expected to produce more than 28 heads?

Answers
to
Odd-Numbered
Exercises

1 An Introduction to Problem Solving

1-1 Problem-Solving Strategies: page 7

1. (a) 12
   (b) Only one, if it's long enough.
   (c) Only halfway; then you start walking out.
   (d) *One* of them is not a nickel, but the other one is.
   (e) There is no dirt in a hole.
   (f) Brother-sister.

3. There are 11 trips needed. First one cannibal and one missionary go over; the missionary returns. Then two cannibals go over and one returns. Then two missionaries go over; one missionary and one cannibal return. Two missionaries go over next and one cannibal returns. Then two cannibals go over and one of them returns. Finally, the last two cannibals go over.

5. After 15 days the cat still has 3 feet to go. She does this the next day and is at the top after 16 days.

7.

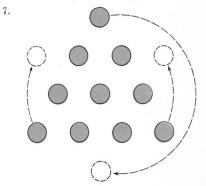

9. If the penny is in the left hand and the dime in the right hand, the computation will give $3 + 60 = 63$, an odd number. If the coins are reversed, we have $30 + 6 = 36$, an even number.

11. There really are no missing dollars. The computation may be done in one of two ways: $(60 - 6) - 4 = 50$, or $50 + 4 + 6 = 60$. In the problem the arithmetic was done in a manner that is not legitimate, that is, $(60 - 6) + 4$.

13. Use the matchsticks to form a triangular pyramid.

15. Both A's and B's will say that they are B's. Therefore, when the second man said that the first man said he was a B, the second man was telling the truth. Thus the first two men told the truth and the third man lied.

17. Eight moves are needed. The coins are identified in the following diagram as well as the squares which may be used. The moves are as follows, where the first numeral indicates the position of the coin and the second one tells you where to move it: D_1: 4–3; P_2: 2–4; P_1: 1–2; D_1: 3–1; D_2: 5–3; P_2: 4–5; P_1: 2–4; D_2: 3–2.

1	2	3	4	5
P_1	P_2		D_1	D_2

19. E, N. These are the first letters of the names (one, two, three, etc.) of the counting numbers.

21. Use H for the half-dollar, Q for the quarter, and N for the nickel. Assume that each coin is on a larger coin. The seven moves may be made in the order listed.

N: A to C,	Q: A to B,
N: C to B,	H: A to C,
N: B to A,	Q: B to C,
N: A to C.	

For the four coins use P for the penny. Then the moves are

P: A to B,	N: A to C,
P: B to C,	Q: A to B,
P: C to A,	N: C to B,
P: A to B,	H: A to C,
P: B to C,	N: B to A,
P: C to A,	Q: B to C,
P: A to B,	N: A to C,
P: B to C.	

For a discussion of the number of moves required for 64 disks see page 171 of *Mathematics and the Imagination* by Edward Kasner and James Newman, Simon & Schuster, 1940. They estimate that it would take more than 58 billion centuries to complete the task.

23. The sum of the values on two opposite faces of any ordinary die is always 7. Therefore for three dice this sum is 21. Thus the difference between 21 and the value on the top face of the top die is the sum of the values on the other specified faces.

25. $\left(\dfrac{1}{1} - \dfrac{1}{2}\right) + \left(\dfrac{1}{2} - \dfrac{1}{3}\right) + \cdots$

$+ \left(\dfrac{1}{9} - \dfrac{1}{10}\right) = \dfrac{1}{1} - \dfrac{1}{10} = \dfrac{9}{10}$

*27. (a) $x, y \rightarrow x + y + 2$
   (b) $x, y \rightarrow xy + 1$
   (c) $x, y \rightarrow$ lesser of x and y
   (d) $x, y \rightarrow x$
   (e) $x, y \rightarrow 10 - (x + y)$, that is, $10 - x - y$

29. Dashed lines show segments removed.

1-2 **Problem Solving with Arithmetic Patterns: page 20**

1. 1×9, first; 2×9, second; 3×9, third; 4×9, fourth; 5×9, fifth; 6×9, sixth; 7×9, seventh; 8×9, eighth; 9×9, ninth.

3. $6^2 = 1 + 2 + 3 + 4 + 5 + 6 + 5$
   $\qquad + 4 + 3 + 2 + 1$

$7^2 = 1 + 2 + 3 + 4 + 5 + 6 + 7$
$\qquad + 6 + 5 + 4 + 3 + 2 + 1$
$8^2 = 1 + 2 + 3 + 4 + 5 + 6 + 7$
$\qquad + 8 + 7 + 6 + 5 + 4 + 3$
$\qquad\qquad\qquad\qquad + 2 + 1$
$9^2 = 1 + 2 + 3 + 4 + 5 + 6 + 7$
$\qquad + 8 + 9 + 8 + 7 + 6 + 5$
$\qquad\qquad + 4 + 3 + 2 + 1$

7. **(a)** $9 \times 47 = 423$

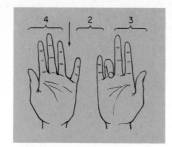

(d) $9 \times 27 = 243$

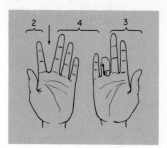

(b) $9 \times 39 = 351$

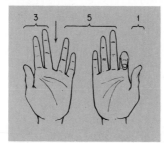

(c) $9 \times 18 = 162$

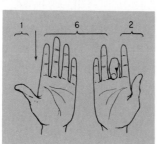

9. **(a)** 45×91, that is, 4095.
    (b) 150×301, that is, 45,150.
    (c) 10×40, that is, 400.
    (d) 75×300, that is, 22,500.
    (e) 125×502, that is, 62,750.

11. The sum appears to be nine times the number in the center of the array.

13. The sum of the two outer numbers appears to be equal to the sum of the two inner numbers.

15. **(a)** The conjecture is false.
     (b) The conjecture appears to be true.

17. There are only three such ways: $1 + 1 + 1 + 7$, $1 + 1 + 3 + 5$, and $1 + 3 + 3 + 3$.

*19. Impossible. The sum of three odd numbers must be an odd number.

1-3 Problem Solving with Geometric Patterns: page 29

1. There will be 4 holes produced by 4 folds, and 2^{n-2} holes with n folds.

3. $V + R = A + 2$. The sum of the number of vertices and the number of regions is 2 more than the number of arcs.

5.

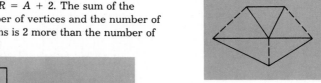

7. Dashed lines show segments removed.

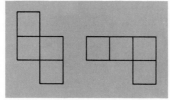

9. Dashed lines show the original outline of the pool.

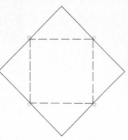

11. Many answers are possible.
(a)

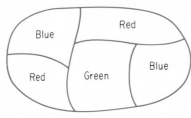

(b)

Red	Blue	Red	Blue	Red

13. Among others:
(a)

(b)

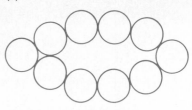

*15. (a) Approximately 7 times.
(b) Approximately 50,000,000 miles.

17. (a) 8
(b) 24
(c) 24
(d) 8

*19. 10; the faces would be all red, 1 red and 5 blue, 1 blue and 5 red, 2 adjacent blue and 4 red, 2 opposite blue and 4 red, 2 adjacent red and 4 blue, 2 opposite red and 4 blue, 3 red including two opposite faces and 3 blue, 3 red with no two opposite and 3 blue, or all blue.

*21. Number of:

chords 1   2   3   4   5
regions 2   4   7   11

(with differences 2, 3, 4 marked between successive region values)

By actual construction we find the answer for four chords to be 11. Now note that the differences between successive number of regions appears to be increasing by 1. The next difference is thus expected to be 5, thus suggesting the number of regions for five chords to be $11 + 5$, that is, 16.

1-4 Problem Solving with Algebraic Patterns: page 39

1. $(n - 11) + (n - 10) + (n - 9) + (n - 1) + (n) + (n + 1) + (n + 9) + (n + 10) + (n + 11) = 9n$, which is nine times the center number n.

3. The numbers in the shaded regions can be represented as shown. The sum of the quantities shown is $9n$. Note that $-20 + 20 = 0$, $-11 + 11 = 0$, $-9 + 9 = 0$, and $-1 + 1 = 0$.

	$n-20$	
$n-11$		$n-9$
$n-1$		$n+1$
$n-11$		$n+11$
	$n+20$	

5. The array has the form shown below. Each sum of numbers in opposite corners is $2a + 16$.

a	$a+1$	$a+2$
$a+7$	$a+8$	$a+9$
$a+14$	$a+15$	$a+16$

7. Use the array for Exercise 5. The products are $a(a + 16) = a^2 + 16a$ and $(a + 2)(a + 14) = a^2 + 16a + 28$. The constant difference is 28.

9. 0; n, $n + 7$, $3n + 21$, $3n$, n, 0.

*11. Represent the four cyclic numbers as follows and add:

$$1000a + 100b + 10c + d$$
$$1000b + 100c + 10d + a$$
$$1000c + 100d + 10a + b$$
$$\underline{1000d + 100a + 10b + c}$$

The sum is $1000(a + b + c + d) +$
$$100(a + b + c + d) +$$
$$10(a + b + c + d) +$$
$$(a + b + c + d) =$$
$$(1000 + 100 + 10 + 1)(a + b + c + d)$$
$= 1111(a + b + c + d)$. Dividing this sum by $a + b + c + d$ gives 1111.

*13. From the top of the deck n cards were removed by your friend and 20 more cards were removed to be spread out. Then your friend counted backward (toward the top of the deck) starting with the displayed card that had been furthest from the top of the deck. Since $n + 20 - n = 20$, the last card counted will be the predetermined card that had been placed 21st from the top of the deck.

2 Calculators and Computers

2-1 Calculators: page 50

Among others:

1. Instructions: 2 ⊠ 21 ⊞ 7 ⊟(=)
   Displays: 2   2   21   42   7   49

3. Instructions: 18 ÷ 3 + 12 =
   Displays: 18   18   3   6   12   18

5. Instructions: 3 × 5 +/- + 20 =
   Displays: 3   3   5   −5   −15   20   5

7. Instruction: 11 × 11 + 5 =
   Displays: 11   11   11   121   5   126

9. Instructions: 64 √ + 36 =
   Displays: 64   8   8   36   44

11. Instructions: 121 √ + 7 − 15 =
    Displays: 121   11   11   7   18   15   3

13. 500; 492
15. 900; 885
17. 3100; 3090
19. 400; 399
21. 300; 343
23. 100; ≈99.6

25. 9
27. 110
29. 39
31. 5
33. 60
35. 91

2-2 Problem Solving with Calculators: page 53

1. 10 miles per hour
3. 4 miles per hour
5. 1500 miles
7. (a) 70¢;
   (b) 20¢
9. $9000
11. ≈4,500,000,000

13. Yes; $1,000,000 < 60 \times 60 \times 24 \times 7 \times 2 = 1,209,600$
15. ≈18 days
17. ≈32 pounds
19. ≈12.4 orbits
*21. ≈4.4 years
*23. 13 years

2-3 Flowcharts: page 59

1. (a) 570;
   (b) 698;
   (c) $n + 483$
3. (a) 2175;
   (b) 5375;
   (c) $25n$

5. 24
7. 31
9. 28
11. 21
13. 30

ANSWERS TO ODD-NUMBERED EXERCISES

15. **(a)** 110;
    (b) $5(n + 7)$

17. **(a)** 8;
    (b) $(n + 9)/3$

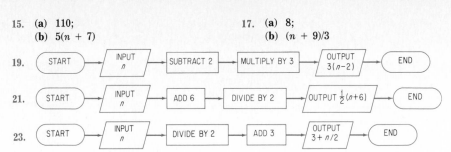

19. START → INPUT n → SUBTRACT 2 → MULTIPLY BY 3 → OUTPUT $3(n-2)$ → END

21. START → INPUT n → ADD 6 → DIVIDE BY 2 → OUTPUT $\frac{1}{2}(n+6)$ → END

23. START → INPUT n → DIVIDE BY 2 → ADD 3 → OUTPUT $3 + n/2$ → END

25. 17, 22, 27, 32, 37, 42, 47, 52

27. 1, 3, 4, 7, 11, 18, 29, 47, 76, 123

*29. a, b, $a + b$, $a + 2b$, $2a + 3b$,
     $3a + 5b$, $5a + 8b$, $8a + 13b$,
     $13a + 21b$, $21a + 34b$

*31. $55a + 88b = 11 \times (5a + 8b)$

2-4 Computers: page 66

1. 9
3. 1
5. 4
7. 3
9. -1
11. 27
13. 2
15. 10/3
17. 200
19. **5*6/3**; 10

21. **6*SQR (25+24)**; 42
23. **8000/(2 ↑ 5)**; 250
25. $7 \times 2 = 14$, $6 + 14 = 20$; 20
27. $6 \times 7 = 42$; $42 \div 3 = 14$; 14
29. $12^2 = 144$, $144 - 100 = 44$; 44
31. $3^2 = 9$, $7 \times 9 = 63$, $5 + 63 = 68$; 68
33. $\sqrt{169} = 13$, $11 + 13 = 24$; 24
35. $12^2 = 144$, $5^2 = 25$, $144 + 25 = 169$,
    $\sqrt{169} = 13$, $17 + 13 = 30$; 30

2-5 BASIC: page 72

1.
2	1
4	2
6	3
8	4

3.
1	1
2	8
3	27
4	64

5.
1	2	3
3	4	7
5	6	11

7.
1	2	3	6
4	5	6	15
7	8	9	24
10	11	12	33

9.
10	100
11	121
12	144
13	169
14	196
15	225

11.
1	2	2
2	3	6
3	4	12
4	5	20
5	6	30
6	7	42

Among others:

13.
```
10   FOR X = 1 TO 20
20   LET Y = X ↑ 5
30   PRINT X, Y
40   NEXT X
50   END
```

15.
```
10   FOR X = 1 TO 1000
20   LET Y = X ↑ 2
30   LET Z = X ↑ 3
40   PRINT X, Y, Z
50   NEXT X
60   END
```

17.
```
10   FOR X = 1 TO 10
20   LET A = X ↑ 2
30   LET B = X ↑ 3
40   LET C = SQR (X)
50   PRINT X, A, B, C
60   NEXT X
70   END
```

*19.
```
10  FOR X = 1 TO 20
20  LET Y = X*0.15
30  LET Z = Y + 1
40  LET W = 1000*Z
50  PRINT X, W
60  NEXT X
70  END
```

*21.
```
10  READ E, G1, G2, G3
20  DATA . . .
30  LET A = E/2 + (G1 + G2 + G3)/6
40  PRINT E, G1, G2, G3, A
50  GO TO 10
60  END
```

3 Mathematics for the Consumer

3-1 Percent: page 83

1. 0.57
3. 0.03
5. 0.0095
7. 2.50
9. 1.00
11. 1/2
13. 19/20
15. 99/100
17. 3/2
19. 2/25
21. 35%
23. 45%
25. 90%
27. 145%
29. 0.1%
31. 9%
33. 92%
35. 150%

37. 76%
39. 170%
41. 0.3; 30%
43. 11/20; 0.55
45. 0.85; 85%
47. 1/20; 0.05
49. 1.1; 110%
51. 0.715; 71.5%
53. 90
55. 60
57. 48
59. $40/100 = n/60$; 24
61. $35/100 = n/80$; 28
63. $20/160 = n/100$; $12\frac{1}{2}\%$
65. $120/160 = n/100$; 75%
67. $25/n = 20/100$; 125
69. $80/n = 125/100$; 64

3-2 Applications of Percent: page 88

1. (b)
3. (b)
5. (a)
7. (b)
9. $180
11. $144
13. $54
15. $44.20
17. $35.88
19. $7.48
21. 25%
23. 24%
25. 17%

27. 24%
29. $33\frac{1}{3}\%$ of $120 = $40; the correct selling price should be $120 − $40, that is, $80. The merchant considered the discount, $30, as $33\frac{1}{3}\%$ of the sale price, $90.
31. $14\frac{2}{7}\%$
*33. $23\frac{1}{2}\%$
*35. In both cases, no action. This is, a 20% cut followed by a 25% increase or a 25% increase followed by a 20% cut returns one to the original salary.

3-3 Simple and Compound Interest: page 96

1. $180
3. $150
5. $630
7. $630

9. $1893.75
11. $1007.93
13. $282.68
15. $6606.60

17. 12

19. (a) 8.16%
    (b) 8.24%

21. (a) $7.20
    (b) $11.00

23. $19,326.12

25. $8269.02

*27. (a) $5512.68
     (b) $4563.92
     (c) 948.76

3-4 Using Tables: page 105

1. $250

3. $1544.50

5. $4177

7. $4891

9. $10,207

11. $1492; 14.2%

*13. $22,421

15. $1408; $1429; $79

17. 17%

19. 17%

21. (a) 98.09%
    (b) 93.63%
    (c) 70.36%
    (d) 42.73%

23. It rises very sharply to reach 1000 at age 99.

25. Through age 40 the number of deaths and the death rate per year are approximately the same. Thereafter, the number of deaths increase and then decrease as fewer people remain alive, whereas the death rate per 1000 continues to climb until it reaches 1000 at age 99.

3-5 Measurement: page 111

1. False

3. False

5. True

7. False

9. True

11. 500

13. 40

15. 8.350

17. 3.58 m

19. 3.785 km

21. 85,000 m

23. 58 mm

3-6 The Metric System: page 116

1. Unlikely

3. Likely

5. Unlikely

7. Likely

9. Likely

11. (b)

13. (c)

15. (b)

17. 8000 m

19. 8.0 cm

21. 15 km

23. 7000 g

25. 2.500 liters

27. (a) 35 mm
    (b) 3.5 cm

29. (a) 80 mm
    (b) 8.0 cm

31. 91

33. 12.7

35. 240

*37. (a) 1,000,000
     (b) 1,000,000
     (c) 1000
     (d) 5

4 Sets and Logic

1. (a) Yes
   (b) Yes
   (c) Yes
3. (a) Yes
   (b) No
   (c) Yes
5. (a) Yes
   (b Yes
   (c) Yes
7. Well-defined
9. Not well-defined
11. (a) Equivalent
    (b) Equal
13. (a) Not equivalent
    (b) Not equal
15. (a) Equivalent
    (b) Not equal
17. Among others: the set of counting numbers less than 100.
19. Among others: the set of positive integral multiples of 5.
21. Among others: the set of perfect squares, 1 through 36.
23. (a) {1, 3, 4, 5, 7}
    (b) {1, 3}
25. (a) {2, 4, 6, 7, 8}
    (b) {4, 6, 8}
27. (a) {1, 2, 3, . . .}
    (b) Ø

29. (a) {3, 4, 5}
    (b) {2, 4}
    (c) {2, 3, 4, 5}
    (d) {4}
31. (a) {2, 4, 6, . . .}
    (b) {1, 3, 5, . . .}
    (c) {1, 2, 3, 4, . . .}
    (d) Ø
33. (a) {2, 3}
    (b) {1, 2}
    (c) {1, 2, 3}
    (d) {2}
35. (a) {b, c}, {a, c}, {a, b}, {a}, {b}, {c}, Ø.
    *(b) {b, c, d, e}, {a, c, d, e}, {a, b, d, e}, {a, b, c, e}, {a, b, c, d}, {a, b, c}, {a, b, d}, {a, b, e}, {a, c, d}, {a, c, e}, {a, d, e}, {b, c,d}, {b, c, e}, {b, d, e}, {c, d, e}, {a, b}, {a, c}, {a, d}, {a, e}, {b, c}, {b, d}, {b, e}, {c, d}, {c, e}, {d, e}, {a}, {b}, {c}, {d}, {e}, Ø.
    *(c) None exist
*37. (a) Not always true; for example, if $A = \{1\}$ and $B = \{1, 2\}$. In general, the statement is false if $A \subseteq B$.
    (b) Always true.
*39. $A \cap B = \emptyset$
*41. $B \subseteq A$
*43. Always true
*45. $A = B$

1. (a) 3
   (b) 7
   (c) 5
   (d) 12
3. A' is shaded with horizontal lines; B is shaded with vertical lines. The union of these two sets is the subset of $\mathfrak{U}$ that is shaded with horizontal lines, vertical lines, or lines in both directions.

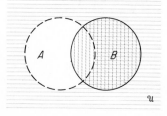

5. A is shaded with vertical lines; B' is shaded with horizontal lines. The intersection of these two sets is the subset of $\mathfrak{U}$ that is shaded with lines in both directions.

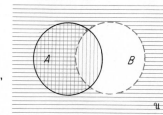

7. In the following pair of diagrams, the final result is the same, showing the equivalence of the statements given.

The set $(A \cap B)$ is shaded with horizontal lines. Its complement, $(A \cap B)'$, is the remaining portion of $\mathfrak{U}$ shaded with vertical lines.

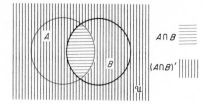

$A \cap B$ ☰

$(A \cap B)'$ ||||||

The set A' is shaded with vertical lines; the set B' is shaded with horizontal lines. Their union, $A' \cup B'$, is the portion of $\mathfrak{U}$ shaded with vertical lines, horizontal lines, or lines in both directions.

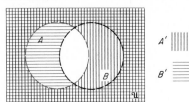

A' ||||||

B' ☰

9. (a) 2
   (b) 4
   (c) 6
   (d) 15
   (e) 28
   (f) 29

11. (a) 33
    (b) 10
    (c) 10
    (d) 39
    (e) 46
    (f) 8

13. (a)

$A \cap B \cap C$

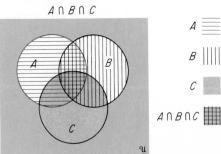

A ☰

B ||||||

C ▦

$A \cap B \cap C$ ▦

(b)

$A \cap B \cap C'$

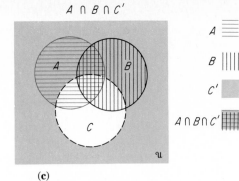

A ☰

B ||||||

C' ▦

$A \cap B \cap C'$ ▦

(c)

$A \cap B' \cap C$

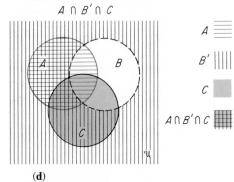

A ☰

B' ||||||

C ▦

$A \cap B' \cap C$ ▦

(d)

$A \cap B' \cap C'$

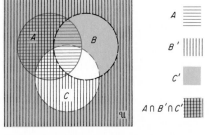

A ☰

B' ||||||

C' ▦

$A \cap B' \cap C'$ ▦

15. $(A \cup B) \cap C'$

17. $A \cap B' \cap C$

19. As shown in the following diagram,
    (a) 12 (b) 9 (c) 3.

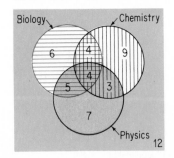

21. As shown in the following diagram, (a) 4 (b) 0.

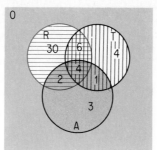

23. As shown in the following diagram, the data would require -2 students in $A' \cap B' \cap C$, which is impossible.

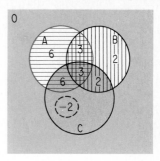

4-3 Simple and Compound Statements: page 145

1. Among others:
(a) $(\sim p) \wedge (\sim q)$
(b) $(\sim p) \wedge q$
(c) $(\sim p) \wedge q$
(d) $\sim[p \wedge (\sim q)]$
(e) $p \vee (\sim q)$.

3. Among others:
(a) $p \wedge (\sim q)$
(b) $p \vee (\sim q)$
(c) $(\sim p) \wedge q$
(d) $\sim[(\sim p) \wedge q]$
(e) $\sim[(\sim p) \wedge q]$.

5. (a) I like this book and I like mathematics.
(b) I do not like mathematics.
(c) It is not the case that I do not like this book, that is, I like this book.
(d) I do not like this book and I do not like mathematics.

9.

p	q	$(\sim p) \wedge q$
T	T	F
T	F	F
F	T	T
F	F	F

11.

p	q	$(\sim p) \vee (\sim q)$
T	T	F
T	F	T
F	T	T
F	F	T

13.

p	q	$\sim(p \wedge q)$
T	T	F
T	F	T
F	T	T
F	F	T

15.

p	q	$\sim$	$[p$	$\vee$	$(\sim q)]$
T	T	F	T	T	F
T	F	F	T	T	T
F	T	T	F	F	F
F	F	F	F	T	T

(d) (a) (c) (b)

17.

p	q	$\sim$	$[(\sim p)$	$\wedge$	$(\sim q)]$
T	T	T	F	F	F
T	F	T	F	F	T
F	T	T	T	F	F
F	F	F	T	T	T

(d) (a) (c) (b)

19. It is not the case that today is Monday. Today is not Monday.

21. It is not the case that these two cars are not made by the same company. These two cars are made by the same company.

23. It is not the case that I am young and I am happy. I am not young or I am not happy.

25. It is not the case the Pedro has $50 and he has two tickets to the game. Pedro does not have $50 or he does not have two tickets to the game.

27. It is not the case that the textbook is expensive or is not used. The textbook is not expensive and it is used.

29. Every college is expensive.

31. There are no cars that are not expensive.

33. (a) ii, iii
    (b) i, iii
    (c) i, iii
    (d) ii, iii

35. (9) I do not like this book and I like mathematics. (12) I do not like this book and I do not like mathematics.

*37. $\sim(p \wedge q)$

4-4 Conditional Statements: page 153

1. If you pass the examination, then you pass the course.

3. If you do not pass the examination, then you do not pass the course.

5. If John drives a red car, then John lives in a red house.

7. If John does not drive a red car, then John does not live in a red house.

9. (a) True
   (b) True
   (c) False
   (d) True

11. (a) True
    (b) False
    (c) True
    (d) True

13. Negation: $x = 1$ and $x = 2$.
    Converse: If $x \neq 2$, then $x = 1$.
    Inverse: If $x \neq 1$, then $x = 2$.
    Contrapositive: If $x = 2$, then $x \neq 1$.

15. Negation: We can afford a new car and we shall not buy it.
    Converse: If we buy a new car, then we can afford it.
    Inverse: If we cannot afford it, then we do not buy a new car.
    Contrapositive: If we do not buy a new car, then we cannot afford it.

17. (13) True   (14) True

19. (13) False   (14) True

21.

p	q	$\sim$	$(p \wedge q)$	$(\sim p)$	$\vee$	$(\sim q)$
T	T	F	T	F	F	F
T	F	T	F	F	T	T
F	T	T	F	T	T	F
F	F	T	F	T	T	T

(b)　(a)　(c)(e)(d)

Columns (b) and (e) are the same.

23.

p	q	p	$\longrightarrow$	q	q	$\vee$	$(\sim p)$
T	T	T	T	T	T	T	F
T	F	T	F	F	F	F	F
F	T	F	T	T	T	T	T
F	F	F	T	F	F	T	T

(a)　(c)　(b)　(d)　(f)　(e)

Columns (c) and (f) are the same.

25.

p	q	r	p	$\wedge$	$(q \vee r)$	$(p \wedge q)$	$\vee$	$(p \wedge r)$
T	T	T	T	T	T	T	T	T
T	T	F	T	T	T	T	T	F
T	F	T	T	T	T	F	T	T
T	F	F	T	F	F	F	F	F
F	T	T	F	F	T	F	F	F
F	T	F	F	F	T	F	F	F
F	F	T	F	F	T	F	F	F
F	F	F	F	F	F	F	F	F

(a)(c)　(b)　(d)　(f)　(e)

Columns (c) and (f) are the same.

27.

p	q	r	p	$\rightarrow$	$(q \wedge r)$	$(p \rightarrow q)$	$\wedge$	$(p \rightarrow r)$
T	T	T	T	T	T	T	T	T
T	T	F	T	F	F	T	F	F
T	F	T	T	F	F	F	F	T
T	F	F	T	F	F	F	F	F
F	T	T	F	T	T	T	T	T
F	T	F	F	T	F	T	T	T
F	F	T	F	T	F	T	T	T
F	F	F	F	T	F	T	T	T

(a)(c)　(b)　(d)　(f)　(e)

Columns (c) and (f) are the same.

4-5 **Forms of Statements: page 159**

Among others:

1. If a piece of fruit is an apple, then it is red.

3. If an animal is a dog, then it is a good watchdog.

5. If a geometric figure is a square, then it is a polygon.

7. If two people are ball players, then they are competitors.

9. If an object is an automobile, then it is expensive.

11. If you like this book, then you like mathematics.

13. If you like mathematics, then you like this book.

15. If you like mathematics, then you like this book.

17. $q \rightarrow p$

19. $q \rightarrow p$

21. $q \rightarrow p$

23. $p \rightarrow (\sim q)$

25. $p \leftrightarrow q$

27. If $12 + 4 = 15$, then $12 - 4 = 7$; true.

29. If $7 + 4 = 11$, then $7 \times 4 = 20$; false.

31. If $7 \times 5 = 75$, then $15 \times 5 \neq 75$; true.

33. I am happy and I pass the test. I am not happy and I do not pass the test.

35. You pass the test and you study hard. You do not pass the test and you study hard. You do not pass the test and you do not study hard.

4-6 **Mathematical Proofs: page 166**

1. For p: Elliot is a freshman.
   q: Elliot takes mathematics.
   The argument has the form $[(p \rightarrow q) \land p] \rightarrow q$ and is valid.

3. For p: The Braves win the game.
   q: The Braves win the pennant.
   The argument has the form $[(p \rightarrow q) \land (\sim q)] \rightarrow (\sim p)$ and is valid.

5. For p: You work hard. q: You are a success. The argument has the form $[(p \rightarrow q) \land (\sim q)] \rightarrow (\sim p)$ and is valid.

7. For p: You are reading this book. q: You like mathematics. The argument has the form $[(p \rightarrow q) \land (\sim p)] \rightarrow (\sim q)$ and is not valid.

9. This argument is of the form $[(p \rightarrow q) \land (q \rightarrow r)] \rightarrow (r \rightarrow p)$ and is not valid.

11. You do not drink milk.

13. If you like to fish, then you are a mathematician.

15. If you like this book, then you will become a mathematician.

17. The argument has the form $[(p \rightarrow q) \land (\sim p)] \rightarrow (\sim q)$ and is not valid. You may still have a headache from some other source.

19.

p	q	$[(p \rightarrow q)$	$\land$	$q]$	$\longrightarrow$	p
T	T	T	T	T	T	T
T	F	F	F	F	T	T
F	T	T	T	T	F	F
F	F	T	F	F	T	F
		(a)	(c)	(b)	(e)	(d)

The argument is not valid since the statement is not true in all possible cases.

21.

p	q	r	$[(p \rightarrow q)$	$\land$	$(q \rightarrow r)]$	$\longrightarrow$	$(p \rightarrow r)$
T	T	T	T	T	T	T	T
T	T	F	T	F	F	T	F
T	F	T	F	F	T	T	T
T	F	F	F	F	T	T	F
F	T	T	T	T	T	T	T
F	T	F	T	F	F	T	T
F	F	T	T	T	T	T	T
F	F	F	T	T	T	T	T
			(a)	(c)	(b)	(e)	(d)

The argument is valid since the statement is true in all possible cases.

*23. If I were to ask you, "Is this the way to the river?" would you say "Yes"?

5 Sets of Numbers

5-1 Numbers and Numerals: page 179

1.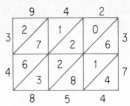

3. ...

5. ...

7. 22
9. 1102
11. 1324
13. $(2 \times 10^3) + (5 \times 10^2) + (0 \times 10^1) + (4 \times 10^0)$
15. $(2 \times 10^5) + (4 \times 10^4) + (5 \times 10^3) + (6 \times 10^2) + (0 \times 10^1) + (0 \times 10^0)$
17. $(5 \times 10^5) + (0 \times 10^4) + (1 \times 10^3) + (2 \times 10^2) + (0 \times 10^1) + (0 \times 10^0)$
19. 8165
21. 60,000,000
23. 700,000,000
25. 37,019
27. 300,023
29. 1,000,305,000
31. Five thousand three hundred seventy
33. Two hundred five thousand thirty
35. Twenty-five million two hundred three thousand five hundred

37.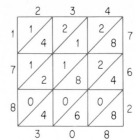
Answer: 34,854

39.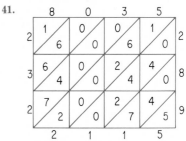
Answer: 178,308

41.
Answer: 2,322,115

5-2 Binary Notation: page 183

1. $100\ 110_2$
3. $11\ 101_2$
5. $1\ 011\ 101_2$
7. $10\ 011\ 100_2$
9. $11\ 001\ 000_2$
11. $110\ 110\ 101_2$
13. 14
15. 27
17. 59
19. 43
21. 106
23. 185
25. $11\ 010_2$
27. $101\ 000_2$
29. 100_2
31. 1110_2

33. 101_2
35. $1\ 110\ 011_2$
37. (a) $1\ 000\ 010_2$
    (b) $1\ 000\ 100_2$
    (c) $1\ 000\ 111_2$
39. $234 = 352_8 = 11\ 101\ 010_2$; when placed in groups of three, starting with the units digit, the binary representation can be translated into the octal system, and conversely.
41. 3531_8
43. $22\ 533_8$
45. $1\ 531\ 046_8$
47. $101\ 000\ 010\ 011_2$
49. $100\ 110\ 010\ 100_2$
51. $11\ 110\ 101\ 100\ 011_2$

Counting Numbers and Their Properties: page 190

1. Cardinal number
3. Ordinal number
5. Identification
7. 1
9. 11
11. 10
13. {(1, 1), (1, 2), (1, 3), (1, 4), (2, 1), (2, 2), (2, 3), (2, 4)}
15. Use any counterexample; for instance, $6 \div 2 \neq 2 \div 6$.
17. No, $4 \neq 1$. No.
19. Commutative, ×
21. Identity, ×
23. Commutative, +
25. $92 + (50 + 8)$
   $= 92 + (8 + 50)$ commutative, +
   $= (92 + 8) + 50$ associative, +
27. $37 \times (1 + 100)$
   $= 37 \times (100 + 1)$ commutative, +
   $= (37 \times 100) + (37 \times 1)$ distributive
   $= 37 \times 100 + 37$ identity, ×
   $= 3700 + 37$ closure, ×
29. $(73 + 19) + (7 + 1)$
   $= [(73 + 19) + 7] + 1$ associative, +
   $= [73 + (19 + 7)] + 1$ associative, +
   $= [73 + (7 + 19)] + 1$ commutative, +

$= [(73 + 7) + 19)] + 1$ associative, +
$= (73 + 7) + (19 + 1)$ associative, +
31. No
33. $7 \times (80 - 1) = (7 \times 80) - (7 \times 1) =$
   $560 - 7 = 553$
35. $8 \times (90 + 2) = (8 \times 90) + (8 \times 2) =$
   $720 + 16 = 736$
37. (a) No
   (b) No
   (c) No
   (d) No
39. (a) No
   (b) Yes
   (c) No
   (d) No
41. (a) No
   (b) No
   (c) Yes
   (d) No
43. (a) <
   (b) >
45. (a) >
   (b) <
47. True
49. True
*51. True

5-4 The Set of Integers: page 198

1. Identity, +
3. Zero, ×
5. Identity, +
7. (a) $20 + 20 + 20 + 20 + 20 = 100$
   (b) $15 + 15 + 15 + 15 + 15 + 15 = 90$
   (c) $18 + 18 + 18 + 18 + 18 + 18 + 18 + 18 + 18 = 162$
9.
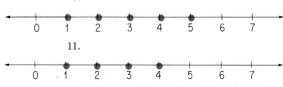
11.

13. The graph is the empty set; there are no points in the graph.

15.

17.

19. True
21. False; for example, -3 is an integer but not a whole number.
23. True
25. False; -0 is not a negative integer.
27. False; for example, $4 \div 3$ is not an integer.
29. False; the set is not closed since the sum of two odd integers is an even integer; also there is no identity element.
31. True   *33. True

5-5 Computation with Integers: page 206

1. $(+5) + (-7) = -2$
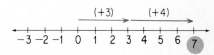

3. $(+3) + (+4) = +7$

5. -4

7. $+4$

9. -27

11. -15

13. -9

15. 0

17. -27

19. -10

21. $+16$

23. -11

25. -11

27. $+7$

29. -45

31. -45

33. -96

35. $+625$

37. -170

39. -105

41. -8

43. -8

45. $+2$

47. $+12$

49. Not defined.

51. -5

53. -20

55. $+40$

57. -4

59. Let $(2k + 1)$ and $(2m + 1)$ represent two odd integers. Then $(2k + 1) + (2m + 1) = 2k + 2m + 1 + 1 = 2k + 2m + 2 = 2(k + m + 1)$ where $(k + m + 1)$ is an integer; thus $2(k + m + 1)$ is an even integer.

61. Let $(2k + 1)$ represent an odd integer. Then $(2k + 1)^2 = 4k^2 + 4k + 1 = 2(2k^2 + 2k) + 1$ where $(2k^2 + 2k)$ is an integer and $2(2k^2 + 2k) + 1$ is an odd integer.

5-6 The Set of Rational Numbers: page 216

1. False

3. False

5. (a) 2
   (b) 1
   (c) 2
   (d) 0

7. 25

9. 8

11. 30

13. 3/10

15. $-3/4$

17. 11/28

19. 5/8

21. $-6/25$

23. $-3/10$

25. $-4/3$

27. $-5/2$

29. 2/5

31. 5/12

33. 7/6

35. $>$

37. $=$

39. $<$

41. Every rational number except 0 has a multiplicative inverse; the set of rational numbers is dense.

43. Among others: 1/101, 1/102, 1/103, 1/200 1/400

45. No, by the density property there is always another rational number between any two rational numbers.

*47. $\left(\dfrac{a}{b} \times \dfrac{c}{d}\right) \div \dfrac{e}{f} = \dfrac{ac}{bd} \times \dfrac{f}{e} = \dfrac{acf}{bde};$

$\dfrac{a}{b} \times \left(\dfrac{c}{d} \div \dfrac{e}{f}\right) = \dfrac{a}{b} \times \left(\dfrac{c}{d} \times \dfrac{f}{e}\right) = \dfrac{acf}{bde}$

	Counting numbers	Whole numbers	Integers	Positive rationals	Rational numbers
49.	✓	✓	✓	✓	✓
51.	x	✓	✓	x	✓
53.	✓	✓	✓	✓	✓
55.	✓	✓	✓	✓	✓
57.	✓	✓	✓	✓	✓
59.	✓	✓	✓	✓	✓
61.	✓	✓	✓	✓	✓

5-7 The Set of Real Numbers: page 227

1. True
3. True
5. True
7. False; 0 is neither positive nor negative.
9. True
11. (a) Irrational
    (b) Irrational
13. (a) Rational
    (b) Irrational
15. (a) Rational
    (b) Rational
17. (a) Rational
    (b) Rational
19. Rational
21. Irrational
23. Rational
25. Terminating; 0.375
27. Nonterminating, nonrepeating
29. Terminating; 10
31. 0.45, $0.\overline{45}$, 0.45455, $0.454554555. . .$, $0.4\overline{5}$
33. 0.06, $0.0\overline{6}$, 0.067, $0.06767767. . .$, $0.06\overline{7}$
35. (a) Not irrational
    (b) Not irrational
    (c) Irrational and between 0.234 and 0.235
    (d) Not between 0.234 and 0.235
37. Among others:
    $0.484484448. . .$,
    $0.486486648666. . .$
39. (a) The set of rational numbers and the set of real numbers (also the set of irrational numbers).
    (b) The set of real numbers.
41. Among others: $\sqrt{2} + (2 - \sqrt{2})$,
    $(1 + \sqrt{3}) + (1 - \sqrt{3})$,
    $(1 + 2\sqrt{5}) + (1 - 2\sqrt{5})$
43. Among others: $\sqrt{2} \times \sqrt{2}$,
    $\sqrt{3} \times 5\sqrt{3}$,
    $(1 - \sqrt{2}) \times (1 + \sqrt{2})$
*45. Among others: $\sqrt{2} \div \sqrt{2}$,
    $5\sqrt{3} \div \sqrt{3}$,
    $(1 + 2\sqrt{3}) \div (1 + 2\sqrt{3})$

6 Elements of Number Theory

6-1 Factors, Multiples, and Divisibility Rules: page 237

1. 2, 4, 6, 8, 10
3. 4, 8, 12, 16, 20
5. 7, 14, 21, 28, 35
7. 10, 20, 30, 40, 50
9. 15, 30, 45, 60, 75
11. 25, 50, 75, 100, 125
13. 1, 2, 3, 5, 6, 10, 15, 30
15. 1, 7, 49
17. 1, 2, 4, 5, 10, 20
19. 1, 2, 4, 8, 16, 32, 64
21. 1, 2, 4, 23, 46, 92
23. 1, 17
25. 1, 2, 4, 5, 8, 10, 20, 40
27. 1, 37
29. 1, 2, 3, 4, 6, 7, 12, 14, 21, 28, 42, 84
31. 1, 5, 19, 95
33. 1, 5, 25, 125
35. 1, 3, 5, 9, 25, 45, 75, 225
37. A whole number is divisible by 5 if its units digit is 0 or 5.
39. A whole number is divisible by 8 if the number formed by the last three digits, in order, of the given number is divisible by 8.
41. (a) Yes
    (b) Yes
    (c) Yes
    (d) Yes
    (e) Yes
    (f) Yes
    (g) No
43. (a) Yes
    (b) Yes
    (c) Yes
    (d) No
    (e) Yes
    (f) Yes
    (g) Yes

45. (a) Yes   (e) No
    (b) No   (f) No
    (c) Yes   (g) No
    (d) Yes

47. (a) Yes   (e) Yes
    (b) Yes   (f) Yes
    (c) Yes   (g) Yes
    (d) Yes

49. (a) Yes   (e) No
    (b) No   (f) No
    (c) Yes   (g) No
    (d) Yes

51. (a) Yes   (e) Yes
    (b) Yes   (f) Yes
    (c) Yes   (g) No
    (d) Yes

6-2 Prime Numbers: page 240

1. Composite
3. Prime
5. Prime
7. Composite
9. Prime
11. Prime
13. $1 \times 16, 2 \times 8, 4 \times 4$
15. $1 \times 21, 3 \times 7$
17. 1×31
19. $1 \times 54, 2 \times 27, 3 \times 18, 6 \times 9$
21. $1 \times 100, 2 \times 50, 4 \times 25, 5 \times 20,$ 10×10
23. $1 \times 125, 5 \times 25$
25. 2, 3, 5, 7, 11, 13, 17, 19
27. 41, 43, 47
29. 83, 89, 97
31. 4, 6, 8, 9
33. 32, 33, 34, 35, 36, 38, 39, 40, 42, 44, 45, 46, 48, 49
35. 81, 82, 84, 85, 86, 87, 88
37. No, for example, 9 is not a prime number. No, 2 is a prime number.

39. There are other possible answers in many cases. $4 = 2 + 2; 6 = 3 + 3;$ $8 = 3 + 5; 10 = 3 + 7; 12 = 5 + 7;$ $14 = 7 + 7; 16 = 3 + 13; 18 = 5 +$ $13; 20 = 7 + 13; 22 = 5 + 17; 24 =$ $7 + 17; 26 = 3 + 23; 28 = 5 + 23;$ $30 = 7 + 23; 32 = 3 + 29; 34 = 5 +$ $29; 36 = 7 + 29; 38 = 7 + 31; 40 =$ $3 + 37.$

41. Any three consecutive odd numbers includes 3 or a multiple of 3. Each multiple of 3 that is greater than 3 is composite, and 1 is by definition not a prime. Therefore 3, 5, 7 is a set of three consecutive odd numbers that are all prime numbers, and, since any other such set contains a composite number that is a multiple of 3, this is the only prime triplet.

43. (a) 13
    (b) 19
    (c) 31

6-3 Prime Factorization: page 245

1. 525
3. 150
5. 864
7. 3375
9. 41,327
11. 21,296
13. $2^5 \times 3$
15. 5×83
17. $2 \times 7 \times 67$
19. 257 is a prime number.
21. $2^3 \times 3 \times 5^3$
23. $5 \times 11 \times 89$
25. Using the factors one at a time: 2, 5, 7; two at a time: $2 \times 5, 2 \times 7, 5 \times 7$; three at a time: $2 \times 5 \times 7$; factors: {2, 5, 7, 10, 14, 35, 70}

27. Using the factors one at a time: 2, 5; two at a time: $2 \times 2, 2 \times 5$; three at a time: $2 \times 2 \times 5$; factors: {2, 5, 4, 10, 20}

29. Using the factors one at a time: 2, 3, 5; two at a time: $2 \times 3, 2 \times 5, 3 \times 3,$ 3×5; three at a time: $2 \times 3 \times 3,$ $2 \times 3 \times 5, 3 \times 3 \times 5$; four at a time: $2 \times 3 \times 3 \times 5$; factors: {2, 3, 5, 6, 9, 10, 15, 18, 30, 45, 90}

*31. 90
*33. 140

6-4 **Greatest Common Factor: page 249**

1. {1, 2}
3. {1, 2, 4, 8}
5. {1}
7. {1, 2, 5, 10}
9. {1, 5, 25}
11. {1, 2, 101, 202}
13. $42 = 2 \times 3 \times 7$; $60 = 2^2 \times 3 \times 5$; the GCF is 2×3, that is, 6.
15. $123 = 3 \times 41$; $287 = 7 \times 41$; the GCF is 41.
17. $123 = 3 \times 41$; $615 = 3 \times 5 \times 41$; the GCF is 3×41, that is, 123.
19. $68 = 2^2 \times 17$; $112 = 2^4 \times 7$; the GCF is 2^2, that is, 4.
21. $600 = 2^3 \times 3 \times 5^2$; $800 = 2^5 \times 5^2$; the GCF is $2^3 \times 5^2$, that is, 200.
23. $2450 = 2 \times 5^2 \times 7^2$; $3500 = 2^2 \times 5^3 \times 7$; the GCF is $2 \times 5^2 \times 7$, that is, 350.
25. $12 = 2^2 \times 3$; $18 = 2 \times 3^2$; $21 = 3 \times 7$; the GCF is 3.
27. $15 = 3 \times 5$; $25 = 5^2$; $40 = 2^3 \times 5$; the GCF is 5.
29. $10 = 2 \times 5$; $20 = 2^2 \times 5$; $35 = 5 \times 7$; the GCF is 5.
31. 2/5
33. 2/5
35. 5/6
37. 17/28
39. 2/5
41. 7/10
43. False, for example, 3 and 10 are relatively prime but 10 is not a prime number.
45. False, for example, 3 and 5 are relatively prime and both numbers are odd.
47. True

6-5 **Least Common Multiple: page 254**

1. 12, 24, 36
3. 15, 30, 45
5. 9, 18, 27
7. 30, 60, 90
9. 140, 280, 420
11. 96, 192, 288
13. $14 = 2 \times 7$; $40 = 2^3 \times 5$; the LCM is $2^3 \times 5 \times 7$, that is, 280.
15. $123 = 3 \times 41$; $287 = 7 \times 41$; the LCM is $3 \times 7 \times 41$, that is, 861.
17. $123 = 3 \times 41$; $615 = 3 \times 5 \times 41$; the LCM is $3 \times 5 \times 41$, that is, 615.
19. $68 = 2^2 \times 17$; $112 = 2^4 \times 7$; the LCM is $2^4 \times 7 \times 17$, that is, 1904.
21. $600 = 2^3 \times 3 \times 5^2$; $800 = 2^5 \times 5^2$; the LCM is $2^5 \times 3 \times 5^2$, that is, 2400.
23. $2450 = 2 \times 5^2 \times 7^2$; $3500 = 2^2 \times 5^3 \times 7$; the LCM is $2^2 \times 5^3 \times 7^2$, that is, 24,500.
25. $12 = 2^2 \times 3$; $18 = 2 \times 3^2$; $21 = 3 \times 7$; the LCM is $2^2 \times 3^2 \times 7$, that is, 252.
27. $15 = 3 \times 5$; $25 = 5^2$; $40 = 2^3 \times 5$; the LCM is $2^3 \times 3 \times 5^2$, that is, 600.
29. $10 = 2 \times 5$; $20 = 2^2 \times 5$; $35 = 5 \times 7$; the LCM is $2^2 \times 5 \times 7$, that is, 140.
31. True
33. True
35. 29/24
37. 29/30
39. 9/20
41. 199/280
43. 38/615
45. $-1/2400$
47. $-13/15$
49. 55/24
51. 5/8
53. 2
55. 4/5
57. 6/5
59. 24/35
*61. $(adf + bcf + bde)/bdf$
*63. $(acf + ade)/bdf$
*65. $(adf + bcf)/bde$

6-6 **Modular Arithmetic: page 260**

1. 5 (mod 12)
3. 4 (mod 12)
5. 0 (mod 12)
7. 2 (mod 5)
9. 3 (mod 5)
11. 3 (mod 5)
13. 4 (mod 5)
15. 3 (mod 5)

17. 2 (mod 5)
19. 4 (mod 5)
21. 4 (mod 5)
23. 3 (mod 5)
25. 3 (mod 5)
27. 3 (mod 5)
29. 2 (mod 5)
31. 2 (mod 7)

33. 5 (mod 7)
35. 3 (mod 9)
37. 3 (mod 6)
39. 4 (mod 8)
41. An impossible equation; that is, there is no value of x for which this equation is true.

43.

×	1	2	3	4	5	6	7	8	9	10	11	12
1	1	2	3	4	5	6	7	8	9	10	11	12
2	2	4	6	8	10	12	2	4	6	8	10	12
3	3	6	9	12	3	6	9	12	3	6	9	12
4	4	8	12	4	8	12	4	8	12	4	8	12
5	5	10	3	8	1	6	11	4	9	2	7	12
6	6	12	6	12	6	12	6	12	6	12	6	12
7	7	2	9	4	11	6	1	8	3	10	5	12
8	8	4	12	8	4	12	8	4	12	8	4	12
9	9	6	3	12	9	6	3	12	9	6	3	12
10	10	8	6	4	2	12	10	8	6	4	2	12
11	11	10	9	8	7	6	5	4	3	2	1	12
12	12	12	12	12	12	12	12	12	12	12	12	12

45. $2 \times 6 \equiv 0 \,(\text{mod } 12)$; $3 \times 4 \equiv 0 \,(\text{mod } 12)$; $3 \times 8 \equiv 0 \,(\text{mod } 12)$; $4 \times 6 \equiv 0 \,(\text{mod } 12)$; $4 \times 9 \equiv 0 \,(\text{mod } 12)$; $6 \times 6 \equiv 0 \,(\text{mod } 12)$; $6 \times 8 \equiv 0 \,(\text{mod } 12)$; $6 \times 10 \equiv 0 \,(\text{mod } 12)$; $8 \times 9 \equiv 0 \,(\text{mod } 12)$.

Thus the zero divisors in arithmetic modulo 12 are: 2 and 6, 3 and 4, 3 and 8, 4 and 6, 4 and 9, 6 and 6, 6 and 8, 6 and 10, and 8 and 9.

7 An Introduction to Algebra

7-1 Sentences and Statements: page 269

1. An open sentence.
3. An open sentence.
5. A true statement.
7. A false statement.
9. A true statement.
11. A false statement.
13. {5}
15. {0, 1, 2, 3}
17. {0, 1, 2, 3, 4, 5, 6}
19. {3, 2, 1, 0, −1, −2, . . .}
21. {2, 1, 0, −1, −2, . . .}

23. {−2, −1, 0, 1, 2, 3, . . .}
25. x is any real number greater than 2.
27. x is any real number less than 6.
29. x is any real number other than 5.
31. A point.
33. A ray.
35. A line segment.
37. A line.
39. A half-line.
41. A ray.
43.

45.  (number line: filled dots at -3 and 4, shaded between, marks -4 -3 -2 -1 0 1 2 3 4 5)

47. (number line: shaded from 2 rightward, marks 0 1 2 3 4 5 6 7 8)

49. (number line: open dot at 2, shaded rightward, marks -1 0 1 2 3 4 5 6 7)

51. 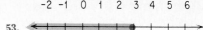 (number line: all shaded, marks -2 -1 0 1 2 3 4 5 6)

53. (number line: filled dot at 3, shaded leftward, marks -2 -1 0 1 2 3 4 5 6)

*55. (number line: filled dot at -2, shaded rightward, marks -3 -2 -1 0 1 2 3 4 5)

*57. (number line: filled dots at -6 and 6, marks -7 -6 -5 -4 -3 -2 -1 0 1 2 3 4 5 6 7)

*59. (number line: filled dots at -3 and 3, shaded between, marks -4 -3 -2 -1 0 1 2 3 4)

7-2 Compound Sentences: page 273

1. $\{1, 2, 3, 4, 5\}$
3. $\{2, 3, 4\}$
5. $\{-1, -2, -3, \ldots\} \cup \{1, 2, 3, \ldots\}$
7. (number line: all shaded, marks 0 1 2 3 4 5 6 7 8)
9. 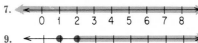 (number line: filled dot at -1, shaded rightward, marks -2 -1 0 1 2 3 4 5 6)
11. $\varnothing$
13. $\varnothing$
15. (number line: filled dots at 2 and 5, shaded between, marks 0 1 2 3 4 5 6 7 8)
17. (number line: filled dot at 0, open dot at 3, shaded between, marks -1 0 1 2 3 4 5 6 7)

19. (number line: open dot at 0, filled dot at 1, shaded between, marks -2 -1 0 1 2 3 4 5 6)
21. $\varnothing$
23. $\varnothing$
25. (number line: all shaded, marks -2 -1 0 1 2 3 4 5 6)
27. (number line: all shaded, marks -2 -1 0 1 2 3 4 5 6)
*29. (number line: filled dots at -3, -1, 1, 3, marks -4 -3 -2 -1 0 1 2 3 4)

7-3 Equations and Inequalities of the First Degree: page 278

1. (a) Given
   (b) Subtract 3
   (c) Divide by 2
3. (a) Given
   (b) Add 4
   (c) Divide by 2.
5. (a) Given
   (b) Subtract 5
   (c) Divide by -3.
7. $x = 4$
9. $x = 5$
11. $x = -3$
13. $x = -4$
15. $x = 2$

17. $x < 4$
19. $x > 3$
21. $x > -4$
23. $x < 3$
25. $x = 22$
27. $x = 12$
29. $x > -8$
*31. Always true.
*33. Not always true; among others:
   $2 - 7 \not< 3 - 10$.
*35. Not always true; among others:
   $2(-4) \not< 3(-4)$.

7-4 Problem Solving: page 283

1. $2n + 3$
3. $5(n - 2)$
5. 33
7. 2
9. 4
11. 4

13. 162
15. Length, 20 cm; width, 10 cm.
17. Length, 10 cm; perimeter, 30 cm.
19. 12
21. 8 people, $39.
23. 525 miles.

7-5 Equations in One Variable: page 289

1. 32 °F
3. 77 °F
5. 140 °F
7. 0 °C
9. 30 °C
11. 50 °C
13. $\{-8, 8\}$
15. $\{-25, 25\}$
17. $\{-r, r\}$
19. (a) 5 seconds
    (b) 25 seconds
21. $\{-5, 5\}$
23. $\{-6, 6\}$
25. $\{-3, 3\}$
27. $\{-3, 3\}$
29. $\{3, 5\}$
31. $\{-5, -1\}$
33. $\{-3, -2, 1\}$
35. $\{-5, 0, 2\}$
37. $\{-5, 1/2, 2\}$
39. $\{-3, 4\}$

41. $\{-5/2, 1\}$
43. $\{-2, 3/5\}$
*45.
     -3 -2 -1 0 1 2 3
*47.
     -1 0 1 2 3 4 5
*49.
     -3 -2 -1 0 1 2 3

*51. (a) Multiplication, $=$
     (b) Zero, $\times$
     (c) Addition, $=$
     (d) Associative, $+$
     (e) Addition
     (f) Zero, $+$
     (g) Addition, $=$
     (h) Commutative, $+$
     (i) Equivalent fractions,
         $-c/a = -4ac/(4a^2)$
     (j) Addition of fractions
     (k) Addition, $=$
     (l) Addition of fractions
     (m) Definition of subtraction

7-6 Sentences in Two Variables: page 299

1. $\{(1, 5), (2, 4), (3, 3), (4, 2), (5, 1)\}$
3. $\{(1, 1), (1, 2), (1, 3), (2, 1), (2, 2), (3, 1)\}$
5. $\{(1, 3), (2, 2), (3, 1)\}$
7. $\{(1, 1), (1, 2), (2, 1)\}$
9. $\{(1, 1)\}$
11. (a) 8
    (b) 8
13. (a) 6
    (b) -4
15. (a) $-2\frac{1}{2}$
    (b) 5
17. (a) -4
    (b) 8
19. (a) 2
    (b) $2\frac{1}{2}$
21. $y = -2x + 9$
23. $y \geq -2x + 8$
25. $y \geq 2x - 8$
27.

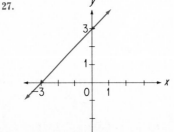

29.

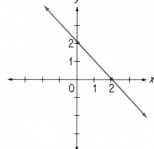

31.

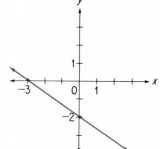

33.

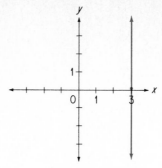

35. (f)
37. (g)
39. (c)
41.

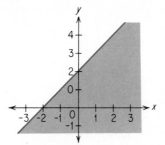

43.

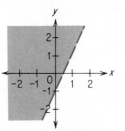

45.

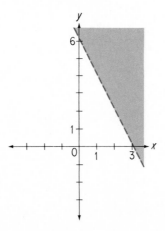

47.

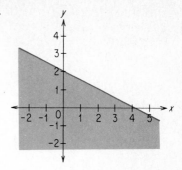

49.

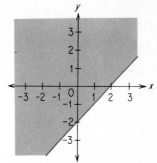

51.

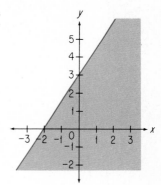

7-7 Linear Systems: page 307

1.

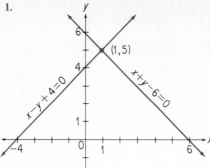

5.

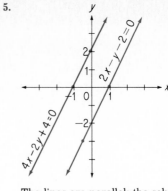

The lines are parallel; the solution set is the empty set.

3.

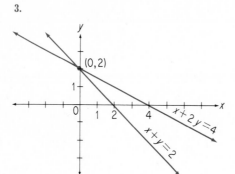

7.

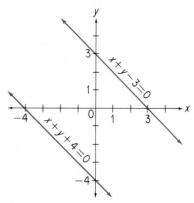

The graph consists of the union of the two lines.

9.

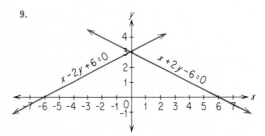

11.

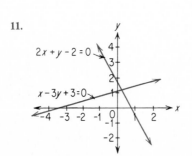

13. $\begin{cases} x + y = 19 \\ x - y = 7 \end{cases}$
$x = 13,\ y = 6$

15. $\begin{cases} x + y = 25 \\ 2x = 3y \end{cases}$
$x = 15,\ y = 10$

17. $\begin{cases} x + 2y = 12 \\ x - y = 3 \end{cases}$
$x = 6,\ y = 3$

ANSWERS TO ODD-NUMBERED EXERCISES

19.

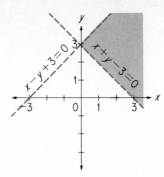

25.

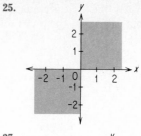

27.

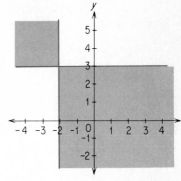

21.

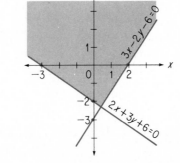

23.

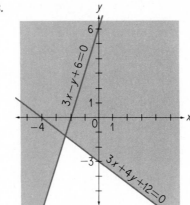

29.

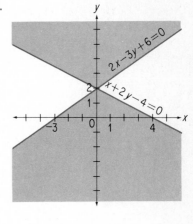

8 Elements of Geometry

8-1 Figures on a Plane: page 319

[Guidelines are provided in the exercises.]

19. (b) The line segments BE and CF are superimposed on each other when the fold is made for the perpendicular bisector of the side BC.

21. (d) $ABCD$ is a parallelogram with at least one (actually all four) of its pairs of adjacent sides perpendicular.

25. n

8-2 Measures of Plane Figures: page 332

1. (a) 2½ inches
    (b) 6 centimeters
3. (a) 1 kilometer
    (b) 1/10 millimeter
5. (a) 1/2 liter
    (b) 1/10 liter
7. (a) Exact
    (b) Approximate
9. (a) Approximate
    (b) Exact
11. Among others:

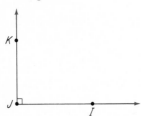

13. Among others:

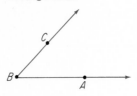

15. Among others:

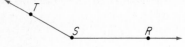

17. 80 cm^2
19. 45 cm^2
21. 68 cm^2
23. 78 cm^2
25. 28.3 cm^2
27. 254 mm^2
29. 117 km^2
31. (a) $60 + 20\pi$ units
     (b) $600 + 100\pi$ square units
33. 184 cm^2
*35. $240 + 96\pi$ m^2

8-3 The Golden Ratio: page 339

1. (a) AB, BC, CD, DE, EA;
    (b) AC, AD, BD, BE, CE;
    (c) $AG, BG, BH, CH, CI, DI, DJ,$
         EJ, EF, AF;
    (d) $GH, HI, IJ, JF, FG.$
3. $2/1 = 2$
         $3/2 = 1.5$
         $5/3 = 1.\overline{6}$
         $8/5 = 1.6$
        $13/8 = 1.625$
       $21/13 = 1.\overline{615384}$
       $34/21 = 1.\overline{619047}$
       $55/34 = 1.61764705 \ldots$
       $89/55 = 1.6\overline{18}$
      $144/89 = 1.61797752 \ldots$
     $233/144 = 1.6180\overline{5}$
     $377/233 = 1.61802575 \ldots$
     approaches
         $G \approx 1.61803398 \ldots$

5. (a) $4b^2$;
    (b) h^2.
7. Since $a^2 = b^2 + h^2$,
    $b^2 + ab = a^2$
    $b(b + a) = a^2$
    $b + a = a^2/b$
    $(b + a)/a = a/b$
    $(b/a) + 1 = a/b$
    $a/b = 1 + 1/(a/b) = G$
    since $G = 1 + (1/G)$.
9. $AE = 1, EB = G - 1 \approx 0.618$
11. (a) 1/2
     (b) $\sqrt{5}/2$;
     (c) $G - (1/2) = [(1/2) + (\sqrt{5}/2)] - (1/2) = \sqrt{5}/2$

Figures on a Coordinate Plane:
page 350

1. 4; (2, 3)
3. 4; (2, 3)
5. 5; $(-3, 9/2)$
7. $(3, -2)$
9. $(-5, 3)$
11. $(2, 5)$
13. $(3, 7)$
15. $(-1, -8)$
17. (a, b)
19. (a) $(a, 6b)$
    (b) $(a, 4b)$
21. 5
23. 25
25. $3\sqrt{2}$
27. A circle with center $(0, 0)$ and radius 4.
29. The interior points of the circle with center $(0, 0)$ and radius 6.
31. The circular region with center $(1, 2)$ and radius 2.
33. $(x - 2)^2 + (y - 5)^2 = 9$
35. $(x + 3)^2 + (y + 1)^2 \le 16$
37. $(x + 4)^2 + (y - 3)^2 > 16$
39. $(x + 1)^2 + (y - 2)^2 \le 25$
41. $9 \le (x - 1)^2 + (y + 2)^2 \le 16$
43. (a) $|x - 2| = 3$
    (b) $|y + 1| = 4$

45. The line segment has endpoints $(b/2, c/2)$ and $((a + b)/2, c/2)$, is on the line $y = c/2$, which is parallel to AB on the x-axis, and has length $a/2$, which is half the length of AB.

*47. Any parallelogram $ABCD$ has a base b and height h and may be represented on a coordinate plane with vertices $A: (0, 0,)$, $B: (b, 0)$, $C: (b + c, h)$, $D: (c, h)$. The diagonals AC and BD have the same midpoint $((b + c)/2, h/2)$ and thus bisect each other.

*49. Among others:
```
10  READ A, B
20  DATA 7, 11, 75, 123
21  DATA -567, 891, 4357, -5437
30  LET M=(A+B)/2
40  PRINT A, B, M
50  GO TO 10
60  END
```

Figures in Space:
page 359

1. Cube
3. Cube
5. Triangular pyramid
7. True
9. False
11. (a) M, N, O, P
    (b) MN, MO, MP, ON, NP, PO
    (c) The triangular regions MNO, PMN, PNO, PMO
13. (a) DC, EF, HG
    (b) EH, FG, DH, CG
    (c) HG, DC, HD, GC
    (d) AD, BC, EH, FG
15. False
17. False
19. True
21. False

23. True
25. (a) True
    (b) False
    (c) False
27. (a) True
    (b) False
    (c) False
29. 10 cm by 10 cm
31. 10 cm, 20 cm, and 20 cm
33. (a) Rectangles 3 m long and at least 1 m wide
    (b) Circles of diameter 1 m
*35. Yes; 30 cm
37. Among others: a circular region; an ice-cream-cone shape as in the figure.

8-6 Measures of Space Figures: page 367

1. (a) 125 cm^3
   (b) 150 cm^2
3. (a) 343 cm^3
   (b) 294 cm^2
5. (a) 3375 mm^3
   (b) 1350 mm^2
7. (a) 84 cm^3
   (b) 122 cm^2
9. (a) 80 m^3
   (b) 132 m^2
11. 100 cm^3
13. (a) 128π cm^3
    (b) 96π cm^2
15. (a) 4000π/3 cm^3
    (b) 400π cm^2
17. (a) 100π cm^3
    (b) 90π cm^2
19. The volume is multiplied by
    (a) 8
    (b) 27
    (c) 1/8
    (d) k^3
    The surface area is multiplied by
    (a) 4
    (b) 9
    (c) 1/4
    (d) k^2

21. The volume is multiplied by
    (a) 8
    (b) 27
    (c) 1/8
    (d) k^3
    The surface area is multiplied by
    (a) 4
    (b) 9
    (c) 1/4
    (d) k^2
*23. $B_1 = e^2$, $B_2 = e^2$, $M = e^2$, $h = e$; $V = (e/6)(e^2 + 4e^2 + e^2) = e^3$
*25. $B_1 = 0$, $B_2 = 0$, $M = \pi r^2$, $h = 2r$; $V = (2r/6)(0 + 4\pi r^2 + 0) = (4/3)\pi r^3$
*27. $B_1 = \pi r^2$, $B_2 = 0$, $M = (1/4)\pi r^2$; $V = (h/6)[\pi r^2 + 4(1/4\ \pi r^2) + 0] = (1/3)\pi r^2 h$

9 An Introduction to Probability

9-1 Counting Problems: page 379

1. (a) 24
   (b) 120
3. (a) 64
   (b) 125
5. (a) 8
   (b) 15
   (c) 18
7. (a) 24
   (b) 30
   (c) 72
9. (a) 12
   (b) 16
11. (a) 30
    (b) 36
13. (a) 56
    (b) 64

15. (a) 48
    (b) 100
17. (a) 180
    (b) 294
19. (a) 448
    (b) 648
21. (a) 12
    (b) 3
*23. (a) 180
     (b) 36
*25. (a) 72
     (b) 328
*27. (a) 250
     (b) 176

*29. $26 \times 25 \times 9 \times 9 \times 8 \times 7$, that is, 2,948,400.

*31. (a) 205,320
(b) 410,640

9-2 Definition of Probability: page 384

1. 1/2
3. 2/3
5. 5/6
7. 0
9. 1/13
11. 1/4

13. 15
15. 1/5
17. 1/12
19. 7/8
*21. (a) 3/7
(b) 1/7

9-3 Sample Spaces: page 388

1. 1/10
3. 2/5
5. 3/5
7. 1/2
9.

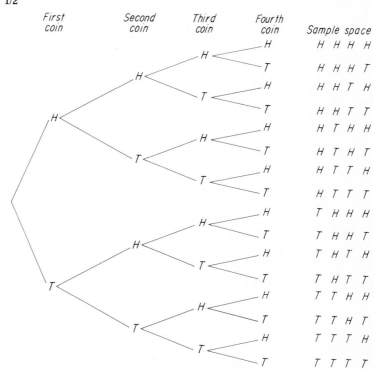

The sample space may also be given as: {HHHH, HHHT, HHTH, HHTT, HTHH, HTHT, HTTH, HTTT, THHH, THHT, THTH, THTT, TTHH, TTHT, TTTH, TTTT}.

11. (a) 5/16
(b) 5/16

13.

	First die	Second die	Sample space

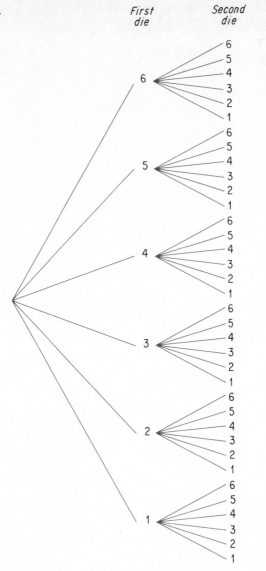

First die: 6, Second die: 6, 5, 4, 3, 2, 1 → (6,6) (6,5) (6,4) (6,3) (6,2) (6,1)

First die: 5, Second die: 6, 5, 4, 3, 2, 1 → (5,6) (5,5) (5,4) (5,3) (5,2) (5,1)

First die: 4, Second die: 6, 5, 4, 3, 2, 1 → (4,6) (4,5) (4,4) (4,3) (4,2) (4,1)

First die: 3, Second die: 6, 5, 4, 3, 2, 1 → (3,6) (3,5) (3,4) (3,3) (3,2) (3,1)

First die: 2, Second die: 6, 5, 4, 3, 2, 1 → (2,6) (2,5) (2,4) (2,3) (2,2) (2,1)

First die: 1, Second die: 6, 5, 4, 3, 2, 1 → (1,6) (1,5) (1,4) (1,3) (1,2) (1,1)

15. (a) 1/36
(b) 35/36

17. (a) 1/18
(b) 17/18

19. (a) 1/12
(b) 1/6

21. R_1R_2, R_2R_1, W_1R_1, W_2R_1, R_1W_1, R_2W_1,
W_1R_2, W_2R_2, R_1W_2, R_2W_2, W_1W_2, W_2W_1;
1/6

23.

R_1R_2	R_2R_1	R_3R_1	R_4R_1	W_1R_1	W_2R_1
R_1R_3	R_2R_3	R_3R_2	R_4R_2	W_1R_2	W_2R_2
R_1R_4	R_2R_4	R_3R_4	R_4R_3	W_1R_3	W_2R_3
R_1W_1	R_2W_1	R_3W_1	R_4W_1	W_1R_4	W_2R_4
R_1W_2	R_2W_2	R_3W_2	R_4W_2	W_1W_2	W_2W_1; 2/5

25.

Coin	Die	Sample space
H	1	H1
	2	H2
	3	H3
	4	H4
	5	H5
	6	H6
T	1	T1
	2	T2
	3	T3
	4	T4
	5	T5
	6	T6

27. **(a)** 1/12
 (b) 5/12

9-4 Computation of Probabilities: page 395

1. 2/3
3. 2/3
5. 1/2
7. 1
9. 1/2
11. 1/52
13. 19/52
15. 0
17. **(a)** 13/204
 (b) 1/663
19. **(a)** 1/16
 (b) 1/2704
21. 1/64
23. **(a)** 1/12
 (b) 1/4
 (c) 7/12
 (d) 3/4
25. **(a)** 7/12
 (b) 7/22

27. 33/66,640
29. **(a)** 27/2197
 (b) 72/2197
 (c) 343/2197
 (d) 307/2197
31. **(a)** 2/143
 (b) 6/143
 (c) 35/286
 (d) 25/286
33. **(a)** 1/8
 (b) 1/8
 (c) 1/8
 (d) 3/8
*35. **(a)** 1/32
 (b) 5/32
 (c) 10/32

9-5 Odds and Mathematical Expectation: page 401

1. 1 to 3
3. 1 to 3
5. 3 to 1
7. 11 to 2
9. 2 to 7
11. $8
13. $(1/6)(10 - 2) + (5/6)(0 - 2) = -1/3$, that is, a loss of $0.33.
15. $2.00
*17. The probability that both of the bills drawn will be tens is 1/10. The probability that both will be fives is

3/10. The probability that one will be a five and one a ten is 3/5. The mathematical expectation is then found to be $14.

*19. The probability of selecting 2¢ is 2/30; the probabilities of 6¢, 11¢, 26¢, and 51¢ are 4/30; and the probabilities of 15¢, 30¢, 55¢, 35¢, 60¢, and 75¢ are each 2/30. Thus the mathematical expectation is $30\frac{2}{3}$¢.

*21. (a) p to $(1 - p)$
(b) $(1 - p)$ to p

*23. $(1/n)(4999) + [1/(n - 1)](999) + \{1 - [1/n + 1/(n - 1)]\}(-1)$, that is, $(4999/n) + [999/(n - 1)] + \{(2n - 1)/[n(n - 1)]\}(-1)$ dollars.

9-6 Permutations: page 407

1. (a) Bat, ball
(b) (bat, ball), (ball, bat) (*Note:* Parentheses are often used for ordered sets, that is, permutations.)
3. (a) A, B, C, D
(b) $AB, AC, AD, BA, BC, BD, CA, CB, CD, DA, DB, DC$
(c) $ABC, ACB, ABD, ADB, ACD, ADC, BAC, BCA, BAD, BDA, BCD, BDC, CAB, CBA, CAD, CDA, CBD, CDB, DAB, DBA, DAC, DCA, DBC, DCB$
(d) (*Note:* Simply add the remaining element at the end of each permutation in part (c).) $ABCD, ACBD, ABDC, ADBC, ACDB, ADCB, BACD, BCAD, BADC, BDAC, BCDA, BDCA, CABD, CBAD, CADB, CDAB, CBDA, CDBA, DABC, DBAC, DACB, DCAB, DBCA, DCBA$.
5. 720

7. 7920
9. 210
11. 10!, that is, 3,628,800
13. 1320
15. 604,800
17. $n!$
19. $n!/2$
21. $n!/n!$, that is, 1
23. 21
25. 5
27. 6
29. (a) 120
(b) 48
(c) 2/5
(d) 3/5
31. 210
33. 6: PAPA, APAP, PAAP, PPAA, APPA, AAPP
35. 34,650
37. 12,600

9-7 Combinations: page 413

1. 35
3. 120
5. 190
7. 28
9. 56
11. 165

13. $_3P_2$: $(r, t)(t, r)$ $(r, s)(s, r)$ $(s, t)(t, s)$
$_3C_2$: $\{r, t\}$ $\{r, s\}$ $\{s, t\}$; 3
15. $_4C_3$: $\{a, b, c\}$ $\{a, b, d\}$ $\{a, c, d\}$ $\{b, c, d\}$
$_4C_1$: $\{d\}$ $\{c\}$ $\{b\}$ $\{a\}$
17. $_nC_0 = 1$; the only possible combination of n things 0 at a time is obtained when none are selected.

19. $_5C_0 + {}_5C_1 + {}_5C_2 + {}_5C_3 + {}_5C_4 + {}_5C_5 =$
    $1 + 5 + 10 + 10 + 5 + 1 = 32 = 2^5$

21. 64

23. 1848

25. 84

27. 80

29. (a) $_{16}C_4$, that is, 1820
    (b) $_{10}C_3 \times {}_6C_1$, that is, 720

31. $({}_4C_4 \times {}_{48}C_2)/{}_{52}C_6$, that is, 3/54,145

33. $_{12}C_4$, that is, 495

35. $_{12}C_2$, that is, 66

37. (a) $_{52}C_4$, that is, 270,725
    (b) $_{52}C_7$, that is, 133,784,560

39. $({}_{13}C_2)({}_{13}C_3) = 22,308$

41. (a) 300
    (b) 325

*43. $(1/2)_8C_4$, that is, 35

10 An Introduction to Statistics

10-1 Uses and Abuses of Statistics
page 426

1. The fact that most accidents occur near home probably means that most of the miles driven are near the home of the driver. It does not necessarily mean that long trips are safer than short trips.

3. The conclusion that mathematics is very popular is not justified by the previous statement even though the conclusion may be true. The fact that more students are studying mathematics reflects the fact that there are more students in college and that colleges are requiring more mathematics.

5. The 100% sale probably indicates that the book was the required text in a course that all students had to take, but the given evidence does not justify the stated conclusion.

7. The survey may not have covered a representative sample of the voting population. Details of the sampling procedure would be needed to supply confidence in the conclusion.

9. One wonders how such a count could possibly have been made. Details of the sampling procedure would be needed to supply confidence in the conclusion.

11. Many people who have asthma go to Arizona because of the climate. Probably a larger proportion of the people in Arizona have asthma than in any other state. Thus the conclusion is not justified.

13. People do not necessarily vote for the candidate of the party in which they are registered.

15. It should be stated what *average* means in this statement. If average denotes the median, then there would always be 50% at or below average.

17. How short? How tall? What is meant by aggressive? How is aggressiveness measured?

19. Effective for what conditions? How does brand A compare to other brands? What is meant by effectiveness? How is effectiveness determined?

21. Fewer than whom? What age groups are under consideration? How much swimming is considered?

23. What is meant by success? What is implied by "student"?

25. Are those who marry during their last year of college included? What is the source of this information?

27. The small changes in temperature appear very large due to the fact that the temperature scale does not start at zero.

29. No; a scale is needed before any such conclusions can be made.

10-2 Descriptive Statistics: page 432

7.
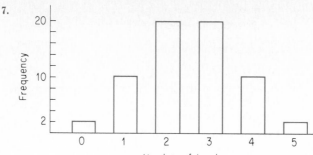

9.
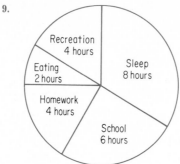

11. 44

13. (a) Yes, 10
     (b) No, percents but not amounts are shown.

15. 158°

17. $100,000

10-3 Measures of Central Tendency: page 438

1. (a) 70   (c) 65, 73
    (b) 70   (d) 72

3. (a) 20
    (b) 20
    (c) There is no mode, all numbers occur with the same frequency (all numbers might be considered modes).
    (d) 18.5

5. (a) 77   (c) 68
    (b) 73   (d) 77

7. (a) 8.2   (c) 8
    (b) 8   (d) 8

9. Median

11. Mode for the waitress; arithmetic mean for the owner.

13. 1125

15. 3440

17. 2065

19. $82\frac{2}{3}$

*21. $\mu = 14$; $(-6) + (-4) + (-1) + (+3) + (+8) = 0$

*23. (Ex. 1) $\mu = 70$;
      $(-10) + (-9) + (-5) + (-5) + (0) + (+3) + (+3) + (+9) + (+14) = 0$
      (Ex. 3) $\mu = 20$;
      $(-10) + (-5) + (-2) + (-1) + (+1) + (+4) + (+6) + (+7) = 0$
      (Ex. 5) $\mu = 77$;
      $(+8) + (-16) + (-9) + (-4) + (+14) + (-9) + (+16) = 0$
      (Ex. 7) $\mu = 8.2$;
      $(+0.8) + (-0.2) + (+0.8) + (-0.2) + (-2.2) + (-1.2) + (-0.2) + (+0.8) + (+1.8) + (-0.2) = 0$

10-4 Measures of Dispersion: page 446

1. 43

3. Among others:
    {51, 70, 75, 87, 89, 97, 98}

5. Among others:
    {55, 70, 80, 90, 91, 97, 98}

7. 50

9. Among others:
    {40, 75, 76, 77, 79, 81, 82, 85, 98}

11. Among others:
     {45, 60, 70, 75, 85, 90, 91, 92, 95}

13.	Approximately normal.	19. ≈3.9
15.	Approximately normal.	21. ≈4.6
17.	(a) 2.5%	*23. ≈9.96
	(b) 16%	
	(c) 95%	

10-5 Measures of Position: page 451

1.	495
3.	440
5.	137.5
7.	99
9.	97
11.	94
13.	Charles
15.	George

17.	Florence
19.	665
21.	(a) 50
	(b) 95
	(c) 40
	(d) 20
	(e) 5
	(f) 99

10-6 Binomial Distributions: page 456

1.	120
3.	150
5.	100 to 140
7.	45 to 55
9.	35 to 65
11.	30
13.	15

15. No; it would appear that her SAT scores should be much closer to 700, that is, about two standard deviations above the mean.

17. Among other things emphasize that *no single class should be expected to have a normal distribution.* Also clarify the meaning of the term *average.*

19. 16 to 24

Index

INDEX

Half-plane(s), 297, 316
  edge of a, 317
  opposite, 316
Hanoi, Tower of, 10
Hecto-, 110
Hectogram, 115
Hectoliter, 115
Hectometer, 114
Heptagon, 318
Herodotus, 340
Hexagon, 26, 318
Hindu-Arabic numerals, 176
Histogram, 430
Hypotenuse, 220, 321

I

Identity, statements of, 265
Identity element:
  for addition, 194
    additive, 194
  for multiplication, 188
    multiplicative, 188
Iff, 158
If-then form, 149
Inclusive or ($\vee$), 140, 142
Inconsistency, 221
Indeterminate, 196
Indirect proof, 162, 221
Induction, reasoning by, 25
Inequality:
  linear, 296
  sense of an, 298
  statements of, 265
  y-form of an, 299
Inferential statistics, 429
Infinite repeating decimal, 223
Infinite set, 125
Initial side of an angle, 328
Insurance, 102
  endowment policy, 105
  term policy, 104
Integer(s), 194, 197
  addition of, 201
  computation with, 201
  congruent modulo m, 259
  division of 203, 205
  even, 205
  multiplication of, 203, 204
  negative, 197
  odd, 206
  opposite of an, 196
  positive, 197
  properties of, 198
  set of, 194
  subtraction of, 203
Intelligence quotient (IQ), 441
Intercepts, 296
Interest:
  compound, 91, 93, 95
  effective annual rate of, 93, 100

Interest *(CONT.)*
  prime rate of, 98
  simple, 90
International System of Units (SI),
    110
Intersecting lines, 313, 358
Intersection of sets, 128, 271
Intervals, open, 271
In the long run, 397
Inverse element(s):
  additive, 197
  multiplicative, 212
  with respect to addition, 197
  with respect to multiplication, 212
Inverse operations, 195
Inverse of a statement, 151
IQ scores, 441
Irrational number, 222
  decimal representation of an, 223
Isomorphic sets, 155
Isosceles trapezoid, 323
Isosceles triangle, 320
  base of an, 320
  base angles of an, 320

K

Kemeny, John, 65
Kepler, Johannes, 337
Kilo, 116
Kilo-, 110
Kilogram, 115, 116
Kiloliter, 115
Kilometer, 110, 114
Kronecker, Leopold, 185
Kurtz, Thomas, 65

L

Large numbers, 199, 371
Last x procedure, 48
Lateral edge of a prism, 359
Law of detachment, 162
Least common denominator (LCD),
    253
Least common multiple (LCM), 251
Length, 196, 326
  unit of, 196, 327
Leonardo of Pisa, 339
Life expectancy, 102
Like fractions, 219, 252
Limits, confidence, 453
Line(s), 313
  broken, 317
  concurrent, 322
  coplanar, 316
  half-, 268, 316
  intersecting, 313, 358
  number, 196
  parallel, 316, 358
  perpendicular, 315
  real number, 314

Line(s) *(CONT.)*
  simple closed broken, 317
  skew, 358
  x-intercept of a, 296
  y-intercept of a, 296
Linear equations, 289
  simultaneous, 303
  systems of, 303
  in two variables, 293
Linear inequalities, 296
Linear measure, 196, 326
Linear system, 303
Line graph, 430
Line, number, 196, 314
Line number, 68
Line segments (*see* Segments, line)
Liter, 109, 115
Logarithmic spiral, 341
Logic, algebraic, 47
Logical arguments, 162, 168
Logical consequence, 162
Logical paradoxes, 161
Lowest terms, for a fraction, 213,
    248

M

Mass, 115
Mathematical expectation, 398, 400
Mathematical proofs, 162
Mathematical tricks, 284
Mean, arithmetic, 435, 453
Measure(s):
  of angles, 328
  area, 329, 330
  of central tendency, 434, 436
  of dispersion, 440, 441
  of length, 196, 326
  of plane figures, 326
  of position, 448
  volume, 364, 365, 366
Measurement, 107, 327
  approximate, 328
  exact, 328
  history of, 107
  metric system and, 109
Median:
  of a set of scores, 435
  of a triangle, 322
Median future lifetime, 106
Mega-, 110
Member of a set, 124
Membership symbol ($\in$), 125
Meter, 109, 114
Method of false position, 270
Metre, 109
Metric system, 110, 112
Micro-, 110
Midpoint formula, 347
Midpoint of a line segment, 320,
    344, 347

Midrange of a set of scores, 436
Milli-, 110
Milligram, 115, 116
Milliliter, 115
Millimeter, 113, 114
Mirror images, 325
Mixed number, 210
Mode of a set of scores, 436
Modular arithmetic, 257
Modus ponens, 162
Monomino, 31
Mortality table, 103
Multiple, 14, 234, 251
  common, 251
  least common, 251
Multiplication:
  of counting numbers, 187
  finger, 16, 21
  of fractions, 210, 253
  galley, 177
  Gelosia, 177
  identity element for, 188
  of integers, 203, 204
  of rational numbers, 210
  of whole numbers, 194
Multiplication property:
  of equality, 210
  of one, 188
  of order, 277
  of zero, 195
Multiplicative identity, 188
Multiplicative inverse, 212
Mutually exclusive events, 383

N

Nano-, 110
Napier, John, 180, 181
Napier's rods (bones), 181
Necessary and sufficient conditions,
    156, 157, 158
Negation of a statement, 141, 152
Negative integers, 197
Negative *n*, 196
n-gon, 26, 318
  number of diagonals of an, 26
Nines, casting out, 21, 261
Nomograph, 208
Nonagon, 318
Nonterminating, nonrepeating
    decimal, 223
Normal curve, 441
Normal distribution, 441
Not ($\sim$), 140
Notation:
  binary, 181
  decimal, 175
  expanded, 176
  factorial, 74, 405
  octal, 184
  place value, 175

Property (CONT.)
  distributive, 189
  of equality, 186, 275
  identity, 188
  of order, 277
  proportion, 80, 211
  reflexive, 186
  of statements of equality, 275
  symmetric, 186
  transitive, 186
Proportion, 80, 211
  cross products of a, 80
Proportion property, 80, 211
Pyramid, 29, 30, 357
  base of a, 356
  face of a, 29, 356
  pentagonal, 30, 357
  rectangular, 30, 357
  square, 357
  surface area of a, 366
  triangular, 30, 353, 357
  volume of a, 366
Pythagoras, 222, 337
Pythagoreans, 219, 350
Pythagorean theorem, 220, 229

Q

Quadrilateral, 318
  adjacent sides of a, 319
  consecutive vertices of a, 319
  opposite sides of a, 319
Quadratic equation, 288
Quadratic formula, 288
Quartiles, 448, 449

R

Radian, 328
Radius:
  of a circle, 348
  of a cone, 367
  of a cylinder, 366
  of a sphere, 356, 367
Random numbers, 402
Range for a set of scores, 441
Rank in class, 448
Rank, percentile, 449
Ratio, golden, 336, 338
Rational number(s), 209
  addition of, 253
  as decimals, 223, 226
  density property of, 215
  difference of, 253
  division of, 212
  equality of, 211
  as a field, 216
  multiplication of, 210
  multiplicative inverse of a, 212
  and the number line, 214
  order relations for, 214
  product of, 210
  properties of, 216

Rational number(s) (CONT.)
  reciprocal of a, 212
  set of, 209
  subtraction of, 253
  sum of, 253
Raw data, 429
Raw scores, 444
Ray(s), 268, 314
  endpoint of a, 268
  opposite, 314
Real number(s), 219, 222
  absolute value of a, 274
  classification of, 226
  completeness of, 222
  decimal representation of a, 223,
    226
  order of, 226
  set of, 219
  trichotomy law of, 190
Real number line, 222, 314
Reasoning by induction, 25
Reciprocal, 212
Recreational uses of binary
    notation, 184
Rectangle, 319
  area of a, 329
  golden, 337
Reduced form, 213
Reflexive property, 186
Region:
  of a circle, 29
  circular, 131
  polygonal, 318
  rectangular, 131
  triangular, 317
Relation(s):
  arrow diagram for a, 192
  equality, 186
  equivalence, 186
  order, 190, 213
Relatively prime, 248
Repeating decimal, 223
  infinite, 223
Replacement set, 265
Review:
  for Chapter 1, 43
  for Chapter 2, 74
  for Chapter 3, 119
  for Chapter 4, 169
  for Chapter 5, 230
  for Chapter 6, 261
  for Chapter 7, 309
  for Chapter 8, 371
  for Chapter 9, 418
  for Chapter 10, 459
Rhind papyrus, 270
Rhombus, 319
Right triangle, 220
  area of a, 329
  hypotenuse of a, 220, 321
  legs of a, 221